Vehicle System Dynamics

Youqun Zhao · Fen Lin

Vehicle System Dynamics

Theoretical Modeling and Application

Science Press
Beijing

Springer

Youqun Zhao
College of Energy and Power Engineering
Nanjing University of Aeronautics
and Astronautics
Nanjing, Jiangsu, China

Fen Lin
College of Energy and Power Engineering
Nanjing University of Aeronautics
and Astronautics
Nanjing, Jiangsu, China

ISBN 978-981-97-2018-7 ISBN 978-981-97-2019-4 (eBook)
https://doi.org/10.1007/978-981-97-2019-4

Jointly published with Science Press
The print edition is not for sale in China (Mainland). Customers from China (Mainland) please order the print book from: Science Press.

This Springer imprint is published by the registered company Springer Nature Singapore Pte Ltd.
The registered company address is: 152 Beach Road, #21-01/04 Gateway East, Singapore 189721, Singapore

Preface

Vehicle system dynamics is an emerging discipline that has developed in modern times, encompassing all the sciences related to automotive motion, with rich and extensive content. Specifically, vehicle system dynamics studies the mechanisms of longitudinal, lateral, and vertical motion of vehicles, as well as the dynamic characteristics and control technologies of various subsystems and assemblies of vehicles derived from these three movements.

With the rapid development of automotive electronic control technology, people are more and more aware of the importance of vehicle system dynamics. The current advanced active safety systems in automobiles, such as anti-lock braking system (ABS), electronic stability program (ESP), adaptive cruise control (ACC), acceleration slip regulation (ASR), electronic brake force distribution (EBD), etc., are based on vehicle dynamics as a theoretical basis for research and development and design. Therefore, for the undergraduate and graduate students majoring in vehicle engineering and the related engineering technicians who will be engaged in the research and development of vehicle engineering technology in the future, it is necessary to study the vehicle system dynamics systematically and deeply.

With the development of modern vehicle performance and speed, and the application of chassis control technology, the research content of vehicle system dynamics is more and more abundant. In the process of writing this book, a large amount of research and material collection are conducted to track the latest industry development status. Combined with the author's years of teaching and research experience in the industry, the latest vehicle handling inverse dynamics, non-pneumatic tire dynamics, vehicle state and parameter estimation methods, and driver-vehicle closed-loop control system dynamics methods are taken as the key chapters of this book. It fully reflects the current advanced technology status and development trend of vehicle system dynamics. In addition, this book systematically combs and classifies the knowledge points of vehicle system dynamics, forming a highly systematic framework structure, aiming to help readers deeply understand the knowledge points of vehicle system dynamics.

The research work of this book has been supported by the National Natural Science Foundation of China (52272397, 11672127, 11072106, 10902049), the

Doctoral Special Research Fund of Chinese Universities (20040287004), the Open Fund of China National Engineering Laboratory for High-Mobility Anti-riot Vehicle Technology (B20210017), the Fundamental Research Funds for the Central Universities (NP20220408, NP2020407, NS2020016, NP2018403, NT2018002, NP2016412, NS2015015), China Postdoctoral Foundation Special Funding Project (2017T100365), China Postdoctoral Foundation superficial Funding Project (2016M601799, 2012M521073), the Key Research & Development and Achievement Transformation Program of Wuhu (2023yf010), and China Jiangsu Postdoctoral Fund Project (1302020C). Here we express our sincere gratitude.

Chapter 7 and Sects. 3.3, 4.3, 6.3 and 6.4 were written by Prof. Lin Fen, and the rest of the book was written by Prof. Zhao Youqun.

In order to present a more complete and rich theoretical system to readers, some domestic and foreign literatures have been quoted to enrich the content of this book during the writing process, and we would like to thank the authors of these literatures.

Due to the limited level of the authors, there are unavoidable omissions in the book, please criticize and correct the majority of readers.

Nanjing, China

Youqun Zhao
Fen Lin

Introduction

This book covers the basic theoretical systems of vehicle system dynamics, based on the modeling and theoretical methods of longitudinal, lateral, and vertical system dynamics. The research results of pneumatic/non-pneumatic tire dynamics, vehicle handling inverse dynamics, vehicle state and parameter estimation method, new evaluation method of driver-vehicle closed-loop system handling dynamics, four-wheel steering vehicle motion stability and vehicle suspension control, and so on are introduced in detail. In the process of reading this book, readers will not only have a comprehensive understanding of the framework system of vehicle system dynamics, but also have a deeper grasp of the connotation and denotation of the discipline.

This book can be used as a course material for undergraduate and graduate students in the field of vehicle engineering. In addition, this book for the vehicle professional teaching and research personnel also has a good reference value.

Contents

Chapter 1
Introduction

In the real world, there are many entities that are composed of a group of objects and connected with each other by certain laws, which is called a system. There are large systems in nature, such as the solar system and the Milky way. Such systems exist naturally without the influence of human beings and are called natural systems. For example, the systems composed of organisms and atoms are all natural systems. The system formed by human design is called an artificial system, such as a production system, transportation system and communication system. There are also systems that combine artificial systems with natural systems, called integrated systems, such as navigation systems.

This book mainly discusses artificial systems and their characteristics [1, 2].

1.1 System and System Dynamics

The term "system" has a wide meaning, so there are many definitions of it. Qian Xuesen, a famous scientist in the world, defined the system as follows: "the extremely complex research object is called a system, that is, an organic whole with specific functions formed by combining several components of interaction and interdependence, and the system itself is a component of a larger system to which it belongs."

This definition indicates that the system has the following characteristics [1].

1. The system is hierarchical

A system is a thing composed of two or more elements. A large system can often be divided into several subsystems, and each subsystem is composed of smaller subsystems (called secondary systems). Each subsystem has its attributes to distinguish it from other systems. Therefore, if a large system is decomposed, many hierarchical structures can be formed, which is the system hierarchy.

Y. Zhao and F. Lin, *Vehicle System Dynamics*,
https://doi.org/10.1007/978-981-97-2019-4_1

2. The system has integrity

Although the system is composed of various elements, the nature of the system is not a simple combination of the performance of each element, but mutual influence. Therefore, this combination makes the overall performance of the system obtain new content and have a higher value. For example, a vehicle is composed of the engine, transmission system, wheels, body, control system, etc. The engine alone can only generate power and cannot walk by itself. However, when the engine is mounted on the chassis of a vehicle with wheels, it becomes a traveling vehicle and a means of transportation. Its function is very different from that of an engine. It can be seen that the study of system characteristics should be viewed from the overall perspective. The performance of a system is represented by its overall performance, not by a certain element.

3. The system is purposeful

This feature is mainly aimed at artificial systems, not natural systems (such as the Milky Way). The purposefulness of the system means that the artificial system is formed for a certain purpose. The purpose is different, and the structure of the system is different. For example, a truck is formed to transport goods. It must have a cargo compartment to carry goods; The bus is designed to transport passengers. Therefore, there must be passenger compartments and seats for passengers in the bus, while the cargo transport equipment (luggage compartment or luggage rack) is secondary, or even not. Therefore, the overall purpose of the system must be studied in the design to correctly select the composition of each element.

4. The system has functional commonality

There is a flow of matter, energy and information in the system, and it exchanges matter, energy and information with the outside (environment), that is, it can input or output matter, energy and information from the outside. This is the function of any system, which is called system function commonality. For example, in the automobile system, the stored energy is converted into the kinetic energy of the automobile due to the combustion of fuel (substance). This is the conversion of energy, and the well generates exhaust gases to be discharged. When driving, the driver obtains information from the environment (road and traffic density), judges it, and then sends out necessary instruction information to ensure that the vehicle runs reasonably and safely. This is the change of information.

The exchanges of materials, energy and information also exist between the system and the outside world. The flow from the environment to the system is called the input of the system, and the flow from the system to the environment is called the output of the system. It can be represented by Fig. 1.1.

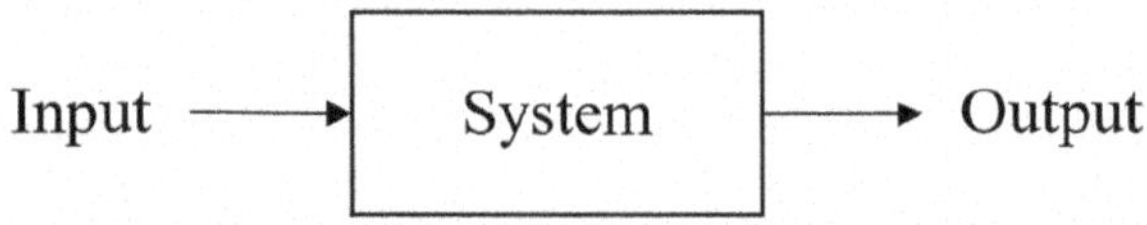

Fig. 1.1 Relationship between the system and input and output

In the study of system dynamics, different research tasks can be derived from the relationship among system, input and output [2].

(1) If the input and the characteristics of the system are known to make its output meet certain requirements, such a task can be called system design (or synthesis). The so-called optimal design is to turn a certain input into an optimal output by selecting system characteristics.
(2) If the characteristics of the input and output research system are known, such research task is called system identification. For example, it is required to judge the advantages and disadvantages of the engine environmental protection system. On the one hand, the input factors such as fuel and lubricating oil can be understood; On the other hand, the exhaust gases, smoke and dust can be collected to understand the output characteristics. According to the comparison of the two, the characteristics of the engine environmental protection system can be identified.
(3) If the characteristics and output of the system are known, the research input is called environmental prediction. If the vibration characteristics of the vehicle are known, the input characteristics of the vehicle on the road surface can be judged by measuring the vibration response value (such as the vibration acceleration on the vehicle body) when it is running on a certain road surface, to understand the uneven characteristics of the road surface.

The systems discussed above all refer to artificial systems. The principles and concepts of various physical and life-related scientific knowledge are applied to the research, creation and design of various systems, which is the content of system dynamics research.

So, what is system dynamics? According to the definition of the famous American scholar Katsuhiko Ogata, system dynamics is "a subject that discusses the mathematical model and response of dynamic systems".

When the system is in a static state, that is, when the system variables remain constant over time, it is called a static system. Strictly speaking, there is no real static system, that is, there will always be slow changes in the system, and judging the system characteristics according to the static analysis results will get incomplete or wrong conclusions. Therefore, the research object of system dynamics should be placed on the dynamic system; The behavior of the dynamic system changes with time and is a function of time. The dynamic system analysis is more complicated than the static system analysis, but it is more necessary because many external disturbances or instabilities can be considered in the analysis process of the dynamic system.

1.2 Structure and Composition of Vehicle System

Chinese national standard Terms and Definitions of Types of Motor Vehicles and Trailers (GB/T 3730.1—2001) has the following definitions for automobiles: powered, non-track carrying vehicles with 4 or more wheels, mainly used to carry people and/or goods; Towing vehicles carrying personnel and/or goods; Special purpose. Modern vehicles are assembled from at least tens of thousands of parts, with many models and different uses and structures. However, from the overall structure of the vehicle, any vehicle includes four major components: engine, chassis, body and electrical equipment [3].

1. Engine

The engine is a machine that can convert other forms of energy into mechanical energy, including an internal combustion engine (gasoline engine, etc.), external combustion engine (Stirling engine, steam engine, etc.), electric motor, etc. The engine is the power device of the automobile, which converts the fuel supplied into heat energy through combustion and into kinetic energy, and drives the automobile through the transmission system of the chassis. It consists of two major mechanisms and five major systems: crank connecting rod mechanism, valve train, fuel supply system, cooling system, lubrication system, ignition system and starting system. The layout of the engine can be divided into three types: front, middle and rear. It can also be divided into horizontal and vertical engines according to the placement form of the engine.

2. Chassis

Chassis refers to the combination of four parts on the vehicle, including the transmission system, driving system, steering system and braking system. It supports and installs the vehicle engine and its components and assemblies to form the overall shape of the vehicle, bear the engine power and ensure the normal running of the vehicle. The transmission system is mainly composed of the clutch, transmission, universal joint, transmission shaft and drive axle. The driving system is composed of the frame, axle, suspension and wheel, which is used to support the quality of the whole vehicle and ensure the driving of the vehicle. The steering system is composed of the steering wheel, steering gear, steering knuckle, steering knuckle arm, tie rod, straight rod, etc. its function is to control the driving direction of the vehicle according to the wishes of the driver. The braking system is generally divided into hydraulic braking and pneumatic braking. Hydraulic braking is composed of a brake pedal, brake master cylinder, sub-cylinder, drum (wheel) brake and oil pipe. Pneumatic braking is composed of a brake pedal, air compressor, air pressure gauge, brake valve, brake air chamber, drum (wheel) brake and air pipe.

3. Body

The vehicle body is installed on the frame of the chassis for the driver and passengers to ride or load goods. The body of vehicles is generally an integral structure, and the body of trucks is generally composed of two parts: the cab and the cargo box. The automobile body structure mainly includes body shell (body in white), door, window, front sheet metal parts, interior and exterior trim parts and body accessories, seats, ventilation, heating, air conditioning devices, etc., and also includes trunk and other equipment on trucks and special vehicles.

4. Electrical system

The on-board electrical equipment is composed of the power supply and electrical equipment. The power supply includes the battery and generator, and the electric equipment includes the starting system of the engine, the ignition system of the gasoline engine and other electric devices.

The above is the general structure of most current vehicles. To adapt to different service requirements and improve the service performance of the vehicle in some aspects, the overall structure and layout of the vehicle can be changed. The development process of vehicle structure is a process of constantly emerging contradictions and solving contradictions. Therefore, when studying the structure of the whole and parts of the vehicle, we should see that they are only the result of solving a series of contradictions in the use and manufacturing process of the vehicle, and their structural forms are not static.

1.3 Research Contents and Methods of Vehicle System Dynamics

1.3.1 Research Contents

According to the above definition of system dynamics, it is not difficult to deduce that, vehicle system dynamics is to regard the vehicle as a dynamic system, study its behavior, and discuss its mathematical model and response [1].

The dynamic characteristics of the vehicle have been studied in automobile dynamics in the early stage. Classical vehicle dynamics mainly studies the interaction of a vehicle under various forces and various dynamic working conditions resulting therefrom and discusses the impact of these dynamic working conditions and their changes on service performance. The shortcoming of this study is that the external forces discussed are all idealized; The model is too simple. While vehicle system dynamics also studies the force and motion of the vehicle, it regards the vehicle as a system placed in the real environment and studies the effects of the environment, such as road surface roughness, soil physical properties, vehicle density, air flow and wind direction, on the vehicle system. It regards them as the input of the system and regards the response of the vehicle system to these inputs as the output. In order to determine the relationship between input and output, it is necessary to study the transmission characteristics determined by the vehicle structure and design parameters, also known as dynamic characteristics, and make these dynamic characteristics adapt to the predetermined requirements. Therefore, the difference between system dynamics and classical dynamics lies in that system dynamics needs to study the environment of the system and find out its characteristics, such as road roughness and aerodynamic characteristics. On this basis, the dynamic analysis of the system in the real environment is carried out, which is the first difference.

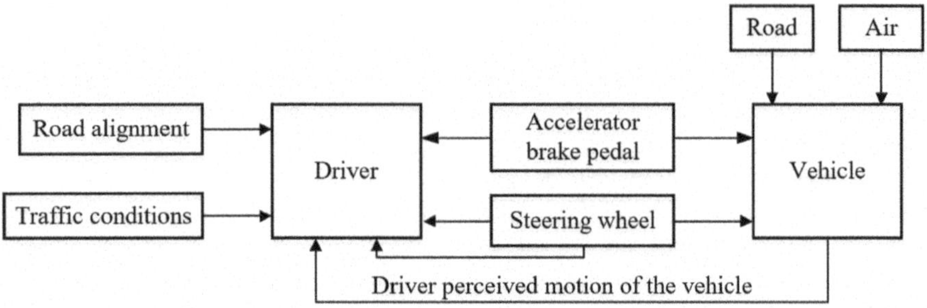

Fig. 1.2 Feedback control system with anthropomorphic transfer function

As a modern means of transportation, with the development of modern science and technology, the function of the automobile is constantly expanding, the amount of social ownership is increasing, and the speed is increasing. Under the high-speed working condition, the requirements for the operation and control of the vehicle are also getting higher and higher. Because under the high-speed working condition, if the driver is slightly careless, the vehicle will deviate from the track, which is very obvious and causes traffic accidents easily. Therefore, the controllability of the vehicle is almost close to the requirements of a control system. Therefore, in system dynamics, the vehicle is analyzed as a control system, which is the second difference from classical vehicle dynamics. The power steering system, active suspension system and anti-lock braking system used in modern vehicles are all typical control systems, so we should use modern control system theory to study their characteristics.

The third characteristic of automotive system dynamics is to consider the driver as an active factor in the automotive system and form a personal computer system to study. The adjustment function of the driver in the movement of the vehicle system can be illustrated in Fig. 1.2.

It can be seen from Fig. 1.2 that the driver controls the driving direction of the vehicle through the action of the brain and muscles. When the actual output of the vehicle deviates from the original route, it will be fed back. The error size will be judged through the detection of the driver's eyes. This information will be transmitted to the brain, and the brain will command the muscle to correct it after processing. The driver's task is to make the deviation between the output and the input as small as possible.

Vehicle system dynamics is to study the interaction between humans and vehicles in the man–machine system. Therefore, it is necessary to study the model of the human-vehicle (or driver–vehicle) system firstly. Only by establishing the model of the human-vehicle system correctly can the system analysis be carried out effectively and correctly. On this basis, the knowledge of Ergonomics is used to study how the engineering technology design of a vehicle system adapts to the role of humans, so as to make the system work most efficiently.

The fourth difference of vehicle system dynamics is to emphasize the relationship between systems and study the interaction between systems. As we all know, a vehicle can be divided into several subsystems, such as the transmission system, steering system, suspension system, etc. These systems affect and restrict each other in the process of vehicle movement. However, in classical dynamics, there are few studies and expositions in this respect, and the performance of each subsystem is often studied in isolation. System dynamics not only studies the above contents but also gradually reveals the internal relations and interactions among these subsystems. For example, after the research on tire lateral elasticity has achieved results, the close relationship between tire and steering system, tire and brake system, and suspension system performance is immediately revealed. Recent research also shows that the kinematics of the steering mechanism of the suspension system has an important influence on the performance of the steering system and the braking system; There is also an interaction between the torsional vibration of the drive train and the vibration of the suspension system. The characteristic of system dynamics is to actively seek and explore the dynamic relationship between each system itself and other systems so that when designing this system or solving the dynamic problems of this system, we can stand on the height of the whole, consider the interaction between each subsystem, deal with the contradictions that arise, and achieve the best goal of the whole.

To sum up, the research contents of vehicle system dynamics can be summarized into the following four points:

(1) Pavement characteristic analysis, environmental analysis and the effect of environment and road surface on vehicles;
(2) Kinematics and dynamics of vehicle system and its components; Interaction of subsystems in the vehicle;
(3) Optimal control and application of vehicle system;
(4) The mutual matching and model research of the driver-vehicle system and the driver model make the engineering technology design of the vehicle suitable for the use of people so that the work efficiency of the driver-vehicle system is the highest.

1.3.2 Research Methods

The first step to solve any system problem is to abstract the actual problem and transform it into a simplified model. The abstract is a logical method to analyze and extract the essence of phenomena through a kind of thinking and discard the other nonessential and secondary properties. A philosopher once said: "The abstraction of matter, the abstraction of natural laws, the abstraction of value, and so on. In a word, all scientific abstractions reflect nature more profoundly, correctly, and completely." Therefore, we must learn scientific abstract methods to correctly deal with various problems of system dynamics [1].

1. Modeling

When the vehicle is observed as a natural entity without any processing, it is a continuous vibration system. This system consists of many single parts with infinite degrees of freedom in three-dimensional space. The vibration of each component is also spatial.

However, in physics, only some special and simple analytical solutions are provided for continuous entities, such as bars, beams and thin plates. Therefore, when studying the continuum, it is usually simplified into a discrete model with "concentration" characteristics. This model is composed of a rigid body which is not deformable and has a certain mass and moment of inertia, and a spring and a shock absorber are connected between them. Through transformation, the degree of freedom of the original system is reduced, and the original system is replaced by an ideal system (usually an extremely simple model), while the basic characteristics of the system are retained.

Through transformation, we can calculate very complex systems. The purpose of the whole calculation is to study the natural vibration performance (free vibration) and forced vibration performance (frequency response characteristics or time response characteristics) of the system.

There are two ways to determine and calculate all data.

(1) Traditional method: measure each component and determine data (such as mass, the moment of inertia, elastic coefficient and damping coefficient).
(2) Data confirmation method: measure the whole system and obtain the required data through the mean method. The application example is qualitative analysis.

The finite element method is an analytical method between a continuous system and a discrete mode. Its characteristic is that the whole system is divided into many basic elements, which means rods, beams and thin plates. These basic units are used to describe complex systems. Because the number of basic units is very large, the calculation is cumbersome and requires a lot of time. It is almost impossible to quantitatively analyze the system with this method. Such a huge calculation can only be completed by high-performance computers.

2. Principles for establishing differential equations

It is far from enough to apply Newton's motion theorem to establish motion equations in solving complex systems. Systematic analysis requires other principles.

(1) Newton's law of motion

The tenet of Newton's law of motion is that the inertial force of an object is always in balance with the external force it receives, and its inertial moment and the moment it receives also form a balance. The mathematical expression is the acceleration formula:

$$m\ddot{y} = \sum_{i=1}^{n} F_i \tag{1.1}$$

and the angular acceleration formula:

$$\Theta\ddot{\varphi} = \sum_{i=1}^{n} M_i \tag{1.2}$$

Decompose each rigid body in the system, and any rigid body conforms to these two basic laws.

(2) D'Alembert principle

It is another form of Newton's law of motion. From the mathematical point of view, D'Alembert's principle is only a shifted expression of Newton's second law. However, from the physical point of view, the dynamic problem is transformed into a static problem by applying inertial force. Inertial force and moment are considered static force and moment respectively.

(3) Virtual work principle

The static problem is transformed into the dynamic problem by the principle of d'Alembert. If the imaginary deformation occurs at the force point, the corresponding virtual work is done. However, the system can maintain equilibrium only when the virtual work is zero:

$$\delta W = \sum_{i} F_i \delta_{s_i} = 0 \tag{1.3}$$

Extended to the moment, the so-called imaginary deformation can also be a very small torsional deformation. The deformation mentioned here is the deformation allowed by geometric conditions. The advantage of this theory is that when the particles are rigidly connected, the forces between them need not be taken into account because these forces do not do virtual work. Therefore, the trouble of decomposing each particle is saved.

(4) Lagrange's theorem

The differential equation of motion can be transformed as follows:

$$\frac{\mathrm{d}}{\mathrm{d}t}\left(\frac{\partial L}{\partial \dot{q}_k}\right) - \frac{\partial L}{\partial q_k} = \sum_{i=1}^{n} F_i \frac{\partial s_i}{\partial q_i} \tag{1.4}$$

where $L = T - U$, T is kinetic energy, U is potential energy, q_k is the number of degrees of freedom in the system. The item to the right of the equal sign contains power F_i,, and it has no "potential" (i.e., nonconservative force, such as resistive force). It also contains the deformation of the point of action of the relative force s_i. This item can be omitted when analyzing conservative systems.

3. Solving differential equations

The differential equation has been established and can be simplified in the application range:

(1) Only solve linear differential equations (homogeneous and non-homogeneous equations);
(2) Only simple models and concentrated coefficients are analyzed;
(3) The coefficient does not change with time.

The following concepts will be encountered in the following chapters:

(1) Natural vibration performance (natural frequency, damping ratio, attenuation/non attenuation state);
(2) Excitation [resonance frequency, frequency response (amplitude and phase change)].

When writing differential expressions, the output value is generally set on the left side of the equation, and the input value or vibration excitation value is written on the right side of the equation.

4. Theoretical basis of automotive system dynamics

When studying the dynamics of a vehicle system, when the mechanical model of the system is established, it becomes the primary task to correctly determine the dynamic equation describing the motion of the system. This kind of equation can be established by different mechanical principles, and it varies according to the degree of simplification of the system. For the single degree of freedom, two degrees of freedom or some simple type of multi-degree of freedom vibration problems, the differential equations or equations of vibration can be established by using Newton's law of motion, Da Lambert's principle, momentum theorem or momentum moment theorem. For the general multi-degree of freedom systems, especially for the complex multi-degree of freedom systems, analytical mechanics is often used. Analytical mechanics is established from the viewpoint of energy. It uses generalized coordinates as independent parameters to describe the motion of the system. In addition, the principle of virtual work in statics is extended to the dynamic problem by applying D'Alembert's principle, to establish the general dynamic equation, and then the Lagrange equation which can be widely used is derived to establish the motion equation of the system. The method of analytical mechanics can explain the general law and calculation method of vibration of the finite degree of freedom system strictly, and the obtained law can be extended to an infinite degree of freedom system. This is its great advantage, but it also has a disadvantage, that is, because the objects it studies are universal, it is relatively abstract and the physical concept is not so direct. On the contrary, because the object studied by the classical mechanical method is relatively simple, it has the advantages of a clear concept and easy calculation. These two methods will be applied in vehicle system dynamics. Possible methods for handling the dynamic system are shown in Fig. 1.3.

Linear system theory and modern control system theory are another important theoretical basis of automotive system dynamics.

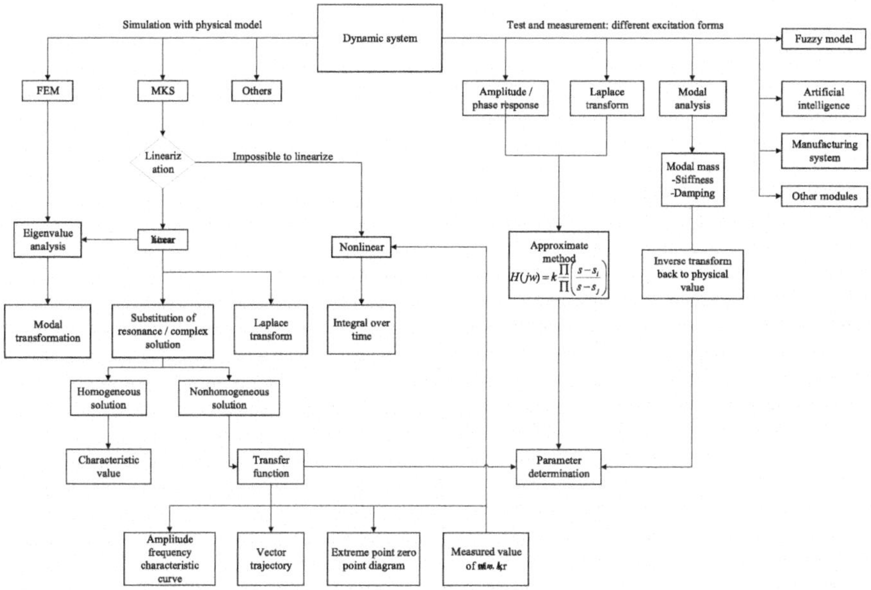

Fig. 1.3 List of possible methods and steps for dealing with the dynamic system

In the past, the study of vehicle dynamics did not consider the factors of environment and people, while the study of vehicle system dynamics also studied the environment and its influence and the role of people on vehicles. Therefore, they were actually combined into a system of human–vehicle–road (or driver–vehicle–road) to study. Study the response of the system under the given input. If human control is taken into account, a closed-circuit control system with feedback will be formed. However, since the input is often changeable and the human control is not perfect, it is necessary to install adjustment devices or control devices in the response system, which are often automatic or semi-automatic devices. Without the knowledge of system analysis and the guidance of modern control theory, designers cannot make scientific analyses of the dynamic characteristics of the system, nor can they design control equipment that can guarantee the best performance. Therefore, system analysis, optimal design and modern control theory (including optimal control theory) should be regarded as important basic theories of vehicle system dynamics.

The vehicle is often affected by the ground when driving. The newly developed vehicle ground mechanics has played a great role in studying the interaction between the ground (hard and soft) and the vehicle. The unevenness of the vehicle passing through the road is random, and the input of the road unevenness makes the response of the vehicle system random. Therefore, statistical laws must be adopted to study the dynamics and output of the system. Therefore, probability theory and its branch of the stochastic process are powerful tools to study this subject, and we must have this knowledge.

Ergonomics is a very useful basic theory for the study of the human-vehicle system. The research object of this discipline is the problems related to the human body in engineering technology design. The purpose is to solve how the engineering technology design adapts to the various requirements of the human body, so as to maximize the working efficiency of the human–machine system. Therefore, it should also be listed as the theoretical basis of vehicle system dynamics.

1.4 Theory and Research Methods of Vehicle Handling Inverse Dynamics

1.4.1 Principle of Handling Inverse Dynamic

The common research methods of vehicle handling dynamics include the open-loop method and the closed-loop method. The open-loop method refers to solving the motion state (vehicle response) of the vehicle with a known vehicle model, road model and driver's manipulation input. The idea is shown in Fig. 1.4. The closed-loop method refers to that, given the driver vehicle closed-loop control system model and the ideal path to be followed by the driver, the driver continuously observes, judges and corrects the control input to find the optimal motion state (vehicle response) of the vehicle following the ideal path. The idea is shown in Fig. 1.5. The above two methods are also called methods to solve the "positive problem" of vehicle handling dynamics.

To fully grasp the dynamic characteristics of vehicle handling, it is necessary to consider the influence of vehicle and driver factors at the same time. When the vehicle is turning at high speed, the handling feeling and subjective initiative of each driver are very different, which makes it difficult to determine the driver model parameters. Therefore, the driver modelling problem is always a "bottleneck" problem that makes

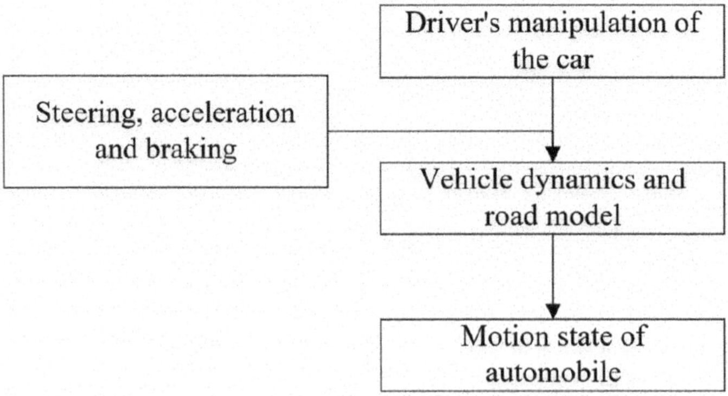

Fig. 1.4 Schematic diagram of driver vehicle open loop method

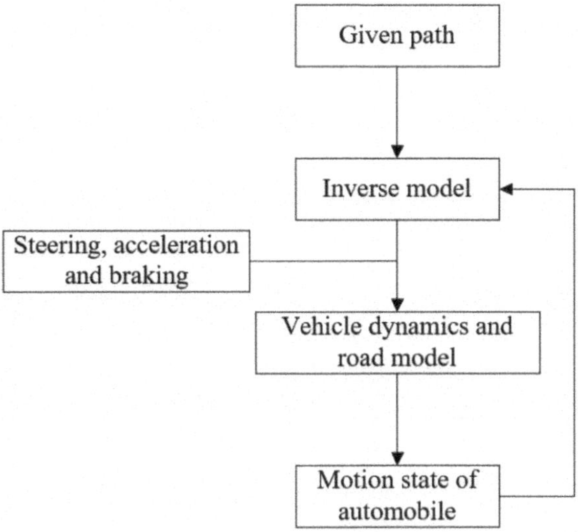

Fig. 1.5 Schematic diagram of driver vehicle closed loop method

the driver vehicle closed-loop system handling dynamics research mature and practical. The research idea of vehicle handling inverse dynamics is different from the "positive problem". Its basic idea is to inversely obtain the allowable driver's manipulation input based on the known vehicle model and vehicle motion state (vehicle response), and then analyze what kind of manipulation is the most acceptable, safest and fastest for most drivers. It can also be called the "negative problem" of vehicle handling dynamics, It belongs to "dynamic load identification" in Mechanics [4], as shown in Fig. 1.6. One of the main advantages of vehicle handling inverse dynamics is the inherent ability to follow a given path, which is conducive to comparing the handling dynamics performance of different vehicles following the same path. This is very different from the closed-loop method. In the closed-loop method, the actual driving path of the vehicle is always different from the ideal following path, while the inverse dynamics method constrains the vehicle to follow the given path. For the closed-loop method, different vehicles follow the same ideal path, their actual driving paths are different, and their steering, braking/driving forces are also different, which is not conducive to comparing the driving performance of different vehicles [5].

Although the development history of vehicle handling dynamics shows that the closed-loop method is the main method to study vehicle handling dynamics, which cannot be replaced by the open-loop method. The vehicle handling inverse dynamics method does not need to establish a driver model. Therefore, it can reverse the driver's manipulation input to the vehicle, and then analyze the vehicle handling dynamics characteristics, which can provide a new research route for the study of vehicle handling dynamics. It can be seen that the development of vehicle handling inverse dynamics may promote the remarkable improvement of vehicle handling dynamics performance and solve some "bottleneck" problems in the current popular vehicle handling dynamics.

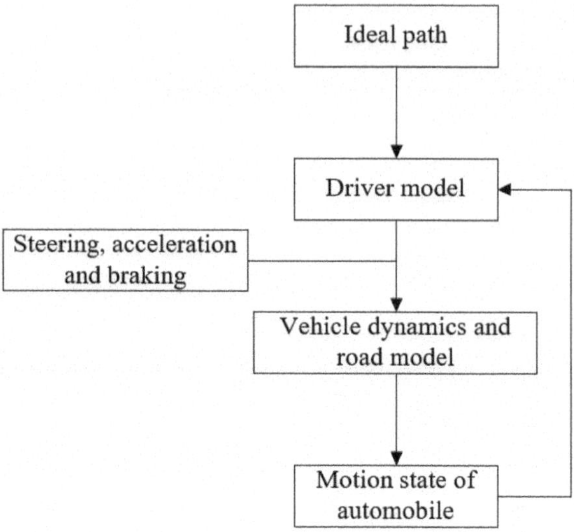

Fig. 1.6 Schematic diagram of inverse dynamics method for vehicle handling

The research of vehicle handling inverse dynamics can solve the driver's manipulation input problem in the path tracking problem and the fastest manipulation problem, and find out the best driving trajectory of the vehicle, which can provide certain guidelines for the actual driver's manipulation process; It also has certain reference value for the research of intelligent vehicles or auto driving of racing vehicles; It provides a theoretical basis for the optimal design of high-speed vehicle handling dynamics; It provides a reference for the analysis and design of vehicle handling stability and further improves the basic theory of vehicle handling dynamics; It provides an important theoretical basis and engineering application guidance for improving vehicle driving stability and safety.

1.4.2 Modeling and Solution of Handling Inverse Dynamics

Research method and application of linear vehicle handling inverse dynamics.

If the vehicle does not perform violent motion with large lateral acceleration or directional control motion on the road surface with a low friction coefficient, the motion of the vehicle is within the linear range. There are two methods to study the inverse dynamics of linear vehicle handling: the frequency domain method and the time domain method.

1. Frequency domain method

(1) Inverse method of frequency response function matrix

In the frequency domain, the relationship between the input vector of linear vehicle motion $\boldsymbol{f}$ and frequency response function matrix $\boldsymbol{H}$, response vector $\boldsymbol{x}$ is as follow:

$$x = \boldsymbol{H}f \tag{1.5}$$

so

$$f = \begin{cases} \boldsymbol{H}^{-1}x & (\boldsymbol{H} \text{ is nonsingular}) \\ \left(\boldsymbol{H}^{-H}\boldsymbol{H}\right)^{-1}\boldsymbol{H}^{H}x & (\boldsymbol{H} \text{ has full rank of singular sequence}) \\ \boldsymbol{U}\boldsymbol{\Lambda}^{-1}\boldsymbol{V}^{H}x & (\boldsymbol{H} \text{ has singular column loss}) \end{cases} \tag{1.6}$$

where $\boldsymbol{U}$, $\boldsymbol{V}$ are two unitary matrices of the singular value decomposition of $\boldsymbol{H}$, $\boldsymbol{\Lambda}$ is a diagonal matrix composed of singular values of $\boldsymbol{H}$.

The frequency domain method is a process of establishing the frequency domain function model of the system in the frequency domain and then identifying the dynamic input through the output of the system. The key to the frequency domain method is to obtain the inverse matrix of frequency response function. Equation (1.6) gives the expression of the numerical method. In addition, the inverse matrix of the frequency response function can be obtained by using the direct excitation method, the reciprocal excitation method, or the modal parameters [6].

(2) Modal coordinate transformation method

The equation of linear vehicle handling dynamics system with proportional damping is

$$x = \boldsymbol{\Phi q} \tag{1.7}$$

$$\left(-\omega^2\boldsymbol{M}_{\mathrm{r}} + j\omega\boldsymbol{C}_{\mathrm{r}} + \boldsymbol{K}_{\mathrm{r}}\right)\boldsymbol{q} = \boldsymbol{P} \tag{1.8}$$

$$f = \boldsymbol{\Phi}^{-1}\boldsymbol{P} \tag{1.9}$$

where $\boldsymbol{\Phi}$ is the modal matrix, $\boldsymbol{q}$ is the frequency domain modal coordinate vector, $\boldsymbol{M}_{\mathrm{r}}$ is the modal mass matrix, $\boldsymbol{C}_{\mathrm{r}}$ is the modal damping matrix, $\boldsymbol{K}_{\mathrm{r}}$ is the modal stiffness matrix, $\boldsymbol{P}$ is the frequency domain modal force vector.

For linear vehicle handling dynamics systems with general damping, its mode is a complex mode, which can be described by a state vector. The specific method can be referred to in the literature [6].

The modal coordinate transformation method can avoid instability in the inversion of the frequency response function matrix, and the matrix operation times are less, but the modal matrix of the system must be determined first.

2. Time domain method

(1) Modal method

Based on the modal parameters of the vehicle handling dynamics system, the inverse model of the system in the time domain is established, and then the dynamic input process is identified through the dynamic response of the system. The modal method is based on the modal coordinate transformation method. On this basis, the series method and the precise integration method are developed. The idea and basic formula of the modal coordinate transformation method in the time domain method are similar to those in the frequency domain method.

In 1996, Zhang and Zhu [7] deduced the series coefficient balance method of load identification in real mode and complex mode space and obtained the calculation formula of dynamic load identification based on the power series expansion theory, which approximately simplified the convolution relationship in the time domain to a linear relationship. This method is suitable for all kinds of transient and steady loads.

In 2002, Xu et al. [8] first transformed the modal coordinates of the proportional damping system to obtain the uncoupled motion equation, then constructed an efficient and accurate load identification formula by using the precise integration method and then calculated the time history of the dynamic load from the measured structural dynamic response.

(2) Inverse system method

The inverse system theory in cybernetics is applied to dynamic load identification so that this inverse problem is transformed into the inverse system of the original system, and the output is solved in the inverse system, that is, it is reduced to the processing of the positive problem, to obtain the dynamic load identification mode of the discrete system. Based on the modern control theory, the sufficient and necessary conditions for the reversibility of the linear system, the stability, controllability and observability of the system are studied, and the state equation of the inverse system is formed.

3. Optimal control method

In 1997, Fu et al. [9] proposed a method for the direct identification of dynamic loads by an inverse optimization system. This method avoids the operation error and operation failure caused by the ill condition of the matrix in the inverse operation of general load identification, thus improving the identification accuracy.

4. Study on the application of linear vehicle handling inverse dynamics

In 1986, Bernard and Pickelmann [10] used the basic method of inverse system method, based on the linear 2-DOF vehicle steering dynamics model under the input of steering wheel angle, and given the lateral acceleration of the vehicle's center of mass, inversely calculated the steering wheel angle, and reached the conclusion that the driver's frequency of steering control is large when driving at high speed, and may even exceed the driver's physiological limit. The methods and conclusions in document [10] are not only helpful to the study of vehicle handling dynamics but also helpful to the study of driver handling burden evaluation.

In 1986, Hatwal and Mikulcik [11] used the optimal control method to inversely calculate the steering wheel angle and braking / driving force input under the given vehicle driving path. It is not only proved that the inverse dynamics problem of vehicle handling can be solved by the optimal control method from many feasible solutions, but also compared that the front wheel drive vehicle and the rear wheel drive vehicle travel along the given path and reach the same final speed, and some conclusions that can improve the vehicle design are drawn.

From 1989 to 1992, Sridhar and Hatwal [12], Sridhar [13] compared the steering wheel angle input and driving force/braking force input required by different four-wheel steering models when tracking a single line shift by using the optimal control method and studied the steering wheel angle input and driving force/braking force input required by the vehicle when sliding and tracking a given path and when the front wheel is locked. The results show that the rear wheel steering model with neutral steering can have good steering characteristics.

In 1997, Allen [14] solved the problem of the shortest time optimal manipulation with the direct shooting method. Compared with the traditional method using Pontryagin's minimum principle, this method has higher computational efficiency.

In 2005, Brayshaw and Harrison [15] introduced the shortest time simulation method similar to the steady state (QSS). This method is based on the optimal control theory. Compared with the instantaneous optimal control method, it needs less simulation time. This method is used to study the influence of the change of the center of gravity on the optimal trajectory. The results show that the change of 6% barycenter position has little effect on the optimal trajectory.

Research method and application of nonlinear vehicle handling inverse dynamics.

The driving conditions of the vehicle are diverse and changeable, such as the change of vehicle speed during acceleration and braking, the transfer of axle load, the role of nonlinear tire force, and the combined action of lateral force and longitudinal force of tire. In fact, some problems in the nonlinear inverse dynamics of vehicle handling are difficult to solve, and there is no mature method so far.

1. Complete linearization method for Nonlinear Systems

In 1997, Xie and Wang [16] proposed to transform the original nonlinear system into a controllable observable system with the linear state equation and output equation by using nonlinear state transformation and nonlinear state feedback at the same time. This method has better tracking ability and better parameter robustness, but the application conditions are harsh.

For the transformed linear system, the inverse system method can be used to solve the vehicle handling inverse dynamics.

2. Two-point boundary value method for optimal control

In 1996, Hendriks et al. [17] studied the simple shortest-time optimal control problem by using the optimal control theory and Pontryagin's minimum principle [18, 19]. This method requires the establishment of adjoint equations and then becomes a nonlinear two-point boundary value problem. The results show that this method can evaluate the best transient handling performance of the vehicle, and the optimal handling strategy is different for different vehicle configurations. The main drawback of this method is that the gradient descent method is used to solve the optimal control problem. It has good robustness to imprecise initial points, but its convergence speed is very slow. In 2000, Bestaoui [20] also studied the speed optimal control problem by using Pontryagin's minimum principle.

In 1998 and 1999, Cossalter et al. [21, 22] took motorcycles as research objects. Under certain constraints on tire grounding and road boundary, motorcycles were required to pass through the path between two fixed points at the fastest speed. By solving the two-point boundary value problem of optimal control, the maneuverability of the motorcycle is evaluated and good results are obtained.

3. Optimization methods for engineering

In 1990, American scholar Trom et al. [23] took the lateral and longitudinal acceleration at the given vehicle center of mass as the objective function. They repeatedly applied Newton Raphson method at each time step until the steering wheel angle and braking/driving force meeting the requirements of the objective function were inversely calculated. The operating results of different vehicles under the same motion state are compared, and the optimal steering hysteresis time of the rear wheel relative to the front wheel of the all-wheel steering vehicle is found.

In 2000, the British scholar Casanova et al. [24] studied the fastest maneuvering problem. Although they did not propose a general solution, they tried to solve the constrained minimum value problem of multivariable function with a sequential quadratic programming algorithm. When studying the influence of yaw inertia on the performance of the vehicle in the process of double-line shifting, it is concluded that the yaw inertia of the vehicle has little influence on the maximum speed of the vehicle. This conclusion is helpful to guide vehicle design.

4. Computational intelligence method

In 1997, Zhang and Zhu [25] deduced the autoregressive function used in the neural network algorithm in the time domain based on the theory of structural dynamics and established the neural network dynamic load identification model with time-delay feedback. The basic learning algorithm and recall algorithm of this network are expounded. The model requires little information and low-test cost in the process of dynamic learning. It is a new dynamic load identification method worthy of popularization and application in engineering. This method can be extended to the inverse dynamics of nonlinear vehicle handling.

In 1998, the American scholar Bernard et al. [26] applied the genetic algorithm to the analysis and evaluation of vehicle handling performance based on the inverse dynamics model of vehicle handling. The problem of how to match the vehicle and the driver to obtain the best driving performance under the condition of a given vehicle motion path is solved. The simulation example proves that the genetic algorithm is a promising method in the study of vehicle handling inverse dynamics.

In 2000, Liang [27] made a good review on several applications of Computational Intelligence in mechanical inverse problems. The artificial neural networks and evolutionary computation mentioned in them have been applied in dynamic load identification and engineering optimization.

In 2006, Wu et al. [28] adopted the objective evaluation method of the handling performance of the human–vehicle–road closed-loop system and obtained the optimal preview time and follow-up order of the driver when the double lane changing road was input by optimizing the evaluation index. For the original and optimized closed-loop systems, the mapping relationship between the lateral displacement and other responses of the vehicle is established by using a radial basis function neural network. The path to be tracked is input into two trained neural networks, and other vehicle responses are inversely solved. This method can conveniently compare the maneuverability of two closed-loop systems tracking the same path.

References

1. Zhang HX. Vehicle system dynamics. Shanghai: Tongji University Press; 1996.
2. Yu F, Lin Y. Vehicle system dynamics. Beijing: Machinery Industry Press; 2005.
3. Chen JR. Automobile construction. Beijing: Machinery Industry Press; 2009.
4. Xu F, Chen HH, Bao M. Present situation and future of research on mechanical vibration load identification. China Mech Eng. 2002;13(6):526–31.
5. Zhao YQ, Yin H, Zhang LX. Present situation and development of vehicle handling inverse dynamics. China Mech Eng. 2005;16(1):77–82.
6. Wu M, Huang M. Load identification method and application of mechanical system. Xuzhou: China University of mining and Technology Press; 1995.
7. Zhang F, Zhu DM. Series method for time domain identification of dynamic loads. J Vibr Eng. 1996;9(1):1–5.
8. Xu Q, Wen XR, Sun SG. Precise stepwise integration method for structural dynamic load identification. J Comput Mech. 2002;19(1):53–7.
9. Fu ZF, Rao ZS, Zhou HT. A method of dynamic load identification. J Shanghai Jiao Tong Univ. 1997;31(3):5–7.
10. Bernard J, Pickelmann M. An inverse linear model of a vehicle. Vehic Syst Dyn. 1986;15(4):179–86.
11. Hatwal H, Mikulcik EC. Some inverse solutions to an automobile path-tracking problem with input control of steering and brakes. Vehic Syst Dyn. 1986;15(2):61–71.
12. Sridhar J, Hatwal H. A comparative study of four-wheel steering models using the inverse solution. Vehic Syst Dyn. 1992;21(1):1–18.
13. Sridhar J. A comparative study of the front wheel steered vehicle and four wheeled steered vehicles using the inverse solution. Kanpur: Indian Institute of Technology, 1989.
14. Allen J. Computer optimisation of cornering line. Cranfield: Cranfield University; 1997.
15. Brayshaw DL, Harrison MF. A quasi steady state approach to race vehicle lap simulation in order to understand the effects of racing line and center of gravity location. Automob Eng. 2005;219(6):725–39.
16. Xie S, Wang Y. Research on complete linearization method for nonlinear systems. Control Theory Appl. 1997;14(1):139–43.
17. Hendriks JPM, Meijlink TJJ, Kriens RFC. Application of optimal control theory to inverse simulation of vehicle handling. Vehic Syst Dyn. 1996;26(6):449–61.
18. Pvtlak R. Numerical methods for optimal control problems with state constraints. Berlin: Springer; 1999.
19. Sethi S, Thompson G. Optimal control theory. New York: Springer; 2000.
20. Bestaoui Y. An optimal velocity generation of a rear wheel drive tricycle along a specified path. In: Proceedings of the American control conference, Chicago; 2000. pp. 2907–2911.
21. Cossalter V, Da Lio M, Biral F. Evaluation of motorcycle maneuverability with the optimal maneuver method. In: SAE paper 983022; 1998.
22. Cossalter V, Da Lio M, Lot R. A general method for the evaluation of vehicle maneuverability with special emphasis on motorcycles. Vehic Syst Dyn. 1999;31(2):113–35.
23. Trom J, Vanderploeg M, Bernard J. Application of inverse models to vehicle optimization problems. Vehic Syst Dyn. 1990;19(2):97–110.

24. Casanova D, Sharp RS, Symonds P. Minimum time manoeuvring: the significance of yaw inertia. Vehic Syst Dyn. 2000;34(2):77–115.
25. Zhang F, Zhu DM. Dynamic load identification based on neural network model. J Vibr Eng. 1997;10(2):156–62.
26. Bernard J, Gruening J, Hoffmeister K. Evaluation of vehicle/driver performance using genetic algorithms. In: SAE Paper 980227; 1998. pp. 73–85
27. Liang YC. Some problems in computational intelligence and mechanical inverse problems. J Mech. 2000;30(3):321–31.
28. Wu J, Zhao YQ, Wu K. Analysis of vehicle handling performance based on inverse problem solving. China Mech Eng. 2006;17(4):435–9.

Chapter 2
Mechanical Characteristics of Pneumatic and Non-pneumatic Tires

2.1 Construction of Tires

2.1.1 Construction of the Pneumatic Tire

According to compositions and structures, pneumatic tires can be divided into two types: Tube tires and Tubeless tires.

1. Tube tire

The tire consists of an outer tube, an inner tube, and a blanket. The inner tube is filled with compressed air; the outer tube is a high-strength and elastic shell for protecting the inner tube without external damage; the blanket is placed between the inner tube and the rim to prevent the inner tube from being scraped and worn by the rim and the bead of the outer tire.

The outer tire mainly includes the following parts: the tread, the shoulder, the side wall and the bead. If more detailed, there are tread patterns, anti-chafing line, bead heel, bead bottom, and bead toe.

2. Tubeless tire

The tubeless tire is similar in appearance and structure to the tube tire, the difference is that the inner wall of the inner tube is added with a layer of rubber seal about 2 ~ 3 mm thickness to use to seal, which is bonded by vulcanization. A self-adhesive layer made of a special mixture of unvulcanized rubber is affixed directly under the seal. When the tire is punctured, the self-adhesive layer can glue the punctured hole on its own, so it is called a tubeless tire with a self-adhesive layer.

Some concentric annular grooves are engraved on the tire bead. Under the air pressure in the tire, the grooves can make the tire bead reliably close to the rim edge to ensure air impermeability between the tire and the rim.

Y. Zhao and F. Lin, *Vehicle System Dynamics*,
https://doi.org/10.1007/978-981-97-2019-4_2

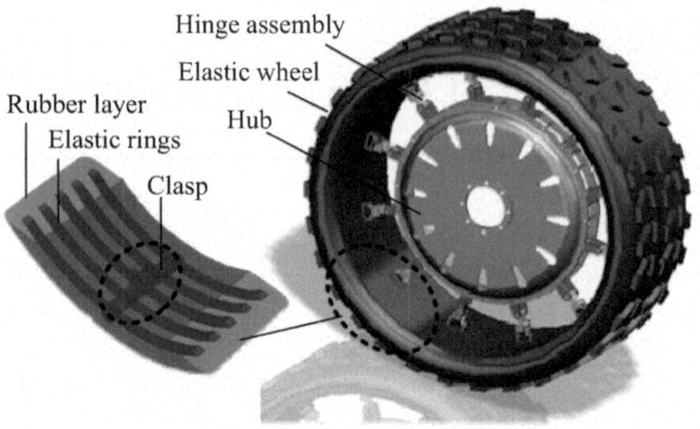

Fig. 2.1 The structural composition of NMESW

2.1.2 Construction of the Non-pneumatic Mechanical Elastic Safe Wheel

The non-pneumatic mechanical elastic safe wheel (NMESW) integrates the traditional wheel and tire into an integrated structure, with main components include the elastic wheel, hinge assembly, and hub (Fig. 2.1). Multiple groups of elastic rings are fixed by the clasp and buried in rubber layer and curtain layer, and then become the elastic wheel after vulcanization. The elastic wheel and hub are connected by a plurality of three-link hinge groups evenly distributed along the wheel circumferential direction. The non-pneumatic, hub-hung, hinged structural designs of the NMESW are free of potential hazards such as air leaks, unstable tire pressure and even blown tires [1–3].

2.2 Mechanical Characteristics of the Pneumatic Tire [4]

2.2.1 Coordinate System of the Tire

To discuss the mechanical properties of the tire, a coordinate system needs to be established, as shown in Fig. 2.2. The tire's central plane perpendicular to the axis of rotation of the wheel is called the wheel plane. The origin O of the coordinate system is the intersection point between the intersection line of the wheel plane and the ground plane and the projection of the wheel rotation axis on the ground plane. The intersection line between the wheel plane and the ground plane is taken as the X-axis, and the forward direction is the positive direction. The Z-axis is perpendicular to the ground plane, and the upward direction is the positive direction. The Y-axis is on the ground plane, and it is the positive direction to point to the left when facing the direction of the wheel running. The forces and torques from the ground acting on the tire are also shown in Fig. 2.2, namely, the tangential ground reaction force F_X, the lateral ground reaction force F_Y, the normal ground reaction force F_Z, and

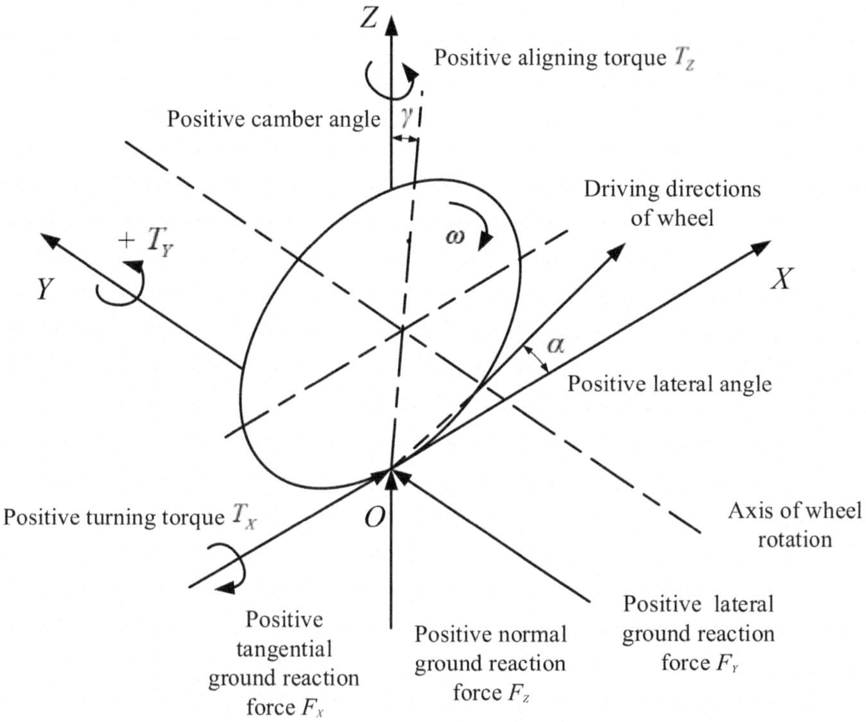

Fig. 2.2 The coordinate system of the tire and the forces and torques from the ground acting on the tire

the aligning torque T_Z by the ground reaction force around the Z-axis. They are determined in positive and negative directions according to the direction specified by the tire coordinate system. The lateral angle α and camber angle γ are also plotted. The lateral angle is the angle between the displacement direction of the tire contact patch center (i.e., the origin of the coordinate system) and the X-axis, the direction shown in the figure is positive. The camber angle is the angle between the vertical plane (XOZ plane) and the wheel plane, the direction shown in the figure is positive.

2.2.2 *Tire Cornering Phenomenon and Lateral Force-Lateral Angle Curve*

In the process of driving, due to the lateral inclination of the road surface, lateral wind or centrifugal force in curve driving, the wheel center will have the lateral force F_y along the direction of the Y-axis, and accordingly, generate lateral reaction force F_Y on the ground, F_Y is also known as side lateral force. When there is a lateral reaction force on the ground, if the wheel is rigid, two conditions will happen:

Fig. 2.3 The rolling of a rigid wheel with lateral forces

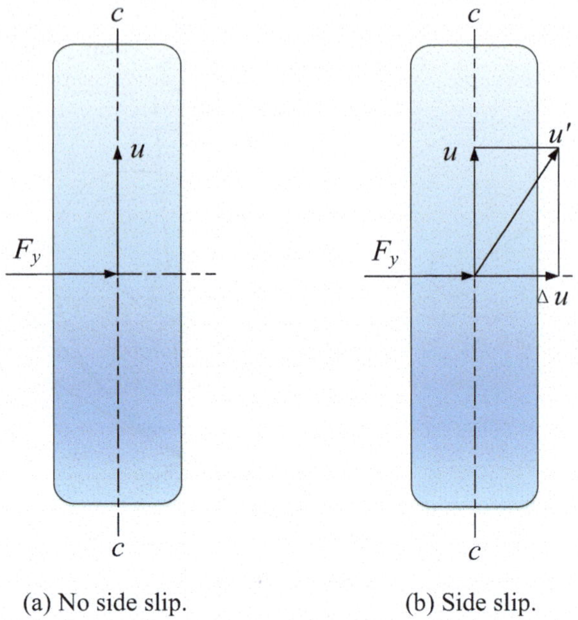

1. When the lateral reaction force F_Y does not exceed the attachment limit between the wheel and the ground, there is no sliding between the wheel and the ground and the wheel is still moving in its plane *cc* (Fig. 2.3).
2. When the lateral reaction force F_Y reaches the attachment limit between the wheel and the ground, the wheel will slide laterally. If the sliding speed is Δu, the wheel will travel in the direction of the resultant speed u', deviating from the wheel plane *cc*.

When the wheel has lateral elasticity, even if F_Y does not reach the attachment limit, the wheel driving direction will deviate from the wheel plane *cc*, which is the cornering phenomenon of the tire. In order to illustrate the cornering phenomenon, we discuss two cases when the wheel with lateral elastic is under the condition of vertical load W, the center of the wheel is subjected to lateral force F_y, and the ground has corresponding lateral reaction force F_Y. The one is that the wheel is stationary and does not roll. Due to the lateral elasticity of the wheel, the tire has lateral deformation. The center line *aa* of the tire contact patch does not coincide with the wheel plane *cc* and is staggered by Δh, but *aa* is still parallel to *cc* (Fig. 2.4a). Another one is the wheel rolling. The center line *aa* of the contact patch is not only staggered with the wheel plane by a certain distance, but is no longer parallel to the wheel plane *cc*. The angle α between *aa* and *cc* is the lateral angle. At this point, the wheel is rolling along the *aa* direction (Fig. 2.4b).

In order to clarify the cause of the lateral angle α, the following specific analysis of the wheel rolling process. Mark points $A_1, A_2, A_3, \ldots$ on the center line of the tread (Fig. 2.4b and c), as the wheel rolls forward, each point will in turn fall on the

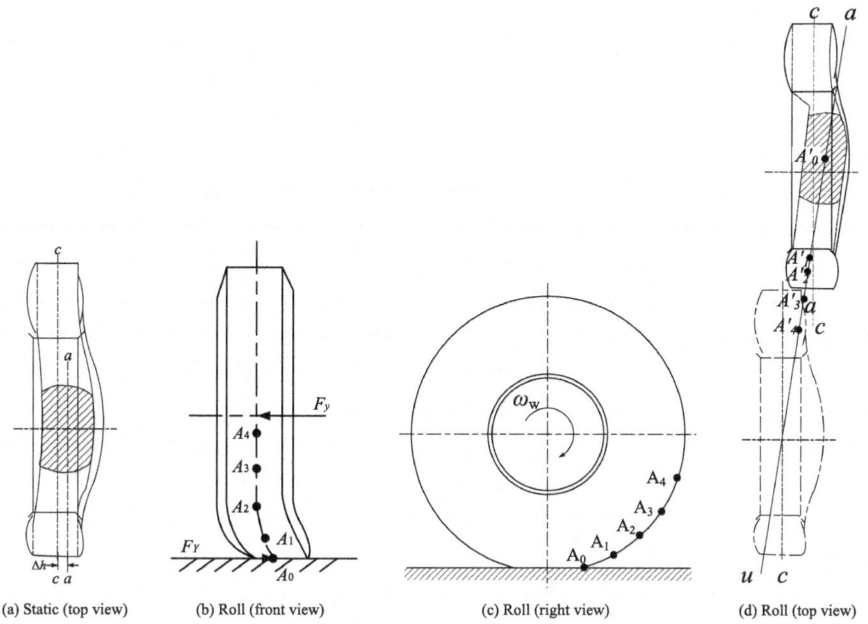

Fig. 2.4 Lateral phenomenon of tire

corresponding points $A'_1, A'_2, A'_3, \ldots$ on the ground (Fig. 2.4d). It can be seen from the main view that on the tread near the ground, the lines of points $A'_1, A'_2, A'_3, \ldots$ gradually become an oblique line when approaching the ground. Therefore, the lines of points $A'_1, A'_2, A'_3, \ldots$ on the ground are not perpendicular to the axis of wheel rotation, that is, there is an angle α with the wheel plane *cc*. When there is no lateral sliding between the tire and the ground, the lines of $A'_1, A'_2, A'_3, \ldots$ are the center lines of the contact patch, which are also the traces left on the ground when the wheel rolls, that is, the wheel does not roll forward in the wheel plane *cc*, but along the direction of the lateral angle α. Obviously, the value of the lateral angle α is related to the value of the lateral force F_Y.

Figure 2.5 shows a curve of lateral force and lateral angle measured by the test. The curve shows that the relationship between F_Y and α is linear when the lateral angle does not exceed 5°. When the vehicle is driving normally, the lateral acceleration does not exceed 0.4 g, and the side slip angle does not exceed 4°–5°. It can be considered that the side slip angle has a linear relationship with the lateral force. The slope of $F_Y - \alpha$ curve at $\alpha = 0°$ is called the lateral stiffness k and the unit is N/rad or N/(°). According to the relevant symbol of the tire coordinate system, negative lateral force produces a positive lateral angle, so the lateral stiffness is negative. The relation between F_Y and α can be written as follows

$$F_Y = k\alpha$$

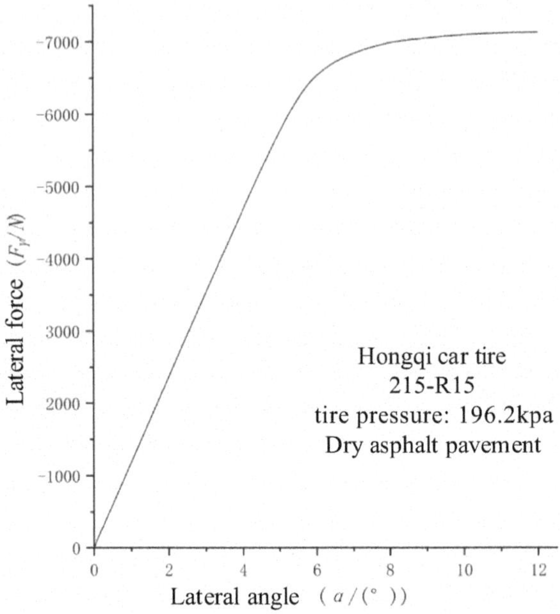

Fig. 2.5 Cornering characteristics of tires

The k value of tires for small vehicles ranges from $-28{,}000$ to $-80{,}000$ N/rad. The cornering stiffness is an important tire parameter that determines the handling stability. Tires should have high cornering stiffness (absolute value) to ensure good handling stability of the vehicle.

When the lateral force is larger, the side slip angle increases at a larger rate, that is, the slope of the $F_Y - \alpha$ curve decreases gradually, and the tire has a partially sideslip at the contact patch. Finally, when the lateral force reaches the limit of attachment, the whole tire slides sideways. Obviously, the maximum lateral force of the tire is determined by the attachment conditions, that is the vertical load, tire tread pattern, material, structure, inflation pressure, road material, structure, moisture degree and wheel camber. Generally speaking, the greater the maximum lateral force, the better the ultimate performance of the vehicle, and the higher the ultimate lateral acceleration in circular driving.

2.2.3 *Aligning Torque*

A torque T_z acting on the tire around the OZ axis is also generated when the tire is sideslip. In circumferential driving, T_z is one of the main recovery torques that restore the steering wheel to the straight driving position, which is called the aligning torque.

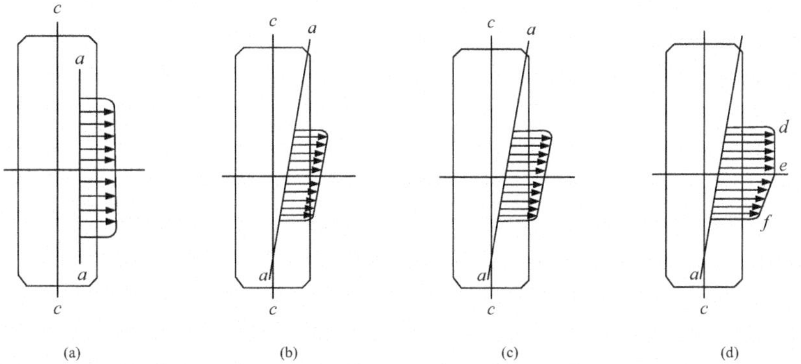

Fig. 2.6 Distribution of lateral reaction force and generation of aligning torque in contact patch

The aligning torque is generated by the micro-element lateral reaction force distributed in the ground. It can be seen from Fig. 2.6a that when the wheel is under lateral force at rest, the long axis *aa* of the contact patch is parallel to the wheel plane *cc*, and the *h* is staggered, that is, the lateral deformation of each point on the long axis *aa* of the contact patch (relative to the *cc* plane) is Δh. Therefore, it can be considered that the lateral reaction force on the ground is uniformly distributed along the *aa* line. When the wheel is rolling, the long axis *aa* of the patch is staggered from the wheel plane by a certain distance and rotates by an α angle, so the front patch is close to the wheel plane and the lateral deformation is small. The rear patch is far from the wheel plane, and the lateral deformation is large. It can be considered that the distribution of the lateral reaction force of the ground element is proportional to the formation of deformation, so the distribution of the lateral reaction force of the ground element is shown in Fig. 2.6b. The resultant force is the lateral force, but its point of action must be behind the geometric center of the contact patch and offset by some distance. *e* is the pneumatic trail and F_{ye} is the aligning torque.

When F_y increases, the distribution of lateral reaction force of the ground element in the contact patch is shown in Fig. 2.6c. When F_y increases to a certain extent, as shown in Fig. 2.6d, some parts of the rear of the contact patch reach the attachment limit, and the lateral reaction force will be distributed along the *d–e–f* line. As F_y increases further, more parts will reach the attachment limit until the entire contact patch is sideslip, so the pneumatic trail will gradually decrease as the cornering force increases.

Figure 2.7 shows the side slip angle curves of the aligning torque under different vertical loads obtained from the experiment. The tire model is 8.00–14, the speed is 8.4 m/s, and the tire pressure is 140 kPa. It can be seen that the aligning torque gradually increases at the beginning and reaches the maximum when the side slip angle is between 4° and 6°. Then, when the side slip angle increases, the aligning torque decreases, and the aligning torque is zero in the range of slip angle of 10° ~ 16°. Furthermore, when the lateral angle increases, the aligning torque becomes

Fig. 2.7 Curves of the aligning torque versus side slip angle under different vertical loads obtained from the experiment

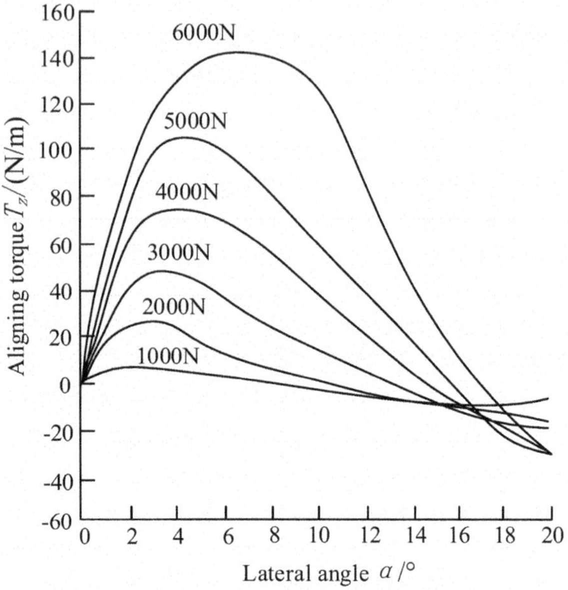

negative. The experiment results also show that the aligning torque increases with the increase of vertical load.

The form and structural parameters of the tire have an important influence on the curves of the aligning torque versus the side slip angle. At the same side slip angle, generally, the size of the tire bigger, the larger aligning torque. The aligning torque of the radial tire is larger than that of the bias tire.

The tire pressure is low, the contact patch is long, the pneumatic trail is large, and the aligning torque is also larger.

The influence of tangential ground reaction force on the aligning torque is shown in Fig. 2.8. As can be seen from the figure, with the increase of the driving force, the aligning torque reaches the maximum value and then decreases. Under the action of braking force, the aligning torque decreases continuously, decreases to zero at a certain braking force, and then becomes negative.

2.2.4 Typical Mechanics Model of the Tire

Tire models are used to describe the mechanical properties of tires. In the aspect of vehicle handling stability, the longitudinal slip characteristics, cornering characteristics, combined slip characteristics and roll characteristics of the tire in a steady state are generally studied. Tire models are divided into theoretical models and empirical models. The theoretical model is a kind of theory to discuss the mechanism of the generation of the lateral force and the aligning torque. Various hypotheses for

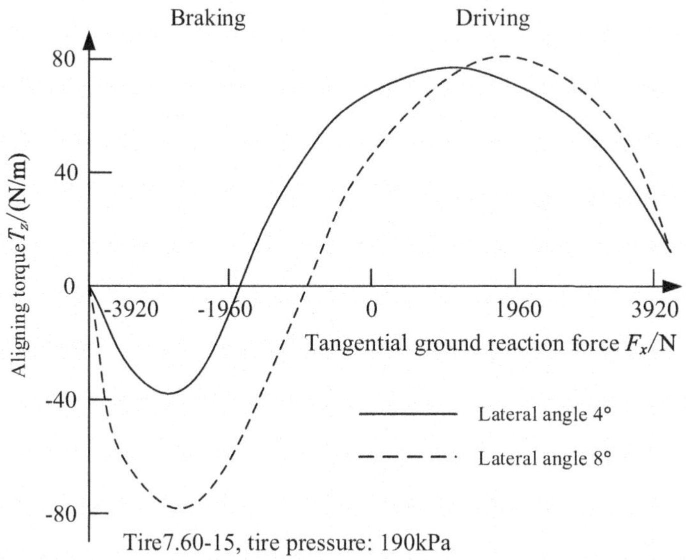

Fig. 2.8 Influence of tangential ground reaction force on the aligning torque

mechanical analysis were put forward. The finite element method is also used for calculation. The empirical model is based on the experiment data of the tire, which is fitted to the calculation formula. For a long time, our automobile industry because of the lack of tire data, vehicle operation stability research has been affected to some extent.

The research on tire mechanical characteristics began in 1930 after Fromm from Germany and Broulhiet from France respectively discovered the lateral phenomenon of automobile pneumatic tires. In 1941, Fromm simplified the tire body into an elastic beam, which is the earliest theoretical model to describe the lateral characteristics of tires. In 1954, based on the elastic beam model, Fiala assumed that the tire only had lateral deformation in the contact area, and the tire could be regarded as an elastic support beam under the action of lateral concentrated force. The relationship between the lateral force and the aligning torque and the lateral angle and camber angle is deduced. In 1966, Pacejka reduced the tire to a stretched string. On this basis, in 1989, Pacejka et al. proposed the magic formula tire model, which is an empirical model that represents the tire's longitudinal force, lateral force and aligning torque by a combination of trigonometric functions. In 1968, Hashishi Tyre Factory in Japan published an experimental modification of Fiala's theoretical expression.

In 1990, Gim et al. derived a complete tire model based on the brush model. This model requires only a few parameters to express tire forces and torques explicitly, but the model accuracy is low. In the same year, Guo Konghui established a general theoretical model of tire lateral characteristics, which is applicable to the assumption of arbitrary tire pressure distribution and tire body deformation. On this basis, he proposed the "unitire model". The model can predict the mechanical properties of slip tires from pure slip test data and the mechanical properties of high-speed tires from low-speed test data.

Among many tire models, the magic formula tire model has high precision, uniform form and strong robustness. Moreover, the team developing the magic formula has a mature business operation mode and will often update the parameters of the magic formula. The magic formula is becoming the standard in the tire industry, where tire factories provide original equipment manufacturers with magic formula coefficients in place of tables or graphs of tire test data. Therefore, many people choose the magic formula tire model as the tire model in the vehicle handling stability model.

The input of the PAC2002 magic formula tire model is the wheel slip rate k and the lateral angle α, and the output is the longitudinal force F_x, lateral force F_y and the aligning torque M_z between the tire and the road. Under pure slip and combined slip, forces and torques have different expressions, so there are six groups of formulas.

The magic formula in sine form is

$$y = D\sin(C\arctan\{Bx - E[Bx - \arctan(Bx)]\})$$

where, y is the longitudinal force F_{x_0} or the lateral force F_{y_0} of pure slip, x is slip rate k or lateral angle α, B, C, D and E are magic formula coefficients.

Wang Dan et al. used the genetic algorithm to identify the results: the relative residuals of longitudinal force and lateral force are about 5%, and the relative residuals of aligning torque are about 10% [5]. Figure 2.9a shows the identification results of the magic formula parameters of pure slip lateral force, and Fig. 2.9b shows the identification results of the magic formula parameters of pure slip aligning torque.

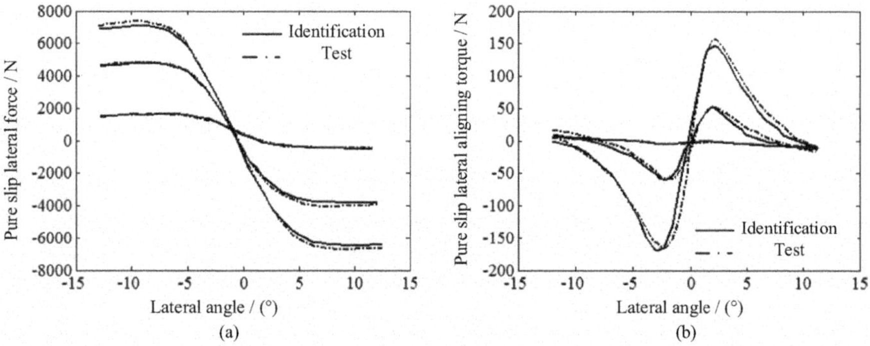

Fig. 2.9 Identification results of magic formula parameters

2.3 Mechanical Characteristics of Non-pneumatic Wheels

2.3.1 NMESW Loading Mode and Working Principle

1. Loading mode

The suspension loading structure of NMESW is different from the traditional bottom loading and top loading. As shown in Fig. 2.10, when the vertical load is applied to the center of the wheel, the hinge assembly in the upper part of the touch-ground area are all in tension with the slight deformation of the elastic wheel, but the hinge assembly in the touch-ground area will bend freely without tension or pressure. The suspension loading structure of NMESW not only ensures that the wheel has high loading efficiency, but also makes the wheel have good touch-ground characteristics [6, 7] and inclusive characteristics [8].

2. Working principle

When the vehicle equipped with NMESW remains stationary, the hinge assembly near the touch-ground area will bend and deform, as shown in Fig. 2.11a. After the

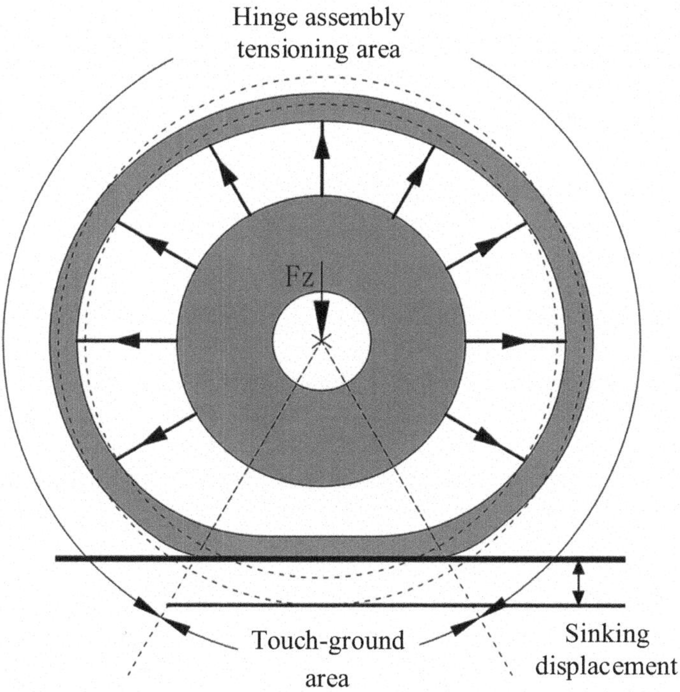

Fig. 2.10 The suspension loading structure of NMESW, F_z is the normal reaction force of the ground to the wheel

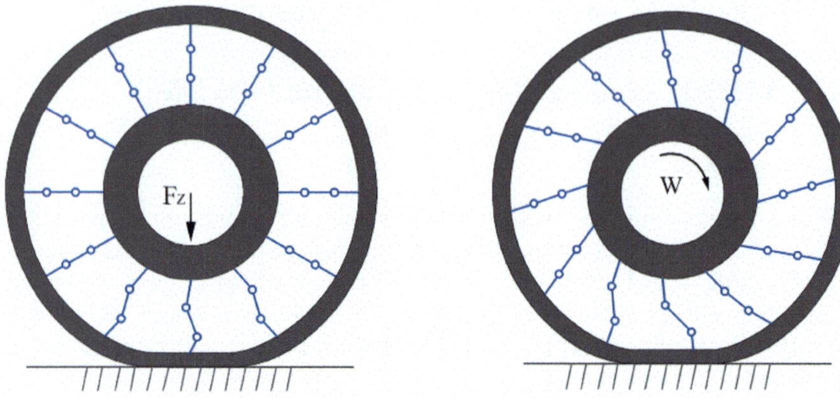

(a) The static state (b) The rolling state

Fig. 2.11 Diagram of static and rolling state of the wheel, W is the vertical load on the wheel, T_t is the driving torque

vehicle starts, the torque of the axle is transmitted to the wheel hub, and the wheel hub rotates and drives the hinge assembly to pull the elastic wheel for rotation, thus making the vehicle run, as shown in Fig. 2.11b. When the wheel is rolling, each hinge assembly will successively undergo a cyclic process of tension, free bending and then tension.

2.3.2 Rolling Mechanism of NMESW

1. The mechanics model of driving

In the driving condition, the driving torque T_t is transferred to the elastic wheel of NMESW through the hinge assembly, and the elastic wheel deforms at the same time. Under steady driving condition, the hinge assembly in the touch-ground area of the wheel are bent and do not bear the tension, while the other hinge assembly are in the tensed state to bear the tension. Moreover, the hub and the elastic wheel have an angle difference δ, namely, the rotation lag angle. The deformation and force of the wheel hinge assembly in the driving condition are shown in Fig. 2.12. In the figure, T_t is the driving torque acting on the wheel, r is the wheel radius, F_t is the tangential reaction force of the ground to the driving wheel.

According to the force exerted by the wheel in the driving condition, the reaction force on the ground to the driving wheel, that is the driving force, is F_t, where F_t can be expressed as

$$F_t = \frac{T_t}{r} \tag{2.1}$$

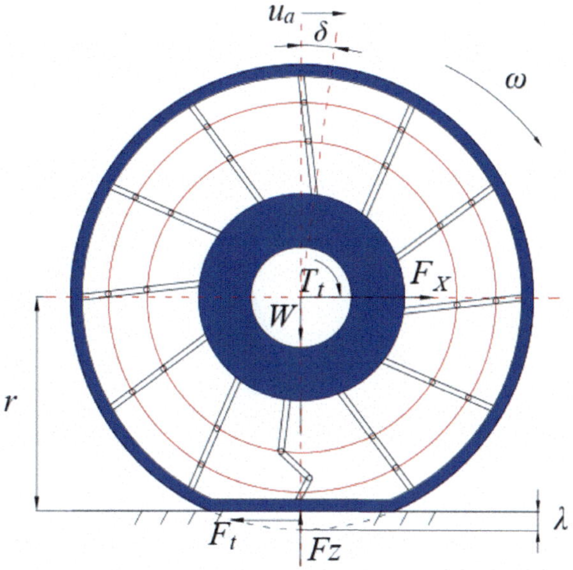

Fig. 2.12 The model of driving force of NMESW, u_a is vehicle speed, ω is wheel speed, λ is the sinking displacement

When calculating the driving force, r is the rolling radius of the wheel, which can be converted according to the relationship between the number of wheel turns and the actual wheel rolling distance, that is

$$r = \frac{S}{2\pi n_W} \tag{2.2}$$

where, n_W is the number of wheel turns, S is the wheel rolling distance when the wheel rolls in n_W turns.

Under the steady driving condition, the force analysis of the elastic wheel and hub of NMESW is shown in Fig. 2.13.

The equilibrium equation of the hub is

$$\begin{cases} \sum_{i=1}^{12} F_{Hir} + \sum_{i=1}^{12} F_{Hi\theta} + W = 0 \\ \sum_{i=1}^{12} F_{Hir} + \sum_{i=1}^{12} F_{Hi\theta} + F_X = 0 \\ \sum_{i=1}^{12} F_{Hi\theta} \times R_H + T_t = 0 \end{cases} \tag{2.3}$$

namely

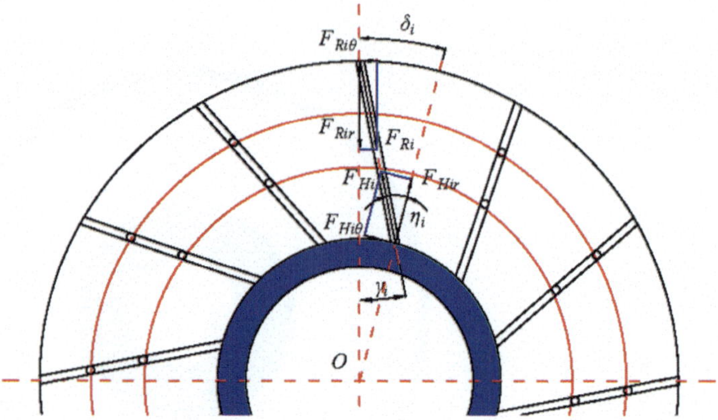

Fig. 2.13 The mechanical model of wheel hinge assembly under steady driving condition

$$\begin{cases} \sum_{i=1}^{12} F_{Hi} \cos \eta_i + \sum_{i=1}^{12} F_{Hi} \sin \eta_i + W = 0 \\ \sum_{i=1}^{12} F_{Hi} \cos \eta_i + \sum_{i=1}^{12} F_{Hi} \sin \eta_i + F_X = 0 \\ \sum_{i=1}^{12} F_{Hi} \sin \eta_i \times R_H + T_t = 0 \end{cases} \tag{2.4}$$

The equilibrium equation of the elastic wheel is

$$\begin{cases} \sum_{i=1}^{12} F_{Rir} + \sum_{i=1}^{12} F_{Ri\theta} + F_z = 0 \\ \sum_{i=1}^{12} F_{Rir} + \sum_{i=1}^{12} F_{Ri\theta} + F_t = 0 \\ \sum_{i=1}^{12} F_{Ri\theta} \times R_R + M_f = 0 \end{cases} \tag{2.5}$$

namely

$$\begin{cases} \sum_{i=1}^{12} F_{Ri} \cos \gamma_i + \sum_{i=1}^{12} F_{Ri} \sin \gamma_i + F_z = 0 \\ \sum_{i=1}^{12} F_{Ri} \cos \gamma_i + \sum_{i=1}^{12} F_{Ri} \sin \gamma_i + F_t = 0 \\ \sum_{i=1}^{12} F_{Ri} \sin \gamma_i \times R_R + M_f = 0 \end{cases} \tag{2.6}$$

where, F_X is the thrust force of the axle to the wheel, i is the hinge assembly number, F_{Hi} is the tensile force of hinge assembly i on hub, $F_{Hi\theta}$ is the tangential component force of the tensile force of hinge assembly i on the hub, F_{Hir} is the normal component force of the tensile force of hinge assembly i on the hub, F_{Ri} is the tensile force of hinge assembly i on the elastic wheel, $F_{Ri\theta}$ is the tangential component force of the tensile force of hinge assembly i on the elastic wheel, F_{Rir} is the normal component force of the tensile force of hinge assembly i on the elastic wheel, R_H is the radius of the hub, R_R is the radius of the elastic wheel, M_f is the drag moment on the wheel from the ground.

2. The mechanics model of braking

In the braking condition, the braking torque T_u is transferred to the elastic wheel of NMESW through the hinge assembly. The deformation and force of the wheel hinge assembly in the braking condition are shown in Fig. 2.14. In the figure, T_u is the braking torque in the brake, F_{Xb} is the ground braking force, W is the wheel vertical load, F_X is the thrust force of the axle to the wheel, F_z is the normal reaction force of the ground to the wheel.

In the analysis of the wheel braking force, the rolling resistance couple moment, the inertia force and the inertia couple moment during deceleration are ignored. From the torque balance relation, we can know

$$F_{Xb} = \frac{T_u}{r} \tag{2.7}$$

With the same method, the force analysis of the elastic wheel and hub of NMESW is carried out under the steady braking condition, as shown in Fig. 2.15.

The equilibrium equations of the hub and the elastic wheel are written as

$$\begin{cases} \sum_{i=1}^{12} F_{Hi} \cos \eta_i + \sum_{i=1}^{12} F_{Hi} \sin \eta_i + W = 0 \\ \sum_{i=1}^{12} F_{Hi} \cos \eta_i + \sum_{i=1}^{12} F_{Hi} \sin \eta_i + F_X = 0 \\ \sum_{i=1}^{12} F_{Hi} \sin \eta_i \times R_H + T_t = 0 \end{cases} \tag{2.8}$$

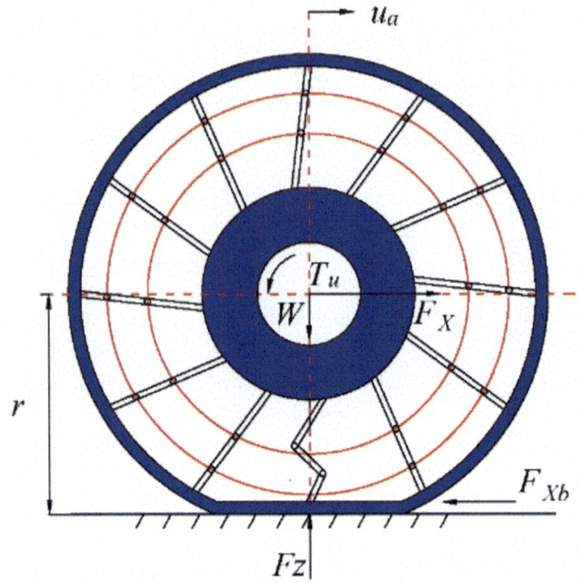

Fig. 2.14 The model of braking force of NMESW

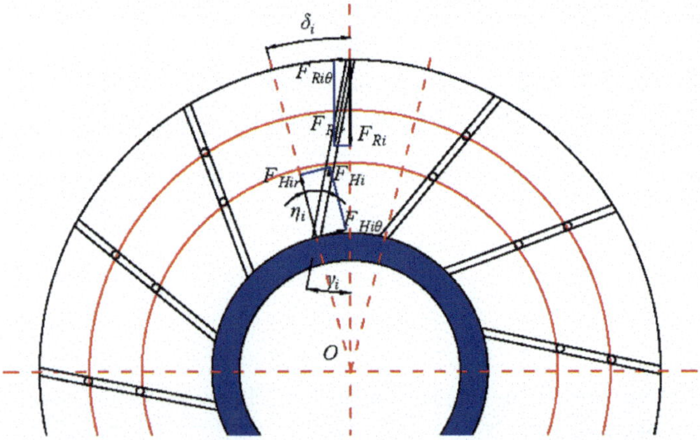

Fig. 2.15 The mechanical model of wheel hinge assembly under braking condition

$$\begin{cases} \sum_{i=1}^{12} F_{Ri} \cos \gamma_i + \sum_{i=1}^{12} F_{Ri} \sin \gamma_i + F_z = 0 \\ \sum_{i=1}^{12} F_{Ri} \cos \gamma_i + \sum_{i=1}^{12} F_{Ri} \sin \gamma_i + F_t = 0 \\ \sum_{i=1}^{12} F_{Ri} \sin \gamma_i \times R_R + M_f = 0 \end{cases} \tag{2.9}$$

3. The model of rolling resistance

When the wheel rolls, similar to a pneumatic tire, the contact area between the wheel and the ground will produce normal and tangential interactions. When the wheel rolls on a hard road, the deformation of the wheel is the main deformation. Due to the internal friction loss, elastic hysteresis loss of the rubber tread of the wheel, and the friction loss between the tread and the touch-ground area, it is shown as the resistance that prevents the wheel rolling, that is the rolling resistance of the wheel.

On a good hard road, the wheel deformation and force considering the wheel rolling resistance are shown in Fig. 2.16.

The equilibrium equations for the wheel are

$$F_{P1} r = T_f \tag{2.10}$$

$$F_{P1} = \frac{T_f}{r} = F_Z \frac{a}{r} = Wf \tag{2.11}$$

where, f is the rolling resistance coefficient.

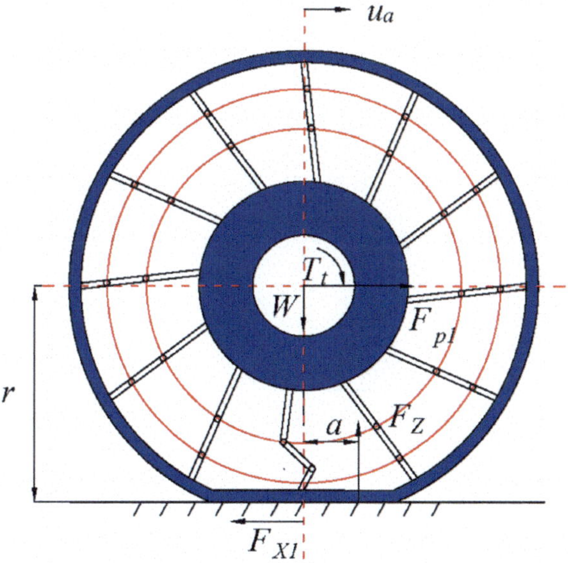

Fig. 2.16 The model of rolling resistance of NMESW

2.3.3 Lateral Characteristic of NMESW

1. The mechanism of lateral characteristic

When NMESW is at rest, the wheel contacts the road under the vertical load F_z, and the center line of the contact patch is in the plane of the wheel. If the lateral force F_y is applied to the wheel on this basis the lateral deformation will occur under the lateral force because of the elasticity of the wheel. Then the center line of the contact patch will have a certain lateral deviation Δh and will no longer be coplanar with the wheel plane as shown in Fig. 2.17a. When NMESW starts to roll forward, the wheel does not roll in the wheel plane, but moves in a straight line along the center line of the contact patch. In this case, the center line of the contact patch is no longer parallel to the wheel plane, and an angle is generated between the center line of the contact patch and the wheel plane, as shown in Fig. 2.17b, and this angle is the lateral angle.

2. The lateral statics characteristics

The lateral static stiffness characteristic curve [9] of NMESW obtained by the test is shown in Fig. 2.18. Figure 2.18a is the curve of lateral deformation and lateral force of NMESW; Fig. 2.18b shows the curve of lateral static stiffness and lateral force of NMESW.

It can be seen from Fig. 2.18a that the lateral deformation of NMESW increases approximately linearly with the increase of the lateral force. Due to the limitation of test conditions, the relevant data of wheel sideslip could not be measured.

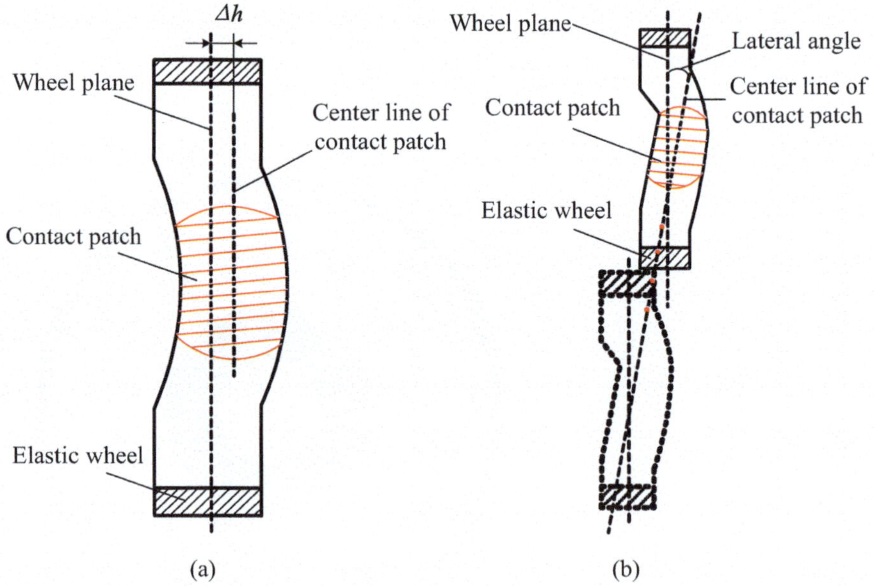

Fig. 2.17 The mechanism of lateral phenomenon of NMESW

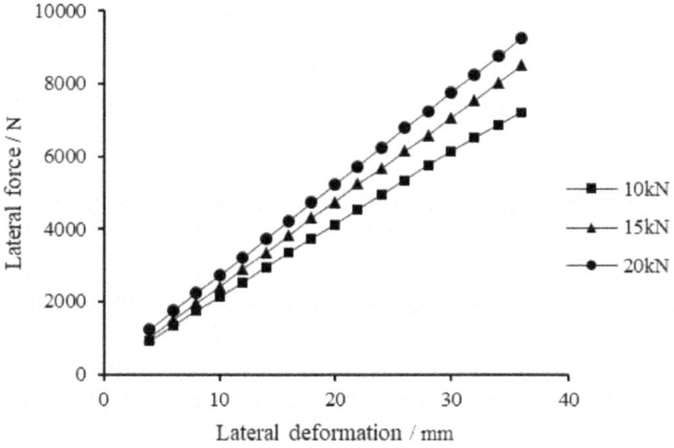

(a) The curve of lateral deformation and lateral force of NMESW

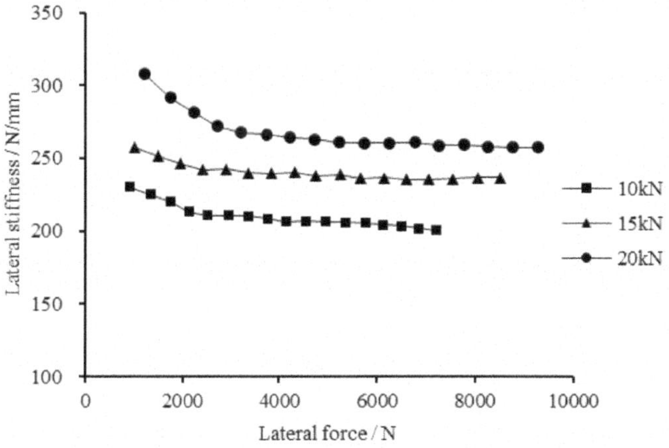

(b) The curve of lateral static stiffness and lateral force of NMESW

Fig. 2.18 The lateral static stiffness characteristic curve of NMESW under different vertical loads

It can be seen from Fig. 2.18b that the lateral static stiffness of the wheel changes nonlinearly with the increase of the lateral force. When the lateral force is small, the lateral static stiffness of the wheel decreases with the increase of the lateral force. Then, with the lateral force gradually increasing, the lateral static stiffness of the wheel remains unchanged. It can also be found from Fig. 2.18b that the vertical load of NMESW has a certain influence on its lateral static stiffness, that is, with the increase of the vertical load, the lateral stiffness of NMESW increases. Compared with the vertical load of 15 kN, the lateral static stiffness of the wheel decreases by 5.6% when the vertical load is 10 kN, and increases by 4% when the vertical load is 20 kN.

3. The comparative analysis of cornering characteristics of pneumatic and non-pneumatic tires

The self-developed tire dynamic mechanical characteristics test bench is used to carry out the steady-state cornering test on the physical prototype of NMESW and the ordinary pneumatic tire. The cornering characteristics of the two tires are compared and analyzed through the obtained lateral test data, as shown in Fig. 2.19. Under the same vertical load (15kN), the variation trend of cornering characteristics of the two kind tires is basically similar, but there are some differences in the evaluation indexes of cornering characteristic. Under the same test conditions, the peak value of lateral stiffness, aligning stiffness and aligning moment of NMESW is larger than that of ordinary pneumatic tire, and the peak value of lateral force is basically the same. When both NMESW and the ordinary pneumatic tire can meet the matching requirements of the whole vehicle, NMESW can make the whole vehicle have better handling stability.

2.3.4 *Camber and Cornering Characteristics of NMESW*

1. The camber characteristics of NMESW

The camber characteristics of the tire have an important influence on the wheel steering and reasonable selection of positioning parameters [10–13]. At the same time, the camber angle will produce a pair of lateral forces in opposite directions on the left and right wheels of the vehicle, as shown in Fig. 2.20, which will directly affect the lateral stability, power performance and abnormal wear of the tread of the vehicle, especially in emergency situations such as high-speed cornering, obstacle avoidance or lane changing.

According to the analysis of the NMESW structure, the elastic wheel is the main elastic support component. Therefore, it is assumed that the elasticity of the elastic wheel is concentrated on the elastic bristles of the tread of the elastic wheel, which is simplified into the brush tire model [14, 15] for the theoretical model of NMESW.

When the width of the elastic wheel is considered, the longitudinal deformation of the wheel is different along the width of the patch. In this case, the longitudinal force on the wheel will produce an aligning torque to the patch center. For the convenience of analysis, the elastic wheel is regarded as composed of multiple discs of equal thickness in the direction of tire width, and the discs roll forward at the same angular velocity ω. At this time, the rolling radius $R(y)$ of each disc can be expressed as

$$R(y) = R_0 + y \sin \gamma \tag{2.12}$$

where R_0 is the rolling radius of the wheel without camber.

The wheel patch half length is

$$a^2(y) = 2R(y) \cdot \Delta z(y) = 2(R_0 + y \sin \gamma)(\Delta z_0 - y \sin \gamma) \tag{2.13}$$

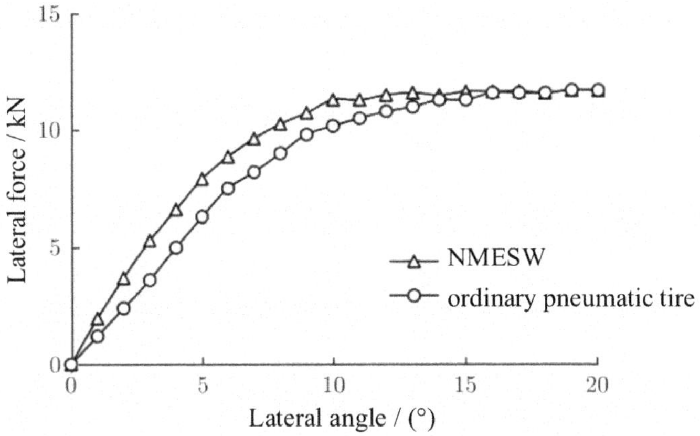

(a) The relation curve between lateral force and slip angle

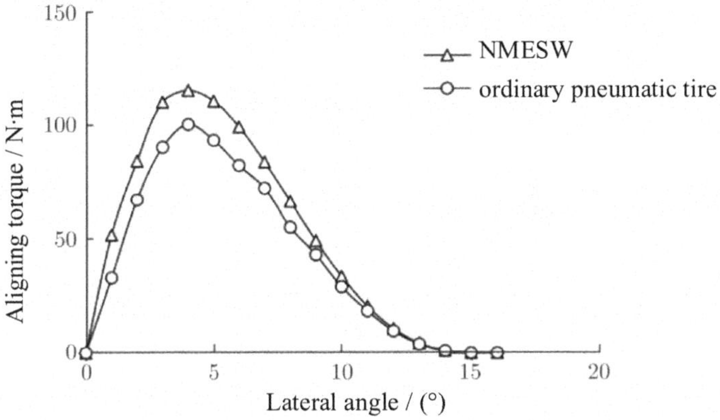

(b) The relation curve between the aligning torque and the slip angle

Fig. 2.19 The comparison curve of lateral characteristics

Fig. 2.20 The schematic diagram of tire camber

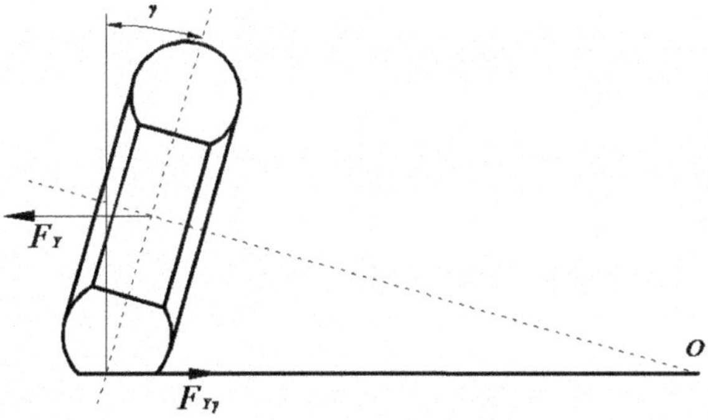

where Δz_0 is the radial sinking displacement of the wheel without camber.

The longitudinal slip rate $S_x(y)$ is defined as

$$S_x(y) = \frac{R_0\omega - R(y)\omega}{R(y)\omega} = \frac{-y\sin\gamma}{R_0 + y\sin\gamma} \tag{2.14}$$

Then, the longitudinal force of each disc in the patch area can be expressed as

$$dF_x = \frac{k_{wx}\cdot a^2(y)\cdot S_x(y)}{b/\cos\gamma}dy = \frac{-k_{wx}\cdot y\cdot a^2(y)\cdot \sin 2\gamma}{2b(R_0 + y\sin\gamma)}dy \tag{2.15}$$

where, k_{wx} is the longitudinal distribution stiffness of the tread, b is the half-width of wheel patch without camber.

The wheel aligning torque M_z can be expressed as

$$M_z = \int_{\frac{-b}{\cos\gamma}}^{\frac{b}{\cos\gamma}} dM_z = \int_{\frac{-b}{\cos\gamma}}^{\frac{b}{\cos\gamma}} y\cdot dF_x = -\frac{4}{3}b^2\cdot k_{wx}\cdot\Delta z_0\frac{\tan\gamma}{\cos\gamma} \tag{2.16}$$

The tangential force of a disc is

$$dF_y = \frac{2}{3}\cdot\frac{k_{wy}}{2b/\cos\gamma}\cdot\frac{a^3(y)}{R}\tan\gamma\cdot dy \tag{2.17}$$

Then the total lateral force with wheel camber is

$$\begin{aligned} F_y &= \int_{\frac{-b}{\cos\gamma}}^{\frac{b}{\cos\gamma}} dF_y \\ &= \frac{k_{wy}\cdot\sin\gamma}{3bR}\cdot\int_{\frac{-b}{\cos\gamma}}^{\frac{b}{\cos\gamma}} a^3(y)\cdot dy \\ &= \frac{4k_{wy}(R_0 - \Delta z_0)}{3R}\cdot\sin\gamma\cdot\tan\gamma\left(b^2\tan^2\gamma - R_0\Delta z_0\right) \end{aligned} \tag{2.18}$$

The wheel camber leads to different vertical deformation in the direction of patch width so that the resultant force of vertical load on the wheel deviates from the wheel center surface. Assuming that the deviation distance is δ_y, it can be obtained

$$\delta_y = \frac{\int_{\frac{-b}{\cos\gamma}}^{\frac{b}{\cos\gamma}} \Delta f_z(y)y \cdot dy}{\int_{\frac{-b}{\cos\gamma}}^{\frac{b}{\cos\gamma}} \Delta f_z(y) \cdot dy} \tag{2.19}$$

$$\Delta f_z(y) = \int_{-a(y)}^{a(y)} \Delta z(y)\left[1 - \frac{x^2}{a^2(y)}\right] K_z(\gamma) \cdot dx \tag{2.20}$$

where $K_z(\gamma)$ is the camber vertical stiffness. So it can get

$$\delta_y = \frac{b^2}{3\Delta z_0} \cdot \frac{\sin\gamma}{\cos^2\gamma} \tag{2.21}$$

Then the turning torque M_x can be expressed as

$$M_x = F_z \cdot \delta_y = \frac{F_z b^2 \sin\gamma}{3\Delta z_0 \cos^2\gamma} \tag{2.22}$$

2. The influence of camber angle on the cornering characteristics

Based on the tire dynamic mechanical characteristics test bench described above, the bench test of NMESW under the combined condition of camber and sideslip is carried out, and the test data are analyzed. Figure 2.21 shows the test results of wheel lateral force changing with slip angle under different camber angles when the load is 12 kN. As can be seen from the figure, when the wheel's camber angle is 0°, the lateral force curve has obvious symmetry, but under the combined action of camber and sideslip, the lateral force produces a certain deviation. When the slip angle is 8°–12°, the lateral force gradually becomes saturated. When the camber angle is 12°, the saturation velocity of wheel slip force at a positive lateral angle is greater than that at a negative slip angle. When the camber angle is −12°, the saturation velocity of wheel lateral force at a positive lateral angle is smaller than that at a negative lateral angle. Therefore, it can be inferred that when the plus-minus signs of camber angle and slip angle are the same, the camber angle will promote the generation of lateral force, so that the wheel is more likely to produce full slip. When the plus-minus signs of camber angle and slip angle are different, the camber angle will inhibit the generation of lateral force, so that the wheel is not easy to produce full slip.

Figure 2.22 shows the relationship between aligning torque and slip angle at different camber angles. By observing the data in the figure, it can be seen that when the lateral angle is large, the aligning torque of the wheel changes sign. When only the camber angle exists, the camber will also produce an aligning torque. The figure also shows that the aligning torque reaches its peak when the slip angle is between 4° and 6°. The influence of the camber angle on the aligning torque also has directivity. When the camber angle and the slip angle have the same plus-minus sign, the camber angle will reduce the aligning torque, while when the sign is different, the aligning torque will increase.

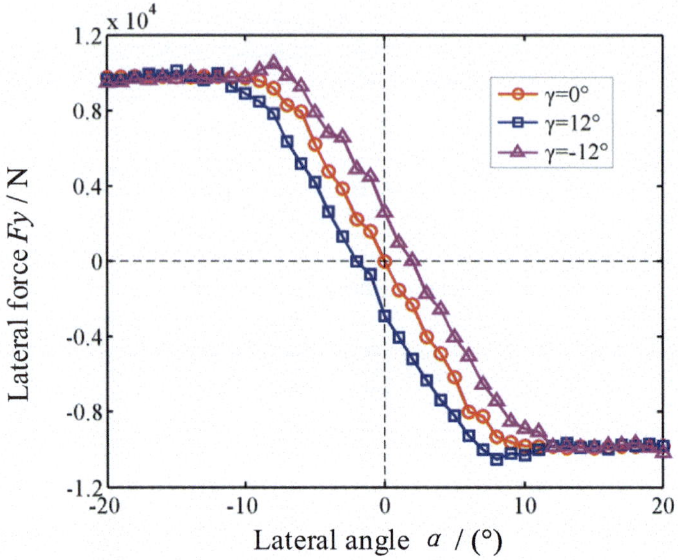

Fig. 2.21 The results of lateral force test under the camber and sideslip conditions

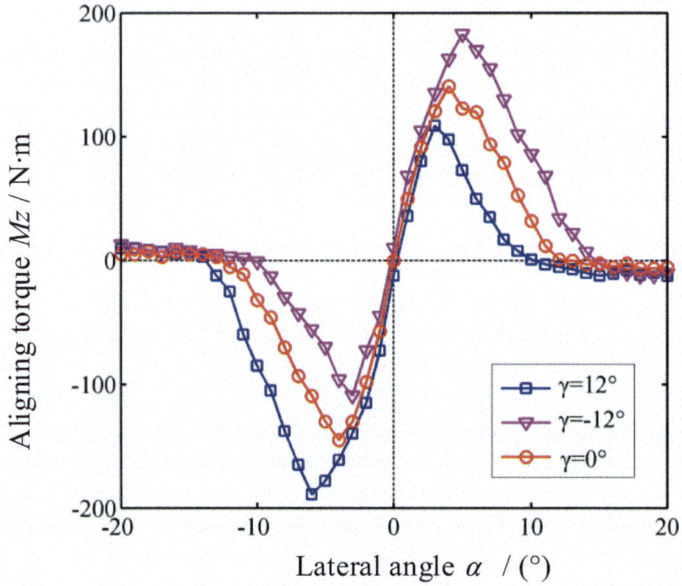

Fig. 2.22 The theory and test results of aligning torque under the camber and sideslip conditions

References

1. Zhao YQ, Zang LG, Chen YQ, et al. Non-pneumatic mechanical elastic wheel natural dynamic characteristics and influencing factors. J Central South Univ. 2015;22(5):1707–15.
2. Wang Q, Zhao YQ, Du XB, et al. Equivalent stiffness and dynamic response of new mechanical elastic wheel. J Vibro Eng. 2016;18(1):431–45.
3. Du XB, Zhao YQ, Wang Q, et al. Numerical analysis of the dynamic interaction between a non-pneumatic mechanical elastic wheel and soil containing an obstacle. Proceed Instit Mech Eng D J Autom Eng. 2017;231(6):731–42.
4. Yu ZS. Vehicle theory. 6th ed. Beijing: China Machine Press; 2018.
5. Wang D, He L, et al. Parameters identification of magic formula of vehicle tire using genetic algorithm. Bus Coach Technol Res. 2015;37(02):1–4.
6. Fu HX, Zhao YQ, Lin F, et al. Influences of bead structure parameters on contact pressure distribution of mechanical elastic wheel. Trans Chin Soc Agricult Eng. 2015;31(17):57–64.
7. Zang LG, Zhao YQ, Li B, et al. Mechanical elastic wheel improving road holding and wear resistance of tire. Trans Chin Soc Agricult Eng. 2014;30(12):56–63.
8. Wang Q, Zhao YQ, Fu HX, et al. Influences of loading and radial stiffness on mechanical elastic wheel enveloping characteristics. Trans Chin Soc Agricult Eng. 2016;32(13):36–42.
9. Fu HX, Zhao YQ, Du XB, et al. Analysis on influencing factors of lateral stiffness of mechanical elastic wheel. J Shanghai Jiaotong Univ. 2017;51(7):863–9.
10. Guo KH, Wang YM, Liu YB, et al. The semi-empirical model for tire lateral properties. Autom Eng. 1986;2:44–54.
11. Pacejka HB, Bakker E. The magic formula tyre model. Vehic Syst Dyn. 1993;21:1–18.
12. Pacejka HB, Besselink IJM. Magic formula tyre model with transient properties. Vehic Syst Dyn. 1997;27:234–49.
13. Oosten IJM, Bakker E. Determination of magic tyre model parameters. Vehic Syst Dyn. 1992;21:19–29.
14. Du XB, Zhao YQ, Lin F, et al. Brush theory modeling of pure camber performance of mechanical elastic wheel. J Harbin Eng Univ. 2018;39(3):569–74.
15. Du XB, Zhao YQ, Xiao Z, et al. Investigation on the camber and cornering characteristics of a catenary non-pneumatic wheel with brush model. J Shanghai Jiaotong Univ. 2018;52(3):305–11.

Chapter 3
Vehicle Stability Control and Estimation of Road Friction Coefficient Based on Vehicle Longitudinal Dynamics

3.1 Review on Vehicle Longitudinal System Dynamics

3.1.1 Significance of Research on Dynamic Characteristics of Vehicle Longitudinal System

The vehicle is a kind of transportation with high running speed, nowadays with the continuous improvement of modern manufacturing technology and road level, the high-speed performance of vehicles has been developed as never before. However, due to the continuous changes in road and traffic conditions, it is often necessary to change the driving speed and even stop instantly. This requires that the vehicle driving at any speed can reach the target speed or stop in the shortest time through longitudinal motion control, and maintain stability while not causing harm to the surrounding environment and the driver [1]. This mainly involves the dynamic characteristics of acceleration and deceleration, which correspond to the dynamic performance and braking performance of the vehicle respectively.

Vehicle dynamic performance refers to the average driving speed that can be achieved when a vehicle is running in a straight line on a good road, which is determined by the longitudinal external force that the vehicle is subjected to. Vehicles are efficient means of transportation, and the efficiency of transportation depends largely on the vehicle dynamic performance. Therefore, vehicle's dynamic performance is one of the most basic and important performances of various vehicles performances.

Vehicle braking performance refers to the ability to artificially force the vehicle to decelerate in a short distance until it stops, maintaining a stable driving direction, and maintaining a certain speed when going down a long slope. Braking is one of the main performances of a vehicle, which is directly related to traffic safety. Major traffic accidents are often related to factors such as too long braking distance, severe sideslip or loss of direction during braking, poor braking stability on long downhill slopes, etc. Therefore, improving the braking performance of vehicles is

Y. Zhao and F. Lin, *Vehicle System Dynamics*,
https://doi.org/10.1007/978-981-97-2019-4_3

always an important task of vehicle design and manufacturing departments. Braking performance also has a certain impact on other performances of the vehicle. If the braking effect of the vehicle is good, the safe driving speed and average driving speed of the vehicle may be higher, and the power and fuel economy of the vehicle will be better.

Longitudinal speed control based on vehicle longitudinal system dynamics can realize active collision avoidance of vehicles. The "2018 Global Status Report on Road Safety" [2] released by the World Health Organization pointed out that the number of road traffic deaths in the world continued to rise. In 2018, the number of people who died in road traffic accidents worldwide reached 1.35 million, which increased by 100,000 people compared with the report in 2017. On average, one person loses his life in a traffic accident every 24 s in the world. Therefore, it is of great significance and value to reduce traffic accidents by researching the vehicle longitudinal dynamics control system. In addition, considering that China has 22% of the world's population, but only 7% of the world's cultivated land, with the acceleration of urbanization, the development of highways will be limited by population, energy and land resources in the future, and will face the challenges of environmental pollution and long-term insufficient capital investment at the same time [3]. Therefore, realizing the automatic control of the longitudinal speed of vehicles, improving driving speed on the basis of safety, and making the most efficient use of existing highway resources has great practical significance for the sustainable development of the Chinese vehicle industry and highway transportation [4].

3.1.2 Basic Problems in the Study of Dynamic Characteristics of Vehicle Longitudinal System

Nowadays, researchers and vehicle manufacturers have carried out research on vehicle longitudinal motion control at different technical levels. Common systems relating to vehicle longitudinal control include cruise control, anti-lock braking systems, and traction control systems. Other advanced longitudinal control systems that are becoming research hotspots include collision avoidance systems, adaptive cruise systems, active differential independent wheel torque control, and longitudinal control systems that control platooning in automated highway systems [5]. However, the research topic of vehicle longitudinal system dynamics still needs to do a lot of in-depth and meticulous work, as follows.

(1) Precise control of vehicle engine: dynamic characteristics of fuel injectors, dynamic characteristics of the intake manifold, torque characteristics generated during combustion, rotational dynamic characteristics of the engine crankshaft, etc.;
(2) Automotive aerodynamics: theoretical research and experimental simulation of computational fluid dynamics, and optimal design of body shape and structure layout;

(3) Advanced vehicle tracking and braking control system: multi-sensor information fusion and identification technology, tracking and braking system integrated control technology, longitudinal dynamics fault-tolerant control and fault detection;
(4) Application of advanced scientific and technological means: including the application of the latest basic scientific achievements, and the application and development of the latest software packages.

3.2 Vehicle Stability Control System [6]

3.2.1 Summarize

Vehicle Stability Control (VSC) or Vehicle Dynamics Control (VDC) system is a relatively new active safety concept, which is used to control the lateral and yaw motion of vehicles in emergencies. The concept was first proposed in 1995 and has been extensively studied as an active safety device to improve stability and maneuverability. For example, when "differential braking" is applied, in which the braking force of the left and right tires is different, the yaw moment is generated. The yaw moment slows down the vehicle and affects the lateral, yaw and roll motion of the vehicle to improve the vehicle's motion. Using the more mature ABS hardware can increase the differential braking function with little cost. Therefore, this chassis control system was quickly accepted by the market [6].

Vehicle handling performance is affected by a variety of excitations. In addition to differential braking or traction, steering is also an incentive. Hac et al. discussed the use of electronic chassis system to improve vehicle stability and emergency handling. Small changes in tire force balance between front and rear axles may affect vehicle yaw moment and stability. They also discussed the use of controllable brakes, steering and suspension to influence vehicle yaw dynamics [7]. Matsumoto proposed a braking force distribution control strategy using steering wheel angle feedforward and yaw rate feedback. The target yaw rate is calculated from the steering wheel angle and actual yaw rate, and left–right brake force distribution control provides a corrected yaw moment as required. The vehicle model has 11° of freedom, and the force generated by the tire model is related to the sideslip angle, normal load, longitudinal slip rate, and road friction coefficient. The simulation results show that yaw rate error decreases significantly when braking during lane change [8]. This work became known as vehicle stability control or vehicle dynamics control.

Over the past 10 years, nearly all major automakers have actively investigated the vehicle control concept, many luxury models have offered vehicle stability control as an option, and more automakers plan to use vehicle stability control systems in future models. The input of the control strategy of the General Motors system is vehicle speed, steering wheel angle, yaw rate, brake pressure and lateral acceleration, and the system has been proven to reduce yaw rate rise time, overshoot time and setup time [9].

3.2.2 Basic Principle

The functions of vehicle stability control may vary slightly in different designs, but essentially include the following tasks: First, the desired vehicle movement (steering angle, accelerator pedal position, brake pressure) is inferred from the driver's action, and then the actual vehicle movement is measured with a set of sensors (yaw rate, lateral acceleration, vehicle speed), compared with the desired movement, and under certain stability constraints (vehicle sideslip angle, roll motion, etc.), the control action (wheel brake pressure) is applied to adjust the vehicle movement to achieve the desired movement.

It is very important to control vehicle stability under extreme conditions. In short, when the vehicle slip sideslip angle is large, due to tire force saturation, the permission of the vehicle's steering angle to generate yaw moment is significantly reduced. This fact was first illustrated by the "β-method" proposed by Shibahata et al. [10]. This method assumes a vehicle running in a straight line, that is, $r = 0$. Using the linear 2-DOF vehicle model, the following equation can be obtained:

$$\begin{aligned} F_{yf} + F_{yr} &= m(\dot{v} + ur) \\ F_{yf} \cdot a - F_{yr} \cdot b &= I_z \cdot \dot{r} \end{aligned} \tag{3.1}$$

where, F_{yf} and F_{yr} are the lateral reaction forces of the ground to the front and rear wheels, m is the total mass of the vehicle, I_z is the moment of inertia of vehicle around z axis, $\dot{r}$ is the yaw angular acceleration, u is the component of the centroid velocity on the ox axis, v is the component of the centroid velocity on the oy axis, a and b are the distances from the vehicle centroid to the center lines of the front and rear axles. Lateral tire force is a function of tire slip angle, where front and rear axle sideslip angles are calculated according to the following formula:

$$\begin{aligned} \alpha_f &= \delta_f - \frac{v + ar}{u} \\ \alpha_r &= -\frac{v - br}{u} \end{aligned} \tag{3.2}$$

The core idea of "β-method" is that when the vehicle sideslip angle $\beta \approx \frac{v}{u}$ is large, the yaw moment $aF_{yf} - bF_{yr}$ and total lateral force $F_{yf} + F_{yr}$ are almost unchanged, regardless of the steering angle δ_f, that is, the driver's authority to control the lateral or yaw movement of the vehicle is significantly reduced.

3.2.3 Control Method

The design of vehicle stability control will be discussed in detail in this section. At the most basic level, the goal of vehicle stability control is to improve vehicle yaw and sometimes even roll stability. Roughly speaking, the design principle of the control algorithm for vehicle stability control can be divided into three parts: ① the estimation and tracking of the desired vehicle (yaw) motion; ② estimation and adjustment of vehicle sideslip; ③ Rollover or wheel lifting prevention.

1. Yaw Rate Following

The desired yaw motion of the vehicle is usually calculated based on the angle at which the person turns the steering wheel, the state of the vehicle and, most importantly, the vehicle's forward speed. In some cases, a human driver may swerve to perform an obstacle avoidance maneuver. In such cases, it is important to use differential braking to help achieve the driver's "desired" yaw rate as quickly as possible. As the sideslip increases, the controller needs to check vehicle motion and achieve a good balance between yaw rate following and limiting sideslip angle.

So how do we determine the desired yaw motion of the driver? The steady-state yaw rate corresponding to the current driver steering input is generally considered to be very close to the yaw rate the driver wishes to achieve. If a linear (for example, bicycle) model is used, the steady-state yaw rate can be calculated from a state-space equation or transfer function.

In general, the bicycle model is used to calculate the desired yaw rate, and the transfer function from the front wheel steering angle to the yaw rate is:

$$G_r = \frac{r}{\delta_f} = \frac{\frac{u}{R}}{\delta_f} = \frac{u}{L + K_{us}u^2} \tag{3.3}$$

where, u is the forward speed of the vehicle (m/s), L is wheelbase (m), $K_{us} = \frac{mb}{LC_{\alpha f}} - \frac{ma}{LC_{\alpha r}}$ is the vehicle understeering coefficient (rad/(m/s^2)).

The usual method is to measure the steering angle and then use this speed-related yaw rate gain to determine the desired yaw rate of the vehicle. The simple control algorithm can help reduce the difference between the ideal and actual yaw rate. Typically, deadzone thresholds are implemented. Here are some examples.

$$r_d = \dot{\psi}_d = G_r \delta_f \tag{3.4}$$

If $\left|\dot{\psi} - \dot{\psi}_d\right| \geq \dot{\psi}_{\text{threshold}}$, and

$$\left|\frac{\dot{\psi} - \dot{\psi}_d}{\dot{\psi}_d}\right| \geq \dot{\psi}_{\text{threshold\_ perct}} \tag{3.5}$$

then

$$M_{\text{VSC\_ yaw}} = K_{\text{yaw}}(\dot{\psi}_d - \dot{\psi})$$

otherwise

$$M_{\text{VSC\_ yaw}} = 0$$

Two thresholds (absolute value and percentage) are used, and $M_{\text{VSC\_yaw}}$ is the vehicle stability control moment, resulting from taking into account yaw rate following.

2. Estimation and Adjustment of Sideslip Angle

While yaw rate following is an important consideration for vehicle handling intentions, vehicle stability is equally or more important. In other words, the vehicle's sideslip and roll angle should not be too large. Because of the lack of economical and practical sensors to measure the lateral velocity of a vehicle, it is not a simple task to accurately estimate the vehicle's sideslip angle. In practice, the sideslip angle of a vehicle is usually estimated from the bicycle model along with available measurements such as steering angle, yaw rate, lateral acceleration and forward speed. If we assume that the sideslip angle can be measured or estimated, a simple control law can also be used. A common algorithm uses the idea of a "sliding mode" type, which is equivalent to PD control for this simple problem: if $|\beta| \geq \beta_{\text{threshold}}$, and $\beta \cdot \Delta\beta > 0$,then

$$M_{\text{VSC\_}\beta} = K_{\beta p} \cdot \beta + K_{\beta d} \cdot \Delta\beta \tag{3.6}$$

otherwise

$$M_{\text{VSC\_}\beta} = 0$$

Among them, constraint $\beta \cdot \Delta\beta > 0$ ensures that the control law only opens when the sideslip angle increases. $M_{\text{VSC\_}\beta}$ is the vehicle stability control moment used to adjust the sideslip angle.

3. Rollover or Wheel Lifting Protection

In addition to the yaw and handling performance considerations discussed above, much recent attention has shifted to the possibility of using vehicle stability control systems for rollover prevention. This function can be achieved by calculating the ideal yaw rate as shown in modified Eq. (3.3), for example:

$$\dot{\psi}_d = \text{sat}(G_r\delta_f, \dot{\psi}_{d\_\text{threshold}}) \tag{3.7}$$

where, sat(·) is the saturation function, and its amplitude limit can be calculated according to the static stability coefficient (SSF) of the vehicle. In other words, the desired yaw rate is finite, so that the corresponding lateral acceleration does not exceed a certain level at which a rollover is likely to occur. However, the idea shown above is essentially "static model-based" and therefore susceptible to external disturbances not considered in the model, such as the road slope angle. A feedback based on the use of vehicle response (roll angle, load transfer, or lateral acceleration) algorithms would be more robust and could be implemented without the use or complement of Eq. (3.7). Here is an example.

If

$$|\phi| \geq \phi_{\text{threshold}} \tag{3.8}$$

then

$$M_{\text{VSC}\_\phi} = -K_{ay} \cdot a_y$$

otherwise

$$M_{\text{VSC}\_\phi} = 0$$

where, a_y is the lateral acceleration of vehicle, and M_{VSC_ϕ} is the vehicle stability control moment used to prevent roll.

Then, the entire vehicle stability control action can be "combined" by adding these three individual control actions together.

$$M_{\text{VSC}} = M_{\text{VSC}\_\text{yaw}} + M_{\text{VSC}\_\beta} + M_{\text{VSC}\_\phi} \tag{3.9}$$

Of course, this is not necessarily the best approach. For instance, when the sideslip or roll angle is large, another option is to emphasize stability. We apply the yaw rate following control behavior only if there is no yaw or roll stability problem. In the following example, Eq. (3.9) is used as a combined vehicle stability control algorithm, which clearly shows the tradeoff between (sometimes) conflicting objectives.

3.2.4 Dynamic Model and Simulation

1. Linear Vehicle Model

Firstly, a vehicle model suitable for vehicle stability control simulation is demonstrated. Obviously, if the vehicle roll motion needs to be simulated (such as studying the effect of a vehicle stability control system to reduce rollover), the minimum vehicle degree of freedom required is 4 (lateral, yaw, roll, and longitudinal). The model can be easily built on the basis of the roll model developed by ME542. Since the longitudinal dynamics of the vehicle are much slower than the rolling dynamics, the simulation model does not need nonlinearity. Instead, assuming a constant vehicle forward speed, the roll model can be updated at each time step. This dynamic separation makes it possible to use the linear simulation command lsim() in MATLAB for vehicle stability control simulation. Another factor to consider is actuator power. It can be assumed that the control signal is a separate brake pressure. However, this obviously adds complexity to the simulation. Assuming that the control signal is the yaw moment M_{VSC}, the following is the vehicle model:

$$\begin{cases} Y_\beta\beta + Y_r r + Y_\phi\phi + Y_\delta\delta_f = m(u\dot{\beta} + ur) + m_R h\dot{p} \\ N_\beta\beta + N_r r + N_\phi\phi + N_\delta\delta_f + M_{\text{VSC}} = I_z\dot{r} + I_{xz}\dot{p} \\ L_\phi\phi + L_p p = m_R h(u\dot{\beta} + ur) + I_x\dot{p} + I_{xz}\dot{r} \end{cases} \tag{3.10}$$

$$-\left|\frac{M_{\text{VSC}}}{T/2}\right| = m\dot{u} \tag{3.11}$$

where, the derivative of vehicle stability is defined as:

$$\begin{cases} Y_\beta \equiv -(C_{\alpha f} + C_{\alpha r}),\ Y_r \equiv \dfrac{bC_{\alpha r} - aC_{\alpha f}}{u} \\ Y_\phi \equiv C_{\alpha r}\dfrac{\partial\delta_r}{\partial\phi} + C_{\gamma f}\dfrac{\partial\gamma_f}{\partial\phi},\ Y_\delta \equiv C_{\alpha f} \\ N_\beta \equiv bC_{\alpha r} - aC_{\alpha f},\ N_r \equiv -\dfrac{a^2C_{\alpha f} + b^2C_{\alpha r}}{u} \\ N_\phi \equiv aC_{\gamma f}\dfrac{\partial\gamma_f}{\partial\phi} - bC_{\alpha r}\dfrac{\partial\delta_r}{\partial\phi},\ N_\delta \equiv aC_{\alpha f} \end{cases} \tag{3.12}$$

The control signal is assumed to be generated by an electronic braking system, so whatever the signal of M_{VSC} is, it always slows the vehicle down. The roll part of the vehicle dynamics [Eq. (3.10)] is not a standard state-space form and therefore requires minor modifications. If we consider the front wheel steering angle as a disturbance input and M_{VSC} as a control signal, Eq. (3.10) can be changed into:

$$\begin{bmatrix} mu & 0 & m_R h & 0 \\ 0 & I_z & I_{xz} & 0 \\ m_R hu & I_{xz} & I_x & 0 \\ 0 & 0 & 0 & 1 \end{bmatrix}\begin{bmatrix} \dot{\beta} \\ \dot{r} \\ \dot{p} \\ \dot{\phi} \end{bmatrix} + \begin{bmatrix} -Y_\beta & mu - Y_r & 0 & -Y_\phi \\ -N_\beta & -N_r & 0 & -N_\phi \\ 0 & m_R hu & -L_p & -L_\phi \\ 0 & 0 & -1 & 0 \end{bmatrix}\begin{bmatrix} \beta \\ r \\ p \\ \phi \end{bmatrix}$$

$$= \begin{bmatrix} 0 \\ 1 \\ 0 \\ 0 \end{bmatrix} M_{\text{VSC}} + \begin{bmatrix} Y_\delta \\ N_\delta \\ 0 \\ 0 \end{bmatrix}\delta_f$$

$$\equiv \boldsymbol{E}\dot{\underline{\boldsymbol{x}}} + \boldsymbol{F}\underline{\boldsymbol{x}} = \boldsymbol{G}M_{\text{VSC}} + \boldsymbol{H}\delta_f \tag{3.13}$$

or

$$\dot{\boldsymbol{x}} = \boldsymbol{A}\boldsymbol{x} + \boldsymbol{B}_u M_{\text{VSC}} + \boldsymbol{B}_w\delta_f \tag{3.14}$$

where, $\boldsymbol{A} = -\boldsymbol{E}^{-1}\boldsymbol{F}, \boldsymbol{B}_u = \boldsymbol{E}^{-1}\boldsymbol{G}, \boldsymbol{B}_w = \boldsymbol{E}^{-1}\boldsymbol{H}$.

To simulate the vehicle model described by Eqs. (3.11) and (3.13), we can assume a simple vehicle stability control algorithm: $M_{\text{VSC}} = \boldsymbol{K}\underline{\boldsymbol{x}} = \begin{bmatrix} k_\beta & 0 & 0 & -k_\phi \end{bmatrix}\underline{\boldsymbol{x}}$, where, Eq. (3.13) defines the state vector $\underline{\boldsymbol{x}}$, and the two control gains k_β and k_ϕ are assumed to be positive. The above control law is a simple proportional control law: when the vehicle sideslip angle β is positive or the vehicle roll angle ϕ is negative, a positive vehicle stability control yaw moment will be required.

As can be seen from Fig. 3.1, when road adhesion is high and a linear tire model is used, the need for vehicle stability control is not obvious. Note that the simulation simulates a violent fish hook steering motion with a yaw rate of up to 20°/s. However, in this case, the vehicle's sideslip angle is quite small and, therefore, it does not appear that vehicle stability control is required. This example illustrates the need to establish a nonlinear simulation model in the simulation and verification of vehicle stability control design.

2. Nonlinear Vehicle Model

Before we present the nonlinear vehicle model, it should be pointed out that since a large tire sideslip angle and large slip rate may occur at the same time, a combined slip tire model is needed. For the combined slip tire model, the tire sideslip angle and slip

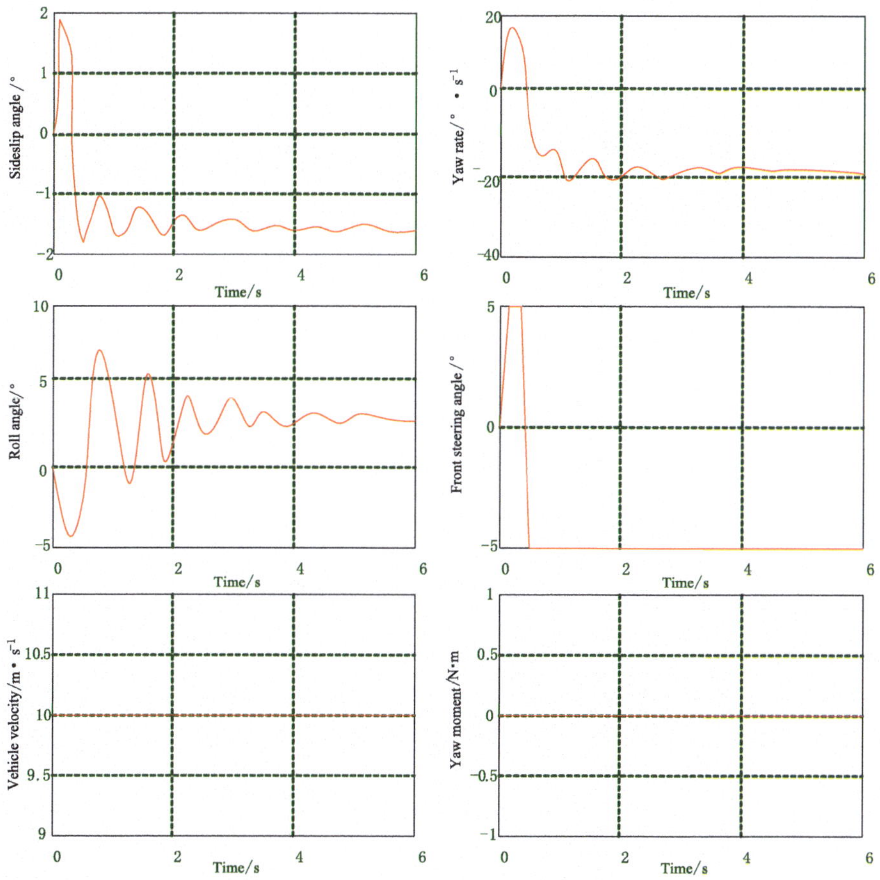

Fig. 3.1 Linear vehicle response (vehicle stability control turned off)

rate need to be tracked to calculate tire forces (lateral and longitudinal). For the sake of simulation, it is assumed that the slip rate of the four tires is directly controlled, or equivalent, the wheel speed is regulated by a servo loop controller. This assumption makes it possible to simulate the vehicle response using a 5-state model. Because wheel power is much faster than vehicle power, this assumption is reasonable. A good servo loop will be able to adjust the wheel slip rate quickly enough around the expected value. The assumption mentioned above has an added benefit: it is consistent with the development of vehicle stability control hardware. Most vehicle stability control is developed based on ABS and traction control systems, so if the basic ABS or TCS does a good job of regulating wheel slip, then the desired wheel slip rate can be set.

On this basis, we assume that the magic tire model with integrated slip correction can simultaneously calculate the transverse and longitudinal forces of the tire, and then the calculated tire forces become the input of the vehicle dynamics equation. When the input of the vehicle roll model is the lateral force of the front and rear axles, the dynamics equation is:

$$\begin{cases} F_{yf} + F_{yr} = m(ur + \dot{v}) + m_R h\dot{p} \\ F_{yf} \cdot a - F_{yr} \cdot b = I_z\dot{r} + I_{xz}\dot{p} + M_{\mathrm{VSC}} \\ L_\phi \phi + L_p p = m_R h(\dot{v} + ur) + I_x\dot{p} + I_{xz}\dot{r} \end{cases} \tag{3.15}$$

or in state space form:

$$\begin{bmatrix} m & 0 & m_R h & 0 \\ 0 & I_z & I_{xz} & 0 \\ m_R h & I_{xz} & I_x & 0 \\ 0 & 0 & 0 & 1 \end{bmatrix} \begin{bmatrix} \dot{v} \\ \dot{r} \\ \dot{p} \\ \dot{\phi} \end{bmatrix} + \begin{bmatrix} 0 & mu & 0 & 0 \\ 0 & 0 & 0 & 0 \\ 0 & m_R hu & -L_p & -L_p \\ 0 & 0 & -1 & 0 \end{bmatrix} \begin{bmatrix} v \\ r \\ p \\ \phi \end{bmatrix}$$

$$= \begin{bmatrix} 0 \\ 1 \\ 0 \\ 0 \end{bmatrix} M_{\mathrm{VSC}} + \begin{bmatrix} 1 & 1 \\ a & -b \\ 0 & 0 \\ 0 & 0 \end{bmatrix} \begin{bmatrix} F_{yf} \\ F_{yr} \end{bmatrix}$$

$$\equiv \boldsymbol{E}\dot{\underline{\boldsymbol{x}}} + \boldsymbol{F}\underline{\boldsymbol{x}} = \boldsymbol{G}M_{\mathrm{VSC}} + \boldsymbol{H}\begin{bmatrix} F_{yf} \\ F_{yr} \end{bmatrix} \tag{3.16}$$

or

$$\dot{\boldsymbol{x}} = \boldsymbol{A}\boldsymbol{x} + \boldsymbol{B}_u M_{\mathrm{VSC}} + \boldsymbol{B}_w \begin{bmatrix} F_{yf} \\ F_{yr} \end{bmatrix} \tag{3.17}$$

where, $\boldsymbol{A} = -\boldsymbol{E}^{-1}\boldsymbol{F}$, $\boldsymbol{B}_u = \boldsymbol{E}^{-1}\boldsymbol{G}$, $\boldsymbol{B}_w = \boldsymbol{E}^{-1}\boldsymbol{H}$. Equation (3.16) indicates that the model has three inputs: lateral forces of front and rear axles, and yaw moments due to longitudinal tire forces. From a servo level perspective, there are five inputs: front axle steering angle (disturbance) and four tire slip rates. Given the steering angle and slip rates, we can calculate the sideslip angles of all four tires and thus the tire forces. Once transverse and longitudinal tire forces are known, three "vehicle dynamics inputs" (M_{VSC},F_{yf},F_{yr}) can be calculated.

The vehicle model described in Eqs. (3.11) and (3.16) are simulated, in which the tire forces are calculated by the combined slip tire magic formula model. In this example, a simple vehicle stability control algorithm, the proportional control law, will be used again. The control action can only be started when the sideslip angle of the vehicle is greater than a certain threshold. Figure 3.2 shows the situation when the vehicle stability control is not turned on. It can be seen that due to the nonlinear tire model used in this simulation, when the yaw rate of the vehicle is much lower than that of the linear vehicle model, a very high vehicle slip angle will occur, and the vehicle will gradually become unstable, the yaw rate slowly increases, and the sideslip angle keeps increasing. The phenomenon is similar to that observed in actual fish hook tests. When the simple proportional algorithm is applied, the vehicle sideslip angle is significantly reduced (Fig. 3.3).

3.3 Estimation of Road Friction Coefficient Based on Longitudinal Dynamics

3.3.1 Slip-Slope Based Method for Estimating Friction Coefficient

The slip-slope method is a method to estimate the road friction coefficient by using the slope of μ-s curve in the linear domain [11]. Here s is the slip rate of the tire. Considering only the longitudinal motion and ignoring the lateral force, μ is defined by the following equation:

$$\mu = \frac{F_x}{F_z} \tag{3.18}$$

where, F_x is the longitudinal force acting on the tire, F_z is the normal load of the tire.

Gustafsson first proposed the slip-slope idea. He designed the Kalman filter to estimate the slope of $\mu - s$ curve in the linear domain. His research shows that when the longitudinal force of the vehicle is sufficient under driving conditions, the estimated slope of $\mu - s$ curve can be used to distinguish the friction levels of different roads. Later, more and more researchers devoted themselves to the slip-slope method to estimate the road friction coefficient: Yi et al. [12] and Hwang and Song [13] also proposed more experimental evidence proving that K_s can be used to distinguish different road conditions in the normal driving process of the vehicle. All of the above methods need to use the speed of the engaged wheel to estimate the reference speed of the vehicle.

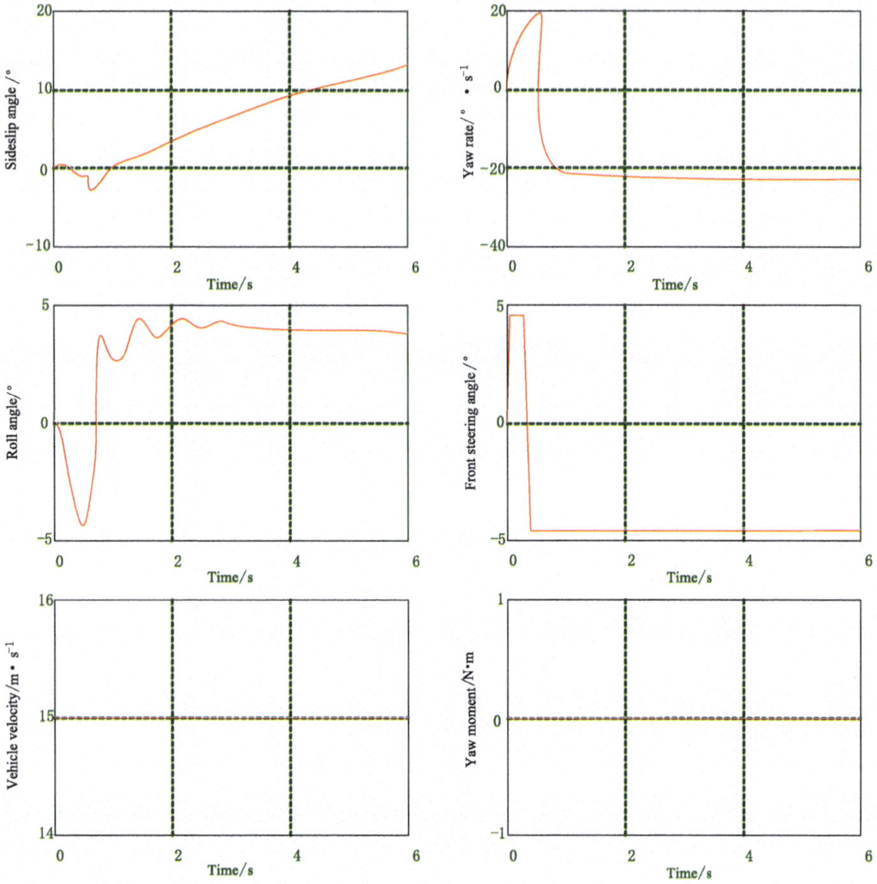

Fig. 3.2 Nonlinear vehicle response (vehicle stability control turned off)

In 2001, Steffen et al. [14] extended the application of slip-slope estimation theory K_s to the braking condition of vehicles. In this thesis, the author turns off the rear brake of the test vehicle so that the rear wheel provides the absolute speed of the reference vehicle and only the front wheel provides the braking force. The front wheel braking force can be measured by the brake pressure sensor. However, it is very expensive to install the brake pressure sensor in the process of real vehicle tests to verify the proposed theory.

All the above slip-slope methods are based on road friction coefficient estimation methods under driving or braking conditions. However, the estimation method combining the two working conditions is rarely seen in the literature. In addition, if virtual experiments can be used to verify the proposed estimation theory, the research cost will be greatly reduced. In this section, the improved slip-slope method is used to estimate the road friction coefficient under driving and braking conditions, and

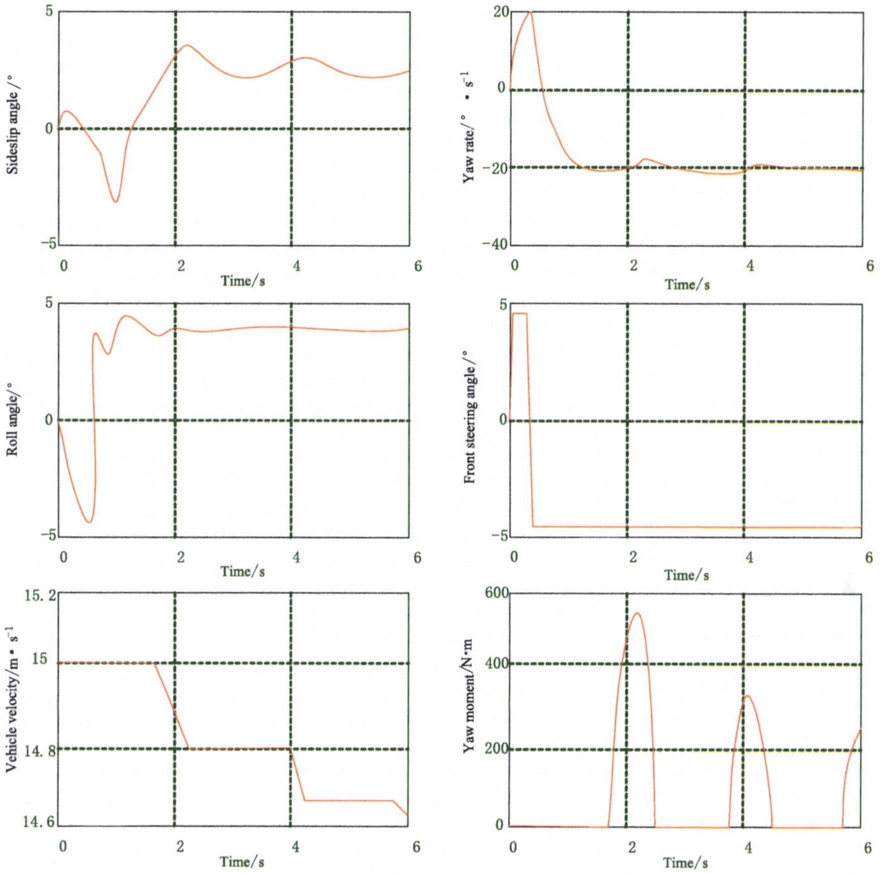

Fig. 3.3 Nonlinear vehicle response (vehicle stability control on front axle turned on)

the proposed estimation method is verified by the virtual test in the ADAMS/Car environment.

3.3.2 Longitudinal Force Estimation

The concept of sliding mode originated in the Soviet Union in the 1960s. At that time, it was studied to introduce discontinuous control rules into dynamic systems, namely to realize variable structure control. It is found that the state of the system can be achieved and maintained in a predefined hyperplane by adopting control rules that switch according to judgment. At this point, the system presents a dimensional reduction motion, called sliding mode motion, which is robust to external interference. It is the inherent robustness of the sliding mode that arouses people's interest in the research of sliding mode, and the concept of sliding mode has also been extended to many other applications [15]. One of the most important areas is the sliding mode observer (SMO) design.

The design of sliding mode observers is one of the important directions of nonlinear observer research. Through sliding mode observer, control can be achieved when the state is partially measurable or completely unmeasurable, and it is robust to parameter uncertainty, model error and external disturbance [16]. As for the sliding mode observer, Utkin [17] first adopted discontinuous switching terms in the design of the observer, that is, the observer with nonlinear terms was designed for a linear system. Later, Slotine et al. [18] adopted a similar method to design a sliding mode observer for nonlinear systems, and the observer also included a linear output feedback term. Walcott and Zak [19] adopted a Lyapunov-based approach to construct an observer with discontinuous terms and proved that its state observation errors converge asymptotically if certain assumptions are satisfied. At present, sliding mode observer is increasingly applied in various fields [20–22]. This section applies the sliding mode observer to the estimation of the longitudinal force of vehicle tires [23].

Different from conventional methods, this chapter takes tire longitudinal force as an unknown variable in the vehicle dynamics model and appears in the sliding mode observer, so as to avoid modeling the tire longitudinal force as a state.

Considering the longitudinal dynamics model of the vehicle shown in Fig. 3.4, ignoring air resistance and rolling resistance, the following equation can be obtained:

$$a_x = \dot{u} = 2\sum_{i=f,r} F_{xi}/m \tag{3.19}$$

Since the virtual test environment in this chapter is linear driving braking and driving conditions without lateral movement freedom, it can be considered that the forces on the left and right tires of the vehicle are the same. Where, F_{xi} is the longitudinal force of the front and rear wheels, $i \in \{f, r\}$ represents the front wheel and rear

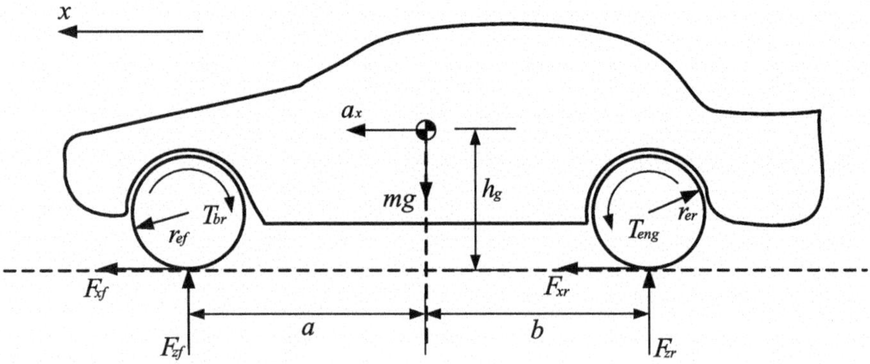

Fig. 3.4 Vehicle longitudinal dynamics model

wheel respectively, m is the mass of the vehicle, a_x is the longitudinal acceleration of the vehicle, and u is the longitudinal velocity of the vehicle.

Considering the tire rotation dynamics model shown in Fig. 3.4, there is:

$$\dot{\omega}_i = \frac{1}{J_\omega}(T_{\text{eng}i} - T_{bri} - r_{ei}F_{xi}) \tag{3.20}$$

where, the tire longitudinal force F_x is positive when driving and negative when braking, $T_{\text{eng}i}$ is the driving torque, J_ω is the moment of inertia of the wheel, r_e is the effective rolling radius of the wheel, and T_{br} is the braking torque.

According to Eqs. (3.19) and (3.20), the state vector $\boldsymbol{x} = [\,u\ \omega_f\ \omega_r\,]^{\text{T}}$, the observation vector $\boldsymbol{z} = \boldsymbol{x}$, the longitudinal force vector of front axis $\boldsymbol{F} = [\,F_{xf}\ F_{xr}\,]^{\text{T}}$, the input vector $\boldsymbol{u} = [\,T_{\text{engr}}\ T_{brf}\ T_{brr}\,]^{\text{T}}$. Here, it is assumed that both driving torque and braking torque can be measured. By combining Eqs. (3.19) and (3.20), it can be obtained:

$$\begin{cases} \dot{\boldsymbol{x}} = \boldsymbol{\Omega F} + \boldsymbol{Bu} \\ \boldsymbol{z} = \boldsymbol{x} \end{cases} \tag{3.21}$$

where, $\boldsymbol{\Omega} = \begin{bmatrix} 2/m & 2/m \\ -r_{ef}/J_\omega & 0 \\ 0 & -r_{er}/J_\omega \end{bmatrix}$, $\boldsymbol{B} = \begin{bmatrix} 0 & 0 & 0 \\ 0 & -1/J_\omega & 0 \\ 1/J_\omega & 0 & -1/J_\omega \end{bmatrix}$.

Let $\boldsymbol{x}$ and $\hat{\boldsymbol{x}}$ represent the actual state value and its estimated value respectively. Let the switching function $\boldsymbol{s} = \tilde{\boldsymbol{x}} = \boldsymbol{x} - \hat{\boldsymbol{x}}$ be the observer error, and construct the following sliding mode observer for $\boldsymbol{x}$:

$$\dot{\hat{\boldsymbol{x}}} = \boldsymbol{Bu} + \boldsymbol{\Lambda}\text{sgn}(\boldsymbol{x} - \hat{\boldsymbol{x}}) \tag{3.22}$$

where, $\boldsymbol{\Lambda}$ represents the observer gain matrix, sgn(·) is a sign function, then

$$\dot{\tilde{\boldsymbol{x}}} = \dot{\boldsymbol{x}} - \dot{\hat{\boldsymbol{x}}} = \boldsymbol{\Omega F} - \boldsymbol{\Lambda}\text{sgn}(\tilde{\boldsymbol{x}}) \tag{3.23}$$

The existence condition of the sliding mode is the prerequisite for the normal operation of the sliding mode observer. If the initial point $\boldsymbol{x}(0)$ of the system is not near $\boldsymbol{s} = 0$, but at any position in the state space, the motion of the system must tend to the switching surface $\boldsymbol{s} = 0$, that is, the reachable condition must be met; otherwise, the system cannot start the sliding mode motion.

The accessibility condition of Lyapunov function is as follows

$$\dot{V}(\boldsymbol{x}) < 0, \quad V(\boldsymbol{x}) = \frac{1}{2}\boldsymbol{s}^2 \tag{3.24}$$

For the sliding mode observer shown in Eq. (3.22):

$$\begin{aligned}\dot{V}(x) &= \tilde{\boldsymbol{x}}^{\mathrm{T}}\dot{\tilde{\boldsymbol{x}}} = \tilde{\boldsymbol{x}}^{\mathrm{T}}(\boldsymbol{\Omega F} - \boldsymbol{\Lambda}\mathrm{sgn}(\tilde{\boldsymbol{x}})) \\ &= \tilde{\boldsymbol{x}}^{\mathrm{T}}\boldsymbol{\Omega F} - \boldsymbol{\Lambda}\left|\tilde{\boldsymbol{x}}^{\mathrm{T}}\right| \le \left|\tilde{\boldsymbol{x}}^{\mathrm{T}}\right|\boldsymbol{\Omega F} - \boldsymbol{\Lambda}\left|\tilde{\boldsymbol{x}}^{\mathrm{T}}\right| \\ &= \left|\tilde{\boldsymbol{x}}^{\mathrm{T}}\right|(\boldsymbol{\Omega F} - \boldsymbol{\Lambda})\end{aligned}$$

Since the output power of the engine is limited, the longitudinal force $\boldsymbol{F}$ of the tire is bounded. Also, it is considered that $\|\boldsymbol{F}\| < \mu$, and μ is a positive constant value. Hence, by choosing a proper positive definite matrix $\boldsymbol{\Lambda}$, such that $\boldsymbol{\Lambda} > \mu\boldsymbol{\Omega}$, we get $\dot{V} < 0$. Thus, it can be proven that the sliding mode exists dynamically with $\tilde{\boldsymbol{x}} = 0$ as the sliding mode surface, so that $\hat{\boldsymbol{x}}$ approaches $\boldsymbol{x}$ within finite time t_0. In addition, from the perspective of statistical average, there are $\dot{\tilde{\boldsymbol{x}}} = 0$, $\forall t > t_0$. And from that we can get

$$\boldsymbol{\Omega F} = \boldsymbol{\Lambda}\mathrm{sgn}(\tilde{\boldsymbol{x}}) \tag{3.25}$$

The estimate can be obtained from Eq. (3.25).

However, the estimate of $\boldsymbol{F}$ obtained above does not have ideal numerical properties. This is because the switching characteristic of $\boldsymbol{\Lambda}\mathrm{sgn}(\tilde{\boldsymbol{x}})$ inevitably brings chattering, causing the estimated $\boldsymbol{F}$ to pass through the surface and bottom of the sliding mode, forming noise interference that covers the truth value, which has an adverse effect on the following friction coefficient estimation. Therefore, in this chapter, a Kalman filtering link is parallel with the sliding mode observation of $\boldsymbol{F}$ to eliminate the influence brought by chattering.

Let $\mathrm{sgn}_{\mathrm{eq}}(\tilde{\boldsymbol{x}})$ be the "equivalent value" obtained through the Kalman filter link, then the estimated tire longitudinal force $\boldsymbol{F}$ can be finally obtained:

$$\hat{\boldsymbol{F}} = \boldsymbol{\Omega}^{+}\boldsymbol{\Lambda}\mathrm{sgn}_{\mathrm{eq}}(\tilde{\boldsymbol{x}}) \tag{3.26}$$

Since $\boldsymbol{\Omega}$ is a non-square matrix, $\boldsymbol{\Omega}^{+}$ in Eq. (3.26) is the generalized inverse of Moore–Penrose. In the process of parallel Kalman filtering, the process noise covariance Q is set as a very small value, and the measurement noise covariance R is adjusted according to the chattering condition.

Figure 3.5 shows an example of a virtual test where the validity of the method proposed in this section is verified by estimating the longitudinal force of the left rear wheel in the process of vehicle linear braking. The braking operation is carried out on the high friction road, the initial speed is 80 km/h, the braking start time is 2 s, the longitudinal deceleration is 0.3 g, the end time is 10 s, and the step length is 0.01 s. In this example, Q is 1×10^{-6} and R is 0.5×10^{-4}.

It can be seen from Fig. 3.5a that when the longitudinal force of the tire is estimated by the sliding mode observer, the estimated initial value is 0 and rapidly approaches the actual value (about 150 N) after a very short time, which shows that the sliding mode observer approaches the sliding surface in a short time and has high robustness to the uncertainty of initial value. However, although the estimated value is in good agreement with the actual value, there is a certain chattering, which can be adjusted by the gain matrix of the sliding mode observer but cannot be completely eliminated.

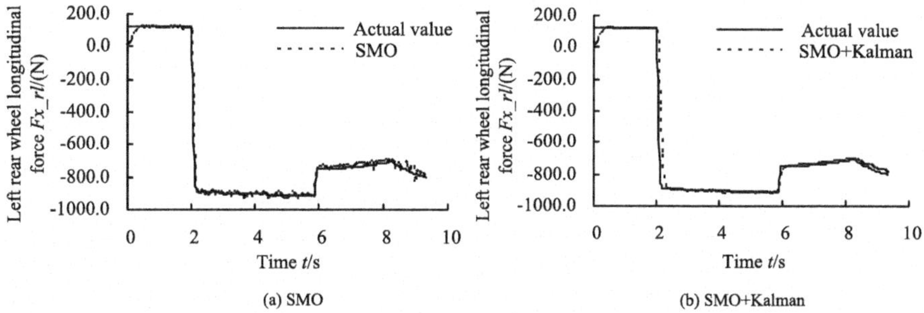

Fig. 3.5 Longitudinal force of rear left wheel estimated by sliding mode observer

Chattering can be effectively eliminated through Kalman filtering, and the filtering value still has a high degree of coincidence with the actual value, as shown in Fig. 3.5b.

3.3.3 Slip Estimation

According to the definition of SAE670e [24], the longitudinal slip rate s of wheels can be solved by the difference value between the peripheral velocity of the driving wheel (or brake wheel) and the absolute speed of the vehicle. The specific equation is as follows:

$$s = \frac{r_e\omega - u}{\max(u, r_e\omega)} \tag{3.27}$$

where, r_e is the effective rolling radius of the wheels, ω is the angular velocity of the tires, and u is the speed of the vehicle. According to the definition of Eq. (3.27), it can be known that when $s = -1$, it is pure holding drag slip; When $s = 0$, it is pure rolling and no slip of tires; When $s = 1$, the tires skid on the extremely low friction road. In the virtual test below, it is assumed that the effective rolling radius r_e of the wheels is known, and the rest is to obtain the angular velocity ω and speed u of the vehicle.

1. Estimation of Slip Rate s in Driving Condition

The estimation of slip rate in the driving condition is relatively simple. In practice, the angular velocity ω of the tire is measured by the wheel speed sensor. The absolute speed of the vehicle, u, is obtained by the angular velocity of the non-driving wheels. It should be noted that the angular velocity measurement signal of the tire often contains high-frequency noise, which is caused by the relatively small rotational inertia of the tire. In the following virtual test, a uniform Gaussian white noise sequence with a certain variance will be added to the tire angular velocity to simulate the actual situation.

2. Estimation of Slip Rate s in Braking Condition

Considering the estimation of the tire-road friction coefficient under braking conditions has two advantages [25].

(1) The system that uses the information under accelerating and braking conditions to estimate the friction coefficient has wider applicability than the system that only uses the information under accelerating conditions.
(2) The order of magnitude of tire longitudinal force under braking conditions is much higher than that under accelerating conditions, which can provide a more accurate estimate for $\mu_{\max}$. Braking maneuvers that are rarely performed use the majority of the maximum friction force (i.e., braking efficiency) and are important for friction estimation system.

During the braking process, because all four wheels brake and slip, there is no non-braking wheel as the absolute speed reference of the vehicle. In practice, the absolute speed can be obtained through the non-contact speed sensor or the GPS-based five wheel device RTK. In order to simplify the study, the speed signal is obtained directly in the braking control of the virtual test below.

3.3.4 Slip-Slope Estimation Method Based on RLS

The two objectives of slip-slope estimation are as follows [11]:

(1) Accurate estimation of K_s can ensure the tracking of the slow change of K_s.
(2) The mutation of K_s can be effectively detected.

1. Time-varying Estimation of Slip-slope

Literature [11] discusses the time-invariant estimation of the friction coefficient under the condition that the road surface condition remains basically unchanged. This off-line estimation method obviously cannot meet the real-time requirements of the vehicle control system.

In the actual situation, we are more concerned about the dangerous working conditions caused by the sudden change in the friction coefficient. For example, in winter, a vehicle suddenly enters a section covered with thin ice from the dry asphalt road section, at this time, the friction condition of the road surface changes abruptly, if the wheel speed is high, it is easy to happen the phenomenon of driving skid or brake lock, resulting in rear-end collisions and other accidents. Therefore, it is of practical significance to study the time-varying estimation of the friction coefficient.

Firstly, a linear recursion model is established. Rewrite $\boldsymbol{\mu}(t) = \boldsymbol{K}_s(t)\boldsymbol{s}(t)$ as:

$$s(t) = \boldsymbol{\mu}(t)\frac{1}{\boldsymbol{K}_s(t)} \tag{3.28}$$

The advantage of the above method is that [11]: the measurement noise covariance matrix of $\boldsymbol{s}$ is much larger than the measurement noise covariance matrix of $\boldsymbol{\mu}$. Compared with the former, we can almost ignore the measurement noise of the latter. Therefore, if written as $\boldsymbol{\mu}(t) = \boldsymbol{K}_s(t)\boldsymbol{s}(t)$, the measurement noise of $\boldsymbol{s}$ is enlarged by $\boldsymbol{K}_s$ times ($\boldsymbol{K}_s > 0$).

There are usually three methods to estimate time-varying parameters [26], which are least mean square (LMS), recursive least squares with forgetting factor (RLS) and Kalman filter. In this chapter, RLS with forgetting factor is used to estimate $\boldsymbol{K}_s$.

For Eq. (3.28), the dynamic system is

$$\boldsymbol{y}(t) = \boldsymbol{\varphi}^{\mathrm{T}}(t)\boldsymbol{\theta}(t) \tag{3.29}$$

where, the system input $\boldsymbol{y}(t)$ is the slip rate of packet content measurement noise, which is assumed to be random Gaussian white noise with variance σ^2 in this chapter. The recursive vector $\boldsymbol{\varphi}(t)$ is $\boldsymbol{\mu}(t)$, and the parameter vector $\boldsymbol{\theta}(t)$ to be estimated is $\frac{1}{\boldsymbol{K}_s(t)}$.

RLS algorithm provides a method to update the unknown parameter vector $\boldsymbol{\theta}(t)$ cyclically. At each sampling time, based on the input and output data contained in the recursive vector $\boldsymbol{\varphi}(t)$ at the previous time, the RLS algorithm updates the unknown parameter vector $\boldsymbol{\theta}(t)$ to minimize the sum of model variance. The steps of the RLS algorithm are as follows [27]:

(1) Measure the output $\boldsymbol{y}(t)$ of the system and calculate the recursive vector $\boldsymbol{\mu}(t)$.
(2) Calculate the estimated residual $\boldsymbol{e}(t)$, which is the difference value between the actual output of the sample system at the current moment and the estimated output including the predicted parameter $\boldsymbol{\theta}(t-1)$ of the previous sample:

$$\boldsymbol{e}(t) = \boldsymbol{y}(t) - \boldsymbol{\varphi}^{\mathrm{T}}(t)\boldsymbol{\theta}(t-1) \tag{3.30}$$

In fact, the residual contains the measurement noise information of the slip rate $\boldsymbol{s}(t)$ and the sliding mode observation deviation information of the longitudinal tire force $\boldsymbol{F}_x$ in $\boldsymbol{\mu}(t)$.

(3) Calculate the update gain $K(t)$ as

$$K(t) = \frac{\boldsymbol{P}(t-1)\boldsymbol{\varphi}(t)}{\lambda + \boldsymbol{\varphi}^{\mathrm{T}}(t)\boldsymbol{P}(t-1)\boldsymbol{\varphi}(t)} \tag{3.31}$$

And calculate the covariance matrix $\boldsymbol{P}(t)$ as

$$\boldsymbol{P}(t) = \frac{1}{\lambda}\left(\boldsymbol{P}(t-1) - \frac{\boldsymbol{P}(t-1)\boldsymbol{\varphi}(t)\boldsymbol{\varphi}^{\mathrm{T}}(t)\boldsymbol{P}(t-1)}{\lambda + \boldsymbol{\varphi}^{\mathrm{T}}(t)\boldsymbol{P}(t-1)\boldsymbol{\varphi}(t)}\right) \tag{3.32}$$

(4) Update the estimated parameter vector $\boldsymbol{\theta}(t)$:

$$\hat{\boldsymbol{\theta}}(t) = \hat{\boldsymbol{\theta}}(t-1) + K(t)\boldsymbol{e}(t) \tag{3.33}$$

where, $\hat{\boldsymbol{\theta}}$ represents the estimated value of $\boldsymbol{\theta}$. λ is defined as the forgetting factor to strengthen the effect of current observed data on parameter estimation, which can effectively reduce the influence of old data unrelated to the model, so as to prevent the oscillation of covariance results. This allows parameter estimation to quickly track changes in the program. The range of λ is $0 < \lambda < 1$, in this chapter, $\lambda = 0.995$. The initial value in the recursion method is chosen as follows: $\hat{\boldsymbol{\theta}}(0) = 0, \boldsymbol{P}(0) = a\boldsymbol{I}$, a is a sufficiently large positive number. In addition, the tire longitudinal force in the recursive vector $\boldsymbol{\mu}(t)$ is estimated in parallel according to the sliding mode observer method proposed above.

2. Application of Change Detection Algorithm

When the slip rate mutates, K_s will also mutate. In this case, for the friction coefficient estimation system with high real-time requirements, the estimation method is particularly important for tracking performance of K_s (especially when road conditions suddenly deteriorate). Therefore, this chapter adopts the CUSUM algorithm [28] to run in parallel with the RLS algorithm. The equation of the CUSUM algorithm is simple and easy to understand and has good robustness to model errors, which is suitable for tracking various mutations [28].

The objective function of the CUSUM algorithm is the residual $e(t)$ of slip rate estimation. When the absolute value of the estimated residual is greater than a given threshold value, the CUSUM algorithm will issue an alarm signal. The equation of the CUSUM algorithm is as follows:

$$\begin{cases} g(t) = \max(g(t-1) + |e(t)| - \gamma, 0) \\ g(0) = 0 \end{cases} \tag{3.34}$$

The input of change detection is the RLS identification error $e(t)$, and the output is the alarm signal $g(t)$. If the output of the CUSUM algorithm is $g(t) > \tau$ (here τ is an adjustable design parameter, which is taken as $\tau = 1 \times 10^{-3}$ in this chapter), the covariance matrix $\boldsymbol{P}(t)$ is immediately multiplied by a gain factor, making $\boldsymbol{P}(t)$ increase rapidly, so as to obtain the good tracking performance of RLS estimator. Thus, the mutation of friction coefficient is rapidly tracked until the absolute value of the estimated residual drops to a certain level and $g(t) = 0$.

γ is defined as the drift parameter, which is used to screen large estimation errors and ignore small estimation errors ($\gamma = 1 \times 10^{-4}$ is taken in this chapter), and the threshold value τ is used to determine when to trigger the alarm signal to activate the gain factor.

Since the improvement of tracking ability comes at the cost of the decrease of anti-interference ability, the selection of gain factor should be decided according to different situations. Fortunately, the time history of K_s mutation is very short, so the CUSUM algorithm is quickly closed after K_s changes gently. In other words, $\boldsymbol{P}(t)$ quickly returns to normal when the error drops to a certain value.

3.3.5 *Virtual Test Verification*

Virtual experiments in ADAMS/Car software are used to verify the effectiveness of the method of road friction coefficient estimation based on the combination of sliding mode observer and recursive least squares. The establishment method of the virtual prototype is the same as that in the previous chapter, which will not be repeated here. The tire subsystem adopts the pac89 tire model inherent in the software.

1. Application of Road Builder

The road surfaces with different friction coefficients in the virtual test are constructed by Road Builder in the ADAMS/Car module. The specific interface of Road Builder is shown in Fig. 3.6. The road generator can define the bank angle and friction coefficient of any three-dimensional point on the pavement in the form of a table. In addition, a roadblock can be defined so that the driving prototype can avoid obstacles by the specified path. Road Builder is a new function that appeared after the ADAMS 2005 version, which provides great convenience for vehicle ride comfort analysis and construction of complex road spectrum. This chapter makes full use of the function that Road Builder can set the friction coefficient at any point on the road.

Figure 3.6 shows the table that defines the characteristics of the pavement coordinate points in detail. Taking the road surface with a sudden change of friction

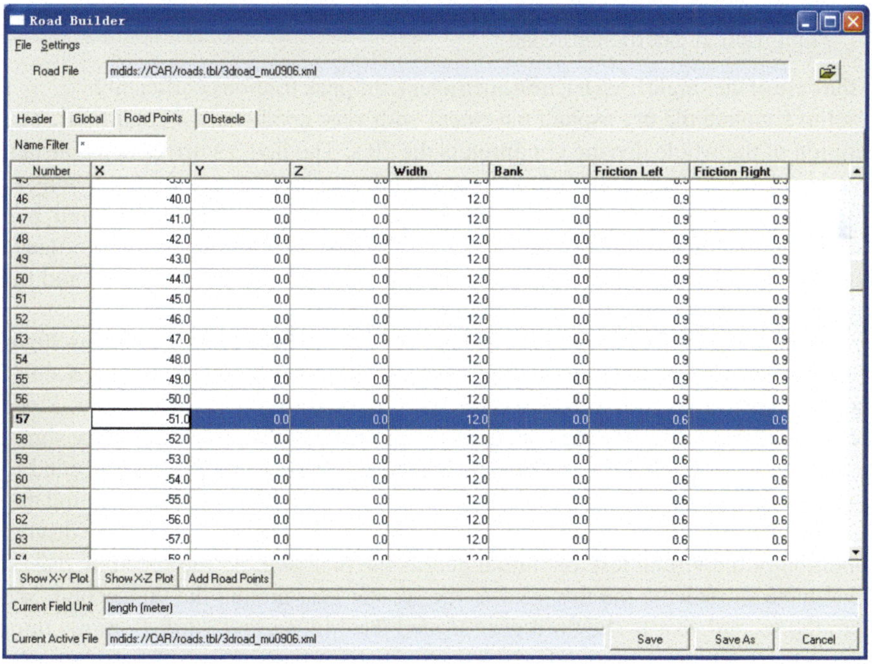

Fig. 3.6 Road Builder interface in ADAMS/Car module

coefficient from 0.9 to 0.6 as an example, a road point coordinate is taken every 1 m in the driving direction of the vehicle, the road width is taken as 12 m, and the lateral slope is set as 0. According to the regulation of the coordinate system in the virtual test, the x coordinate of the starting point of the pavement is 5 m; the x coordinate of the end point is −200 m; the adhesion coefficients of the left and right sides of the road surface centerline are axisymmetric; the friction coefficient of 5–50 m is 0.9; and the friction coefficient of the pavement is set to 0.6 from −51 m (that is, the 57th coordinate point in the pavement editor). In this way, a standard test pavement with a mutant attachment coefficient of 205 m in length and 12 m in width is constructed. After the pavement is defined, it is saved as a custom pavement file, which can be called directly under the driving and braking conditions carried out below.

2. Road Condition Estimation with Single Friction Coefficient while Driving

Firstly, the estimation of road conditions under driving conditions and road friction coefficient is considered. Considering that the tire angular velocity measurement is easily disturbed by noise signals. To simulate the actual test situation, the tire angular velocity measurement value is added with a high-frequency uniform Gaussian white noise sequence with mean 0 and standard deviation of amplitude 2% in the virtual test. A low-frequency uniform Gaussian white noise sequence with a mean of 0 and standard deviation of 1% of amplitude is also added to the driving torque transmitted to the wheel in the estimation of the longitudinal tire force. In addition, a single driving wheel is taken as the research object under the driving condition.

(1) High friction Coefficient Road

In the case of the single high friction coefficient, the peak friction coefficient $\mu_{\max} = 1$ is set to simulate the dry asphalt pavement with very good friction conditions. The duration of the whole driving condition is the 20 s. The first 5 s drives at a constant speed, and the driving control starts from the 5th second. The initial gear is hung in the first gear, allowing shift control. The initial vehicle speed is set to 20 km/h, and the longitudinal acceleration is set to 0.3 g. The pavement file is set in accordance with the above method. The road width is 12 m, the road length is 500 m, and the friction coefficient of the left and right wheels is equal.

As can be seen from Fig. 3.7a, since the tire angular velocity contains high-frequency measurement noise signals, the estimated slip rate also contains noise signals. Figure 3.7b shows the speed signal processed by the Kalman filter. Since the speed signal under driving conditions is obtained by the angular velocity signal of the non-driving wheel with noise interference, filtering processing is needed to maintain the effectiveness of the signal. From Fig. 3.7a and b, it can also be found that there is a sharp mutation in the estimated slip rate. This is because, under the driving condition of the virtual test, the initial gear is the first gear. As the speed increases, the vehicle shifts from the first gear to the second gear around the 9th second, and from the second gear to the third gear around the 14th second. Shift from the third gear to the fourth gear around the 20th second, which causes the vehicle to be in the state of uniform speed in the short instant of shift. At this time, the slip rate of the driving wheel also rapidly reduces to near 0.

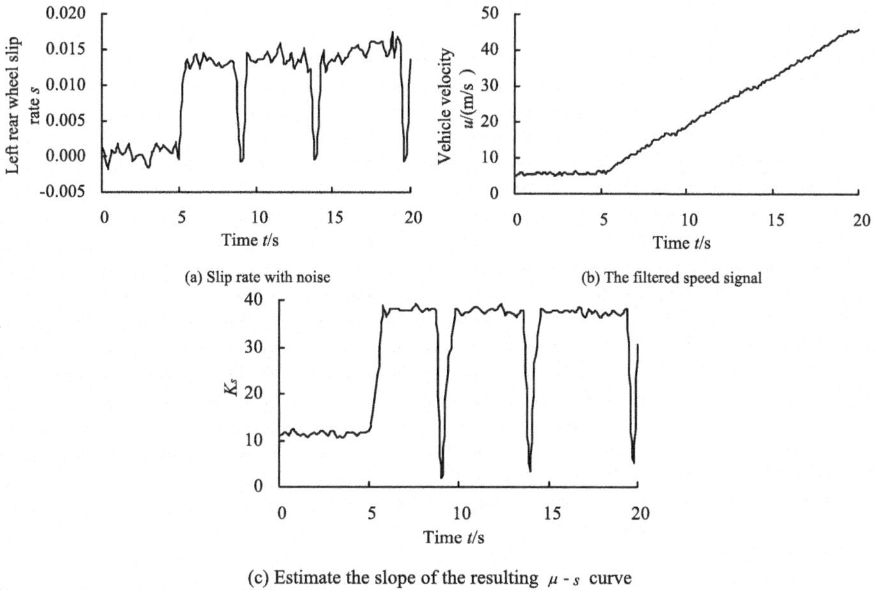

Fig. 3.7 Estimation of road K_s with single high friction coefficient (drive)

Figure 3.7c reflects the law of K_s variation with time obtained by the method estimated in this chapter. It can be clearly seen from the figure that, although the gear shifting operation is performed, K_s basically stays at a constant in the accelerated driving condition, about 38, which also reflects the characteristic that K_s remains unchanged in the straight working condition of the road with a single friction coefficient.

It is also found in the study that as long as the slip rate is kept small (within 0.2), the K_s value of driving conditions with different initial speeds basically keeps around 38 slight fluctuations. Of course, the K_s is also related to many other factors, but reflected in the specified vehicle type, road condition and acceleration in the virtual test, it can be considered that when K_s is around 38, it represents the road surface with good friction performance, such as dry asphalt or concrete road surface.

(2) Low Friction Coefficient Road

Similarly, in the case of a single low adhesion coefficient, the peak friction coefficient $\mu_{\max} = 0.5$ is set to simulate the wet soil pavement with poor friction conditions. The driving conditions and road surface settings are the same as in the case of high friction coefficient.

It can be seen from Fig. 3.8a that the slip rate of the wheel increases significantly when driving on the road surface with poor friction coefficient, and the slip rate increases slowly with the increase of the vehicle speed. As in the case of a high friction coefficient, shifting also causes a sudden change in slip rate. In addition, we can see that no matter on the low friction road surface or the high friction road

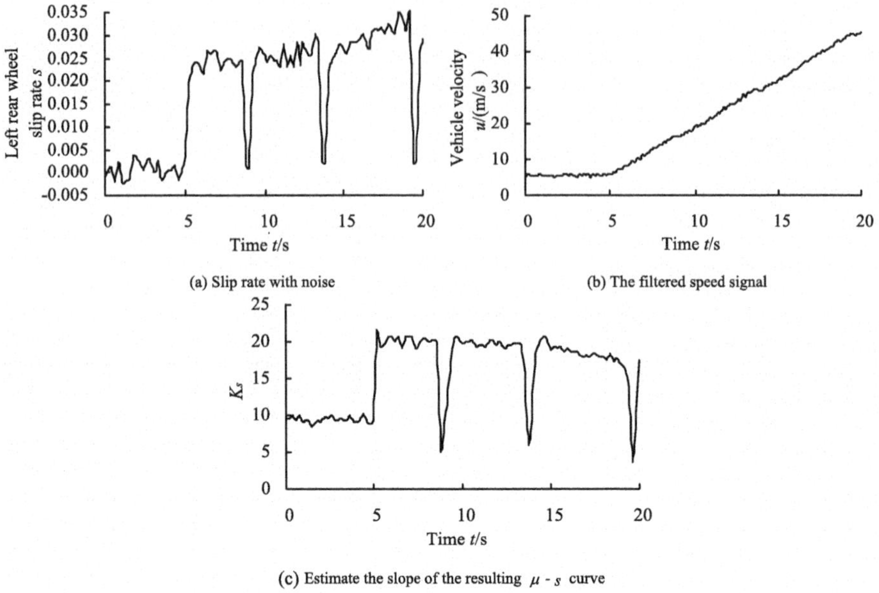

Fig. 3.8 Estimation of road K_s with single low adhesion coefficient (drive)

surface, thanks to the throttle control, the vehicle can achieve the same final speed in the 20 s acceleration time.

From Fig. 3.8c, it can be found that K_s, which represents the road friction, has a significant change with the high friction coefficient. The steady-state value of K_s is basically kept around 20, which is significantly lower than 38 when the friction coefficient is high. Therefore, it can be considered that the size of K_s reflects the quality of road friction. The larger K_s is, the better the road friction coefficient is, and the worse it is otherwise. This also supports the slip-slope idea originally proposed by Gustafsson, and also indicates that this theory can also be reproduced in ADAMS virtual experiment. Therefore, it can be considered that when K_s is around 20, it means that the car passes through the road surface with poor friction performance, such as wet and slippery dirt roads.

3. Road Condition Estimation of Abrupt Friction Coefficient during Driving

As mentioned above, the abrupt change in road friction coefficient is an important factor leading to traffic accidents. Therefore, this section considers the conditions with abrupt changes in road conditions to investigate the ability of the proposed estimation method to sense abrupt changes in road friction level.

In the case of mutated friction coefficient, the road width is set to 12 m, the road length is set to 1000 m, the friction coefficient of left and right wheels is equal, and the friction coefficient of 0–70 m road surface is 0.9, simulating the asphalt pavement in good condition. The friction coefficient of 71–220 m road surface is 0.5, which simulates slippery soil road surface. The recovery friction coefficient of the road surface is 0.9. The duration of the whole driving condition is the 20 s, and the driving control is carried out from the 2.5th second. The initial gear is hung in the first gear, allowing shift control. The initial vehicle speed is set to 20 km/h, and the longitudinal acceleration is set to 0.3 g.

It can be seen from Fig. 3.9a that the slip rate on the road surface with mutated friction coefficient also mutates with the change of road conditions (excluding the slip rate mutation when shifting). The slip rate has an obvious upward mutation near the 7th second and a downward mutation near the 12.5th second, which is obviously because the vehicle passes the road surface with a low friction coefficient between the 7th second and the 12.5th second. In addition, it can be seen from the figure that the slip rate slightly increases in the latter half of the low-friction road surface. This is because the wheel speed also increases with the increase of vehicle speed in the driving process, leading to an increase in the wheel slip rate. If the wheel continues to accelerate on this low-friction road surface, the wheel will quickly slip and idle, and the slip rate will reach 100%.

As can be seen from the Fig. 3.9c, the estimated K_s also mutates with the mutation of pavement conditions, K_s rapidly decreases from 35 to 20 near the 7th second, and K_s rapidly recovers from 20 to about 35 near the 12.5th second, which shows that due to the application of the CUSUM algorithm, the estimation method proposed in this section has a good ability to track the sudden change in pavement conditions.

4. Analysis of Proportional Relation of Braking Force of Front and Rear Brakes

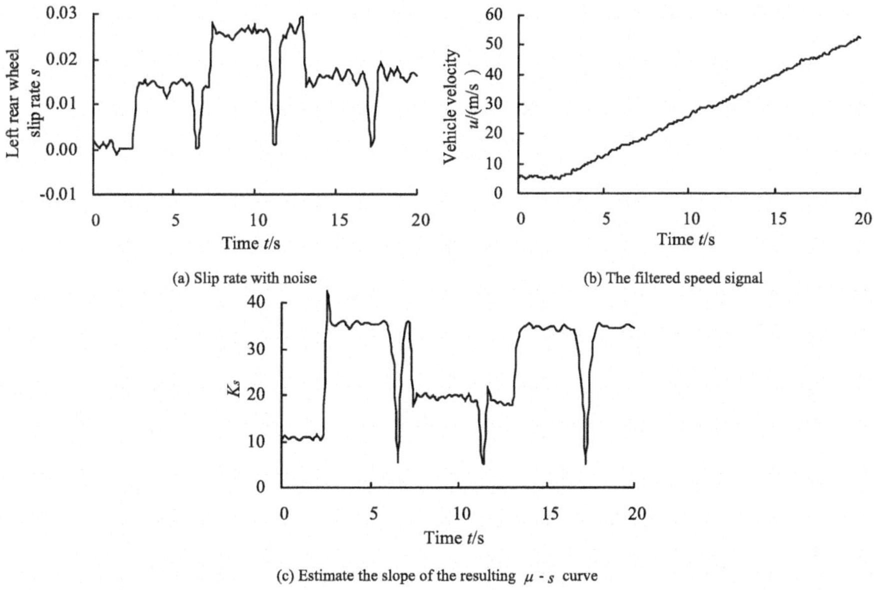

Fig. 3.9 Estimation of road surface K_s with mutation friction coefficient (drive)

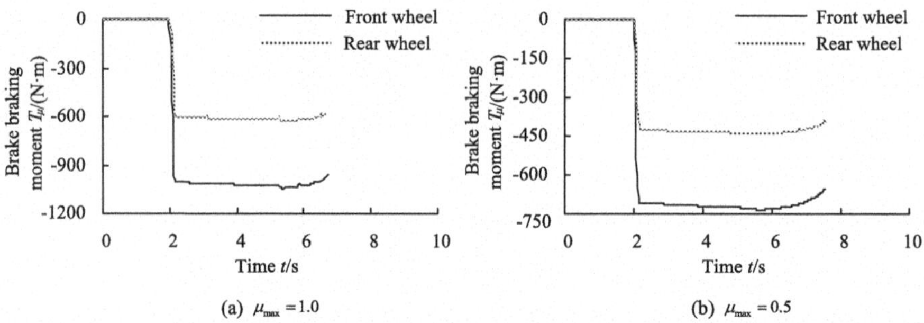

Fig. 3.10 Braking torques of front and rear wheels on roads with different adhesion coefficients

The braking characteristics of the front and rear brakes of the virtual test vehicle are analyzed before the estimation of the adhesion coefficient under braking conditions.

Firstly, a test condition is constructed to obtain the braking force distribution coefficient of the vehicle brake. The condition can be divided into two conditions. The first one is to brake on the road surface with the adhesion coefficient of 1.0 with an initial speed of 120 km/h and a braking deceleration speed of 0.7 g. The second type braking on the road surface with an adhesion coefficient of 0.5 at the initial speed of 100 km/h, braking deceleration speed of 0.5 g. In both cases, the vehicle starts to brake from the 2nd second, and the braking torque T_μ of the front and rear wheel brakes on the two road surfaces is investigated respectively. The results are shown in Fig. 3.10.

It can be seen from Fig. 3.10 that whether the road surface with high adhesion coefficient or a low adhesion coefficient, the braking torque is proportional in the braking stage, so the virtual test vehicle has a fixed braking force distribution coefficient. Because $F_\mu = T_\mu / r$, the slight difference between the front and rear wheel radii is ignored, then

$$\frac{F_{\mu 1}}{F_{\mu 2}} = \frac{T_{\mu 1}}{T_{\mu 2}} = \frac{\beta}{1-\beta} \tag{3.35}$$

According to Eq. (3.35) and Fig. 3.10, the brake force distribution coefficient $\beta = 0.6259$ of the virtual vehicle can be obtained, and the synchronous adhesion coefficient $\varphi_0 = (L\beta - b)/h = 1.21$ of the vehicle can be further calculated. According to the related literature about different adhesion coefficient road vehicle braking process analysis, due to $\varphi_0 > 1$, it can be known that no matter how large the road adhesion coefficient is in this virtual test, there is a $\mu_{\max} < \varphi_0$, so for the virtual prototype in this virtual test, if the wheels are locked, the front wheel is always hugged first, which effectively avoids the phenomenon of rear axle side sliding.

Through the above analysis, it can be seen that under a certain braking intensity, the slip degree of the front wheel is higher than that of the rear wheel, so the front wheel is taken as the research object during braking.

5. Road Condition Estimation with Single Friction Coefficient during Braking

In the braking condition, the speed signal is obtained directly in the virtual test because there is no free rolling wheel as the speed reference. In practice, the speed signal is usually acquired by radar or GPS position signal for difference, with high accuracy, so the speed signal does not add noise interference. In addition, because the estimation of tire longitudinal force in braking operation includes braking torque signal, a low-frequency uniform Gaussian white noise sequence with mean 0 and standard deviation 1% of amplitude is added to the braking torque measurement. In addition, the left front wheel (non-driving wheel) is set as the research object under braking conditions.

(1) High Friction Coefficient Road

Similar to the previous working conditions, the peak friction coefficient of the road is $\mu_{\max} = 1$. Brake control is performed from the second, with the initial gear hanging in the 5th gear, allowing shift control. The initial vehicle speed is set to 120 km/h, and the longitudinal deceleration is set to 0.55 g. The road is 12 m wide, 500 m long, and the left and right wheel friction coefficient is equal.

It can be seen from Fig. 3.11a that the slip rate suddenly increases after the start of braking. Figure 3.11c shows the change law of the slope of $\mu - s$ curve estimated by the method in this chapter with time and K_s increases rapidly at about the 2nd second, which reflects the good tracking performance of the estimation method in this chapter. It can be seen from the figure that the slope of the braking $\mu - s$ curve converges to about 50 on similar dry asphalt roads. It is also found in the study that as long as the slip rate is kept small (within 0.2), the K_s value of braking conditions with different initial speeds basically keeps around 50 with slight fluctuations. Under the specified vehicle type, road condition and braking intensity in this virtual test, it can be considered that K_s around 50 represents the road surface with good adhesion performance, such as a dry asphalt road surface.

(2) Low Friction Coefficient Road

Similarly, in the case of a single low friction coefficient, the peak adhesionfriction coefficient $\mu_{\max} = 0.5$ is set to simulate the wet soil pavement with poor friction conditions. The braking conditions and road surface settings are the same as in the case of high friction coefficient.

It can be seen from Fig. 3.12a that the slip rate of the wheel increases significantly when braking on the road surface with poor friction. After the 6th second, the clutch disconnects and the car slides freely, reducing the slip rate. Figure 3.12c shows that the slope of the $\mu - s$ curve has a significant change compared with the high friction coefficient. The steady-state value of K_s basically stays around 28, which is significantly lower than 50 at high friction coefficient. Similar to the previous situation, it can also be considered that the size of K_s in the braking situation also reflects the quality of road friction. The larger K_s is, the better the road friction is, and the worse it is on the contrary.

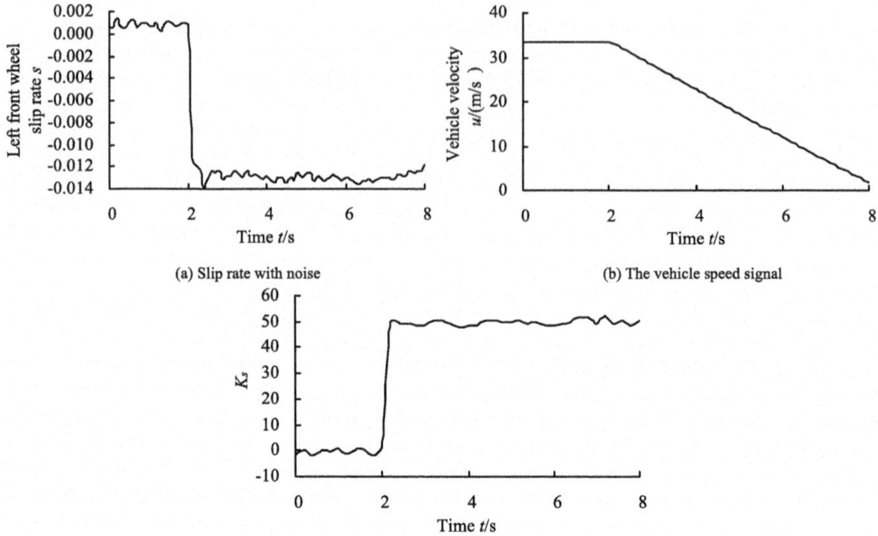

(a) Slip rate with noise

(b) The vehicle speed signal

(c) Estimate the slope of the resulting μ - s curve

Fig. 3.11 Estimation of road K_s with single high adhesion coefficient (braking)

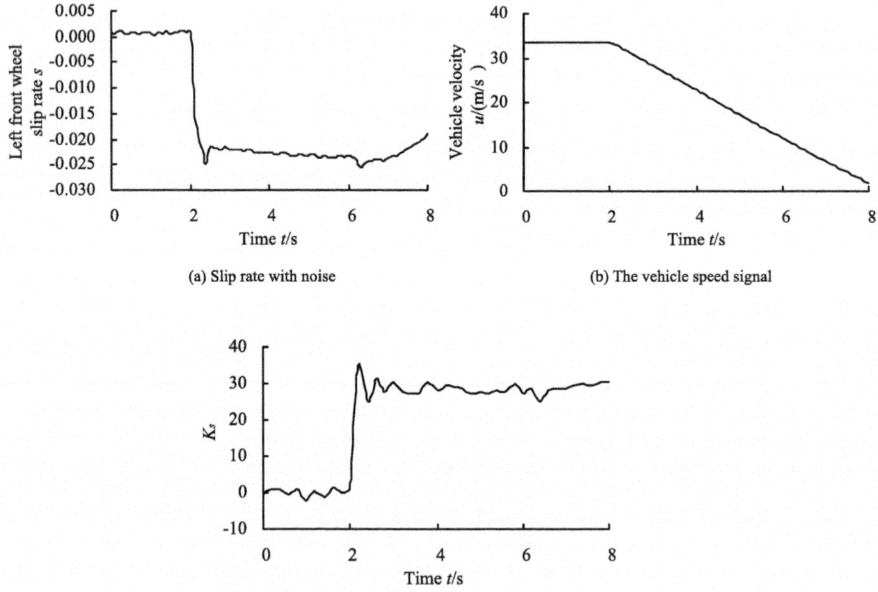

(a) Slip rate with noise

(b) The vehicle speed signal

(c) Estimate the slope of the resulting μ - s curve

Fig. 3.12 Estimation of road K_s with single low friction coefficient (braking)

6. Road Condition Estimation of Abrupt Friction Coefficient during Braking

The sudden deterioration of road conditions during braking may cause the wheel to lock up in a very short time, which leads to the phenomenon of vehicle braking deviation and losing the ability to steer. That is an important factor leading to traffic accidents. Therefore, it is very important to study the abrupt change of road friction coefficient under braking conditions.

In the case of mutated friction coefficient, the road width is 12 m, the road length is 500 m, the friction coefficient of the left and right wheels is equal, the friction coefficient of 0–110 m road surface is 1.0, the simulation condition is very good dry asphalt pavement. The friction coefficient of 111–150 m road surface is 0.5, which simulates the slippery soil road surface. The friction coefficient of 151–500 m road surface is 0.7, which simulates gravel road surface. Brake control is performed from the 2nd second, with the initial gear hanging in the fifth gear, allowing shift control. The initial vehicle speed is set to 120 km/h, and the longitudinal deceleration is set to 0.55 g.

It can be seen from Fig. 3.13 that the estimated slope K_s of $\mu - s$ curve is obviously divided into three stages during the whole braking process. With the sudden change of K_s at the beginning of braking, K_s converges to about 50 from the 2nd second to 3.5th second, reflecting that the dry asphalt pavement with friction coefficient of 1.0 at this stage. K_s converges to about 28 from the 3.5th to the 6th second, which

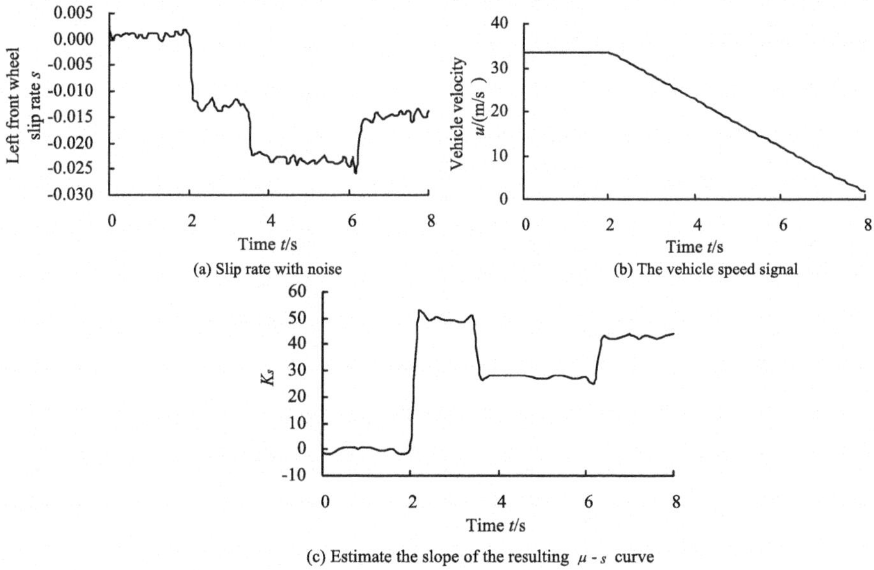

Fig. 3.13 Estimation of abrupt adhesion coefficient road (braking)

reflects the wet soil pavement with friction coefficient of 0.5 at this stage. From the 6th to the 8th second, K_s converges to about 43, reflecting the gravel pavement with an friction coefficient of 0.7 at this stage. For the mutation of K_s among various road surfaces, it can be seen from Fig. 3.13c that the estimation algorithm in this chapter has good tracking performance, and the mutation tracking time of K_s is all within 0.1 s.

Appendix

Procedure 3.1:

```
% λ − μ curve
% Define the λ − μ data
Data = …
[0.0000 0.0000
0.0250 0.2250
0.0500 0.4500
0.1000 0.6500
0.1250 0.6850
0.1500 0.7050
0.1750 0.6900
0.2000 0.6800
0.2500 0.6500
0.3000 0.6350
0.3500 0.6300
0.4000 0.6275
0.4500 0.6250
0.5000 0.6225
0.5500 0.6200
0.6000 0.6175
0.6500 0.6150
0.7000 0.6125
0.7500 0.6100
0.8000 0.6075
0.8500 0.6050
0.9000 0.6000
0.9500 0.5975
1.0000 0.5950];
lambda = Data(:,1);
mu = Data(:,2);
% We would like to fit the function
% mu = c(1)*lambda^n + c(2)*lambda^(n-1)
% + c(3)*lambda^(n-2) + … + c(n)*lambda + c(n + 1);
% to the data.
% In the following, n = 6
[c,S] = polyfit(lambda, mu, 6);
c
pause
t = (0: 0.05: 1)';
y = polyval(c,t);
plot(lambda,mu,'or',t,y,'-b')
title('Polynomial interpolation,
degree n = 6');
```

Procedure 3.2:

```
ti = 0.0; tf = 1.0; xi = [20.0,20.0];
tol = 1.0e−4; trace = 1;
[t,x] = ode45('Ex17_3a', ti, tf, xi, tol,trace);
plot(t,x); title( 'Vehicle Braking');
xlabel (' Time (sec) '); ylabel ( 'x');
grid;
legend( ' Forward speed ',' Wheel
speed ')
% A separate file Ex17_3a.m
function xdot = Ex6_3a(t,x);
% Define the simulation parameters:
m = 1400; rho = 1.202; Cd = 0.5; A =
1.95;
g = 9.81;
Theta = 0.0; bw = 0.0; f = 0.01; uw = 0.0;
fw = 0.0;
Iw = 2.65; rw = 0.31; Nw = 4; Fz =
3560.0;
% Define the mu-lambda polynomial
coefficients:
c = [-68.593, 238.216, -
324.819,219.283,
-75.58, 12.088, -0.0068];
if x(1) > = x(2),
lambda = (x(2)-x(1))/x(1);
else
lambda = (x(2)-x(1))/x(2);
end;
al = abs(lambda);
if al > 1.0, al = 1.0; end;
mu = sign(lambda)*c*[al^6;al^5;
al^4;al^3;al^2;al;1];
% Define the torque input T = Te -Tb;
Te = 0.0; Tb = 1000.0;
T = Te - Tb;
% Define the state equations:
if x(1) < 0.0, x(1) = 0.0; end;
if x(2) < 0.0, x(2) = 0.0; end;
xdot = [(-
(0.5*rho*Cd*A)*(uw + rw*x(1))^2 +
Nw*Fz*mu-f*m*g*cos(Theta)-
m*g*sin(Theta))/(m*rw);
(fw*Fz-bw*x(2)-
Fz*rw*mu + T)/Iw];
if x(1) < = 0.0, xdot(1) = 0.0; end;
if x(2) < = 0.0, xdot(2) = 0.0; end;
```

Procedure 3.3:

```
% Define linear model parameters
u0 = 20;g = 9.81;m = 1000;f =
0.015;Theta = 0;
rho = 1.202;A = 1;Cd = 0.5;uw = 2;
% Calculate the equilibrium force,Fx0:
Fx0 = m*g*sin(Theta) + f*m*g +
0.5*rho*A*Cd*(u0 + uw)^2;
% Time constants and DC gains
Tau = (m/(rho*A*Cd*(u0 + uw)));K =
Tau/m;
% Calculate the Ti gain of PI control(Ki =
Kp/Ti)
Ti = 186.86/10.0;
% The open-loop transfer function of PI
control is
%Kp*(num/den),num/den = (1 + 1/
(Ti*s))*K/(1 + Tau*s)
%
num_o = K*[Ti 1];den_o = [Ti*Tau Ti 0];
% Calculate the open loop zero pole
z = roots(num_o);
p = roots(den_o);
% Get the root locus as Kp changes
K_root = 0:10:300;
R = rlocus(num_o,den_o,K_root);
plot(R,'*');title('RootLocus Plot');
axis([−0.2 0 −0.05 0.05]);
% [K_root,POLES] = rlocfind(num_o,den_
o);
% Simulation of unit step response of
velocity
t = 0:0.5:100;
sys = tf(K,[Tau 1]);
y = step(sys,t);
Kp = 186.86;KI = Kp/Ti;
numc = K*[Kp KI];denc = [Tau K*Kp + 1
K*KI];
% Calculate the zero pole of the closed-loop
transfer function
zc = roots(numc);
pc = roots(denc);
sys_c = tf(numc,denc);
yc = step(sys_c,t);
Fig. (1)
plot(t,y,'-r',t,yc,'–b');grid;
xlabel('Time/s')
title(' Reference velocity unit step response
')
legend('open loop', 'closed loop')
% Simulation of disturbance unit step
response
yd = step(sys,t);
sys_disturbance = tf([K 0],denc);
ycd = step(sys_disturbance,t);
Fig. (2)
plot(t,yd,'-r',t,ycd,'–b');grid;
xlabel('Time/s')
title(' disturbance unit step response ')
legend('open loop', 'closed loop')
```

Procedure 3.4:

```
g = 9.81; m = 1000; f = 0.015; Theta = 0;
rho = 1.202; A = 1; Cd = 0.5; uw = 0;
Ka = 10; TauA = 0.2;T = 0.01;u(1) = 15.0;
error = 0.0;
error_old = 0.0;
int_error = 0.0;
F_dot = 0.0;
F = 0.0;F_vec = F;
t = 0:T:40;
r = [15*ones(1,5/T + 1),15 + (1:30/
T)*0.5*T,30*ones(1,5/T)];
d = [zeros(1,10/T +
1),-0.02*m*g*ones(1,10/T), zeros(1,20/T)];
for i = 1:40/T,
Tau = (m/(rho*A*Cd*(u(i) + uw))); K =
Tau/m;
Kd = (TauA + Tau-0.66*TauA*Tau-1)/
(Ka*K);
Kp = (0.34*(TauA + Tau)-0.3*TauA*Tau)/
(Ka*K);
Ki = (0.04*(TauA +
Tau)-0.04*TauA*Tau)/(Ka*K);
error_old = error;
error = r(i)-u(i);
int_error = int_error + error*T;
c = Kd*(error-error_old)/T + Kp*error +
Ki*int_error;
F_dotdot = (Ka*c-F_dot)/TauA;
F_dot = F_dot + F_dotdot*T;
F = F + F_dot*T;F_vec = [F_vec,F];
u_dot = (K*(F + d(i))-u(i))/Tau;
u(i + 1) = u(i) + u_dot*T;
end
figure();plot(t(1:40/T),u(1:40/T),'r'), grid
xlabel('Time/s')
title(' Actual speed /(m/s)')
figure();plot(t(1:40/T),r(1:40/T)-u(1:40/
T),'r'), grid
xlabel('Time(s)')
title(' Speed error /(m/s)')
figure();plot(t(1:40/T),F_vec(1:40/T),'r'),
grid
xlabel('Time(s)')
title('Tire force/N')
```

Procedure 3.5:

```
g = 9.81; uw = 0.0; u0 = 30.0; rho = 1.202;
Theta = 0.0; ThetaPrime = 0.0;
% Charged vehicle parameters
mc = 1000.0; Cdc = 0.5; Arc = 1.5;
fc = 0.015;
Kc = (1/(rho*Cdc*Arc*(u0 + uw)));Tc =
mc*Kc;
wc0 =
mc*g*(fc*sin(Theta)-cos(Theta))*ThetaPrime;
Fc = (u0/Kc);
% Forward vehicle parameters:
ml = 1500.0;Cdl = 0.6;Arl = 1.95;fl = 0.015;
Kl = (1/(rho*Cdl*Arl*(u0 + uw)));Tl =
ml*Kl;
Fl = (u0/Kl);
wl0 =
(ml*g*(fl*sin(Theta)-cos(Theta))*ThetaPrime);
t = [0:0.1:20]';
U0 = Fc*ones(size(t)); % Nominal control
force
wl = -800*(1 + 0.01*t); % Slope change
function
wc = wc0*ones(size(t));
disturbance = [U0 wc wl];
% 4-state variable system for controller design:
Aa = [0 -1 0 0;0 -1/Tc 0 0;1 0 0 0;0 0 1 0];
Ba = [0;Kc/Tc;0;0];
% Controller design
pc = [roots([1 2*0.9*0.4 0.4^2]); -1.08; -1.18];
K = place(Aa,Ba,pc);
% Closed-loop simulation, tracking v_l(x_5):
Ac = [0 -1 0 0 1;-K(1)*Kc/Tc -(1 + K(2)*Kc)/Tc
-K(3)*Kc/Tc -K(4)*Kc/Tc 0;1 0 0 0 0;0 0 1 0 0;0 0 0 0
-1/Tl];
Bc = [0 0 0;0 Kc/Tc 0;-1 0 0;0 0 0;0 0 Kl/Tl];
% Output: × 1 (range), vc and vl
Cc = [1 0 0 0 0;0 1 0 0 0; 0 0 0 0 1];
Dc = zeros(3,3);
r = 30.0*ones(size(t));
disturbance = [r U0 + wc wl];
xc0 = [30 u0 0 -(u0/Kc + K(1)*30 + K(2)*u0)/K(4) u0];
[yc,xc] = lsim(Ac,Bc,Cc,Dc,disturbance,t,xc0);
subplot(211), plot(t,yc(:,1));
title(' Distance between two vehicles/m');
xlabel('Time/s'); grid;
subplot(212), plot(t,yc(:,3),'r',t,yc(:,2),'-.b');
title('Vehicle speed/(m/s)');
xlabel('Time/s'); grid;
legend('Vl', 'Vc');
figure; subplot(211)
u =
U0-K(1)*xc(:,1)-K(2)*xc(:,2)-K(3)*xc(:,3)-K(4)*xc(:,4);
plot(t, u'); title(' Control force /N');
xlabel('Time/s'); grid
```

References

1. Cui SM. Vehicle system dynamics and simulation. Beijing: Peking University Press; 2014. p. 50–1.
2. Zhang YL. The World Health Organization released its 2018 global status report on road safety. Chin J Disast Med. 2019;7(2):100.
3. Liu YZ. Problems and countermeasures of sustainable development of cross-century highway traffic in China. J Changsha Commun Univ. 1996;4:9–14.
4. Qiao YJ. Research on control method of vehicle longitudinal dynamics system. Chongqi: Chongqi University; 2004.
5. Li L, Wang FY, Guo W, et al. Intelligent vehicle: advanced sensing and control. Beijing: China Machine Press; 2016. p. 106.
6. Ulsoy A, Peng H, Cakmakci M. Automotive control system. Cambridge: Cambridge University Press; 2012.
7. Hac A, Bodie MO. Improvements in vehicle handling through integrated control of chassis systems. Int J Vehic Auton Syst. 2002;1(1):83–4.
8. Matsumoto S. Improvement of vehicle dynamics through braking force distribution control. Soc Autom Eng. 1992;13(2):52–4.

9. Hoffman DD, Rizzo MD. SAE technical paper series. Chevrolet C5 Corvette Vehic Dyn Control Syst. 1998;12(5):1–3.
10. Shibahata Y, Shimada K, Tomari T. Improvement of vehicle maneuverability by direct yaw moment control. Vehic Syst Dyn. 1993;21(13):5–7.
11. Lin F, Huang C. Unscented Kalman filter for road friction coefficient estimation. J Harbin Instit Technol. 2013;45(7):115–20.
12. Yi K, Hedrick K, Lee SC. Estimation of tire-road friction using observer based identifiers. Vehic Syst Dyn. 1999;31(4):233–61.
13. Hwang W, Song B. Road condition monitoring system using tire-road friction estimation. In: Proceedings of AVEC, Michigan; 2000, pp. 437–442.
14. Steffen M, Michael U, Karl H. Slip-based tire-road friction estimation during braking. In: Proceedings of ASME international mechanical engineering congress and exposition, New York; 2001, pp. 213–220.
15. Luan JH. Fault reconstraction technology and its application in satellite attitude control system fault diagnosis. Harbin: Harbin Institute of Technology; 2006.
16. Yu XH, Xu JX. Variable structure systems: towards the 21st century. Berlin: Spring; 2002.
17. Utkin V. Variable structure systems with sliding modes. IEEE Trans Autom Control. 1977;22(2):212–22.
18. Slotine J, Hedrick J, Misawa E. On sliding observers for nonlinear systems. J Dyn Syst Measur Control. 1987;109(3):245–52.
19. Walcott B, Zak S. Combined observer-controller synthesis for uncertain dynamical systems with applications. IEEE Trans Syst Man Cybern. 1988;18(1):88–104.
20. Zheng YH, Fattah HA, Loparo K. Non-linear adaptive sliding mode observer-controller scheme for induction motors. Int J Adapt Control Sig Process. 2000;14(2):245–73.
21. Tan CP, Edwards C. Sliding mode observers for robust detection and reconstruction of actuator and sensor faults. Int J Robust Nonlinear Control. 2003;13(5):443–63.
22. Kim SM, Han WY, Kim SJ. Design of a new adaptive sliding mode observer for sensorless induction motor drive. Electr Power Syst Res. 2004;70(1):16–22.
23. Zhao YQ, Lin F. Estimation of road pavement adhesion factor based on virtual experiment. J Jilin Univ. 2011;41(2):309–15.
24. SAE. Vehicle dynamics terminology. In: SAE J670e, society of automotive engineers; 1976.
25. Steffen M, Michael U, Karl H. Estimation of the maximum tire-road friction coefficient. J Dyn Syst Measur Control. 2003;125:607–17.
26. Ljung L, Soderstrom T. Theory and practice of recursive identification. Cambridge: MIT Press; 1983.
27. Kailath T, Sayed AH, Hassibi B, et al. Linear estimation. London: Prentice Hall; 2000.
28. Rajamani R. Radar health monitoring for highway vehicle applications. Vehic Syst Dyn. 2002;38(1):23–54.
29. Perruquetti W, Barbot JP. Sliding mode control in engineering. New York: Marcel Dekker; 2002.

Chapter 4
Independent All-Wheel Drive Distribution Control and Estimation of Tire Effective Cornering Stiffness Based on Vehicle Lateral Dynamics

With the development of the economy and automobile technology, highway traffic shows the trend of high-speed driving, intensive traffic flow, and non-professional drivers. Frequent traffic accidents make highway traffic safety a social issue of widespread concern. The solution to this security problem is divided into two parts: active safety and passive safety. Active safety refers to the vehicle performance that helps the driver avoid accidents as safely as possible in all situations, such as reliability, environmental visibility, lateral system dynamics, and acceleration or braking. Arranged in descending order via the proportion of inducing factors of automobile accidents, they are lateral system dynamics, acceleration and braking, visibility and reliability. Passive safety refers to vehicle performance to minimize the loss of an unavoidable accident (especially when it comes to casualties). It contains external safety (cyclists and pedestrians) and internal safety (passengers). For decades, how to design automobiles to obtain excellent active safety, especially lateral system dynamics, has been one of the main research directions of scholars and designers from all over the world.

Vehicle lateral system dynamics is an important component of automobile theory, which usually contains two interrelated aspects, namely vehicle handling and stability [1]. Under the condition that the driver does not feel excessive tension or fatigue, maneuverability means that the car can follow the direction commanded by the driver through the steering system and steering wheels. Stability refers to the ability to resist interference and maintain stability when there is external disturbance [2]. The dynamic characteristics of the vehicle lateral system not only affect the maneuvering ease of vehicle driving, but also belong to the main performance for high-speed vehicle safety. Therefore, people call it the lifeline of high-speed vehicles.

For a moving car, the external actions (inputs) consist of three aspects: the driver's operation (from the steering wheel, acceleration, brake pedal), the road force (from the tires) and the aerodynamic force (from the body and tires). The gravitational force is a constant factor and does not belong to external actions. In terms of the vehicle lateral system dynamics, the response caused by the driver's input, where

Y. Zhao and F. Lin, *Vehicle System Dynamics*,
https://doi.org/10.1007/978-981-97-2019-4_4

the steering wheel input is directly affected by the road alignment that the driver is following or the spatial frequency of the radius of curvature of the road alignment, is the vehicle handling problem, that is whether the driver could control the car. The response caused by the road and air inputs belong to the problem of vehicle stability.

4.1 Review of Vehicle Lateral System Dynamics

4.1.1 Introduction of Evaluation Method Development for Vehicle Lateral System Dynamic Characteristics

There are two evaluation methods for vehicle lateral system dynamics: the open-loop method and the closed-loop method. The evaluation method of vehicle motion characteristics, which is called open-loop method, is based on the general requirements for maneuverability, steady-state quality and transient response characteristics. In fact, the pros and cons of the dynamic characteristics of the vehicle lateral system not only depend on the vehicle structural parameters, but also involve subjective and objective factors, such as the driver and the road traffic environment. If the vehicle is regarded as the control objective of a driver–vehicle–environment closed-loop system (Fig. 4.1), the evaluation method based on the whole system characteristics is called the closed-loop method.

Before the 1960s, the methods were basically open-loop methods, and the application was based on classical control theory. According to the steady-state and transient analysis, the under-over-steer characteristics and the step response characteristics of the steering input are used to evaluate the vehicle lateral system dynamics. However, the above-mentioned open-loop method is hard to directly apply in practice. Because the designers couldn't be able to evaluate the steady-state and transient steering characteristics, and cannot figure out whether understeering or oversteering characteristics that meet the stable conditions were better [3].

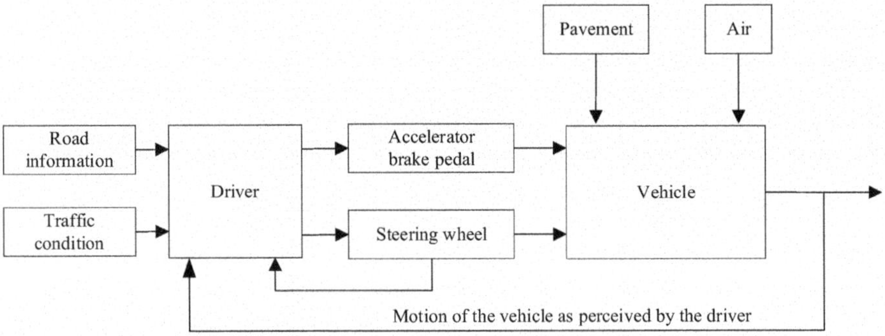

Fig. 4.1 Driver–vehicle–environment closed-loop system

In the early 1970s, the implementation of the ESV research program prompted people to study practical lateral system dynamics design methods. Because the driver model at that time was still at the level of improving the simulation accuracy of the closed-loop tracking response, researchers in various countries mainly adopted the method of systems engineering to explore the evaluation method of the lateral system dynamic characteristics. Based on a large number of experiments and theoretical analysis, the safety allowable limits of steady-state response characteristics, transient response characteristics, self-aligning characteristics and lateral slip characteristics were proposed, and objective evaluation was carried out on the lateral system dynamic characteristics. Since the mid-to-late 1970s, the driver's perception of the vehicle straight-line performance, cornering performance, and steering lightness has been used to conduct subjective evaluations. Although the subjective evaluation method has not been theoretically deduced, it reflects the thought of closed-loop design to a certain extent because the driver factors and the characteristics of the road environment were considered. However, after a large number of experimental design processes, it was found that above methods could effectively obtain stability (asymptotic stability and ability to resist external disturbances), while it was difficult to effectively design the vehicle lateral system dynamics for vehicle transient response characteristics, where objective evaluation and subjective evaluation are even inconsistent. The reason lies in that the objective evaluation method was still proposed by the requirements of the general control system, which was an open-loop method. The subjective evaluation refers to the driver's feeling of the difficulty of the maneuvering action when driver is maneuvering the car according to certain tracking requirements. This feeling is not entirely dependent on the vehicle characteristics, but also on the characteristics of human behavior and the requirements of road tracking.

In response to the above problems, people started from theoretical and experimental aspects in the early 1980s, and began to study the driver–vehicle–road closed-loop system in depth. In terms of theory, many deterministic driver direction control models were established, taking fully into account human learning and adaptability, which effectively simulated the process of tracking a given path in the driver–vehicle–road closed-loop system. In terms of testing, considering that the progress of the driver model cannot meet the requirements for active safety closed-loop design, and that the design method of safety test can only be used after the prototype car was trial-produced and limited by natural conditions, a development-type driving simulator was developed. This kind of simulator adopted advanced real-time simulation, digital imaging, hydraulic control and other technologies to combine real drivers with modeled cars and replaces field tests with indoor computer simulation, which alleviates the contradiction between the development of theoretical research and requirements of vehicle active safety closed-loop design. In terms of simulation evaluation, Japanese scholar Masato Abe proposed a method to theoretically predict the dynamic characteristics of the lateral system based on the preview model. Masato Abe believes that while manipulating a car, people do not completely follow tracking performance objective to adjust their own dynamic characteristics parameters, but also to take into account vehicle movement safety and their own physical factors. The maneuverability and the control burden required by the according driver dynamic parameters mainly depend on the preview distance (the preview time) and the performance index that the entire closed-loop system can achieve, which is called the evaluation index.

In the 1990s, Professor Guo Konghui also studied the driver–vehicle–road closed-loop control system model. And in consideration of many factors affecting the dynamic characteristics of the vehicle lateral system, he proposed various single total variance evaluation indexes with clear physical meanings (trail, lateral acceleration, steering wheel angular velocity, and lateral force coefficients of front and rear wheels), and applied the statistical analysis method of frequency domain to propose a comprehensive evaluation and optimization design method for the active safety of closed-loop systems, which was convenient for engineering applications [2]. Based on constant driver characteristics, Harada established a comprehensive performance index based on three single mean square value indexes (trail error, steering wheel angle, and driver handling burden). Professor Zhao Youqun has compared and studied each individual evaluation index of the active safety of the driver–vehicle–road closed-loop control system, and verified that only a limited number of individual indexes can be used to determine a reasonable comprehensive evaluation index in automobile design. The single evaluation index of the dynamic characteristics of the open-loop lateral system is also supplemented; the comprehensive evaluation index of the dynamic characteristics of the vehicle open-loop lateral system and the comprehensive evaluation index of the active safety of the driver–vehicle–road closed-loop control system are also analyzed.

For the closed-loop design method of vehicle lateral system dynamics, both theoretical research and simulation test design starts from the ease of handling, which is consistent with the subjective evaluation index. The results of the study show that maneuverability can be predicted by the optimal performance index that the driver–vehicle–environment closed-loop system can achieve, and the ideal driver characteristic parameters required to achieve this index. Moreover, this prediction of maneuverability has good consistency with the subjective evaluation results. Professor Zhao Youqun has researched the response and active safety of the stochastic driver–vehicle–road closed-loop system in consideration of the uncertain parameters of the driver dynamic characteristics. It is necessary and valuable for the prediction of the driver handling feeling and the optimal design of the vehicle structure parameters based on active safety to establish a closed-loop evaluation method for the dynamic characteristics of the vehicle lateral system by considering the uncertainty of the driver. Since the 1990s, the application of developmental driving simulators to study the active safety of driver–vehicle–environment closed-loop systems for the improvement of the vehicle performance has been one of the major development directions in the world.

Up to now, the mathematical model of the road input (the road alignment which the driver needs to follow) in the simulation of the driver–vehicle–environment closed-loop system has been mainly deterministic. There are various road alignments in the objective world, that is, the roads encountered by different drivers and the roads encountered by the same driver at different times are random. Professor

Zhao Youqun conducted a response analysis and active safety evaluation of a driver–car–road closed-loop system with stable random road following, of which the results were effective and reasonable. On this basis, the response analysis and active safety evaluation of the driver–vehicle–road stochastic closed-loop system with evolving stochastic road input are also studied. This research involves the condition that vehicle speed varies with time, thereby simulating the real driving process well. This method is not only an accurate algorithm, but also can shorten the evaluation period and avoid the huge economic burden of the real vehicle test.

In this chapter, the roads are all good hard surfaces. Vehicles running on soft ground (such as deserts, fields, ect.) belongs to the research content of soil-vehicle system mechanics, because the main contradiction in the study lies in the relationship between the ground and the vehicle forces.

4.1.2 The Significance of Research on Closed-Loop Lateral System Dynamics

It is well-known that the vehicle safety depends on three factors: the driver, the vehicle and the road. The first task of accident research is to determine which accidents are caused by driving techniques or road traffic systems that are not suitable for the driver. There are generally two ways to improve traffic safety: the first one, which is called the strategy of adapting the human being to vehicles, is to influence drivers to make their behavior in the current road traffic system meet the safety requirements, such as training and traffic education. The second, which is the strategy of adapting vehicles to the human being, is to make the road traffic and vehicle system better adapted to the driver's behavior to avoid too many demands on the driver due to technical reasons. Most traffic accidents are caused by the contradiction between the driving requirements of the driver–vehicle–road closed-loop system and the driving skills of the driver. And the driver is affected by various factors such as physiology, gender, age and so on, which is the most difficult step in the driver–vehicle–road closed-loop system to change. So the strategy of adapting vehicles to human is the fundamental method to solve the problem. The strategy of adapting vehicles to people also embodies the idea of the vehicle's active safety design.

As the control object of the driver with uncertainty, the car cannot be designed in an open loop according to the general control principle. The main reasons are as follows:

(1) The open-loop evaluation of the dynamic characteristics of the vehicle lateral system usually takes the response of the steering wheel angle step and the steering wheel force step input as the analysis object, but the characteristics of the step response only indirectly reveal the characteristics of the vehicle response under the actual manipulation by the driver;
(2) When a car runs in the actual traffic environment, the active safety performance is the most important. In terms of the car, it should specifically include the tracking

track error to avoid obstacles, the lateral acceleration to prevent rollover, and the force on the front and rear wheels to prevent sideslip. The open-loop evaluation cannot fully reveal the performance of these three aspects;

(3) The driver's feeling is a subjective standard for revealing and testing the dynamic characteristics of the vehicle lateral system. In addition to the factors in (2), the factors affecting this subjective evaluation also include fatigue level. This index cannot be replaced by open-loop evaluation;
(4) Not only the vehicle characteristics, but also the driver characteristics affect the dynamic characteristics of the driver–vehicle–road closed-loop control system. The reasonable matching of them can improve the performance of the entire closed-loop system.

Given a car, the driver can improve the performance of the driver–vehicle–road closed-loop system by training, accumulating the experience, and adjusting own characteristic parameters. But this is just for individual drivers. As far as the driver group, the change and adjustment of characteristic parameters are limited. In other words, the adaptation of people to the car is limited. The vehicle design must be based on the conditions of the driver, taking the route of fitting the car to the man. That is to say, the pros and cons of the dynamic characteristics of the vehicle lateral system should regard the ability to allow non-ideal drivers to control under certain safety requirements as the main evaluation index.

4.1.3 Basic Problems in the Vehicle Lateral System Dynamics

At present, the research on the vehicle open-loop system has been quite in-depth. The in-depth study of tire characteristics has made it possible to carry out a fairly comprehensive and realistic simulation of the dynamic response. A lot of research has been done on the theory of the driver model, which has made breakthroughs in the research of driver–vehicle–road closed-loop control system dynamics simulation, vehicle lateral system dynamics and vehicle active safety research. However, the research topic of driver–vehicle–road closed-loop control system evaluation still needs to do a lot of in-depth and meticulous work, as follows:

(1) Rationality and feasibility of evaluation methods and evaluation indicators: the status and relevance of subjective and objective evaluation, the representativeness and controllability of test conditions, the rationality and feasibility of evaluation indicators, etc.
(2) Shorten the evaluation cycle: replace subjective evaluation with objective evaluation, replace real environment with simulation environment, replace experimental evaluation with simulation evaluation, replace old evaluation method with new evaluation method, etc.
(3) Optimal design problems: consideration of multiple objectives (complexity of operating conditions) with numerous performance indicators, driver dynamic

characteristics in closed-loop systems, subjective factors in objective function (quantification of evaluation indicators), time-varying structural parameters, etc.
(4) Application of advanced scientific and technological means: including the application of the latest basic scientific achievements, the application and development of the latest application software packages.

4.2 Independent All-Wheel Drive Distribution Control

4.2.1 Conventional Four-Wheel Drive System

If the differential braking based on the yaw stability control system is used during vehicle acceleration, the acceleration will be reduced and thus will not provide the vehicle longitudinal response required by the driver [4]. The solution that the automotive industry is keen on is the application of independent drive torque control with all-wheel drive control technology to enhance drive performance and handling [5].

In order to make readers understand better, the concepts of four-wheel drive and all-wheel drive are explained as follows. In a four-wheel drive system, the drive torque is transmitted to all four wheels (in contrast, the drive torque is transmitted only to the two front wheels in a front-wheel drive vehicle).

The advantage of a four-wheel drive system is that the longitudinal tire drive forces that propel the vehicle forward are generated on all four wheels. This is beneficial for situations with low adhesion, such as driving in snow and climbing slopes. But all-wheel drive isn't helpful for braking on slippery surfaces, as it's entirely dictated by the braking system, not the drive system.

The main components that ensure the work of the four-wheel drive system are the differentials and transfer cases of the front and rear wheels. Front and rear differentials allow the left and right wheels to spin at different speeds. It is necessary when cornering, as the outer wheel travels a larger radius and rotates faster. The transfer case transfers torque from the transmission to the front and rear axles. Based on this design, the transfer case can provide equal or proportional torque to the front and rear axles. The transfer case distributes torque to the front and rear axles via the center differential.

When the four wheels are combined in the four-wheel drive system, the front and rear drive shafts would be locked so that they must spin at the same speed. The four-wheel drive system is divided into full-time four-wheel drive and time-sharing four-wheel drive system. For the time-sharing four-wheel drive system, the driver can select four-wheel drive or two-wheel drive through a handle or a switch. The driver can make a quick switch (switching between four-wheel drive and two-wheel drive while driving). This allows the vehicle to adopt two-wheel drive on regular dry roads, and four-wheel drive on slippery roads which require more drive.

In addition, the full-time four-wheel drive system can keep the vehicle in two-wheel drive (front or rear) until the system determines that four-wheel drive is needed, and it would automatically transmit torque to the four wheels. If necessary, the torque ratio of the front and rear axles could be adjusted. Normally, if the system detects a wheel slip, the four-wheel drive will be automatically activated. However, the system can switch to four-wheel drive depending on the driving situation in some latest advanced systems, even if there is no slip phenomenon. The full-time four-wheel drive system is also called all-wheel drive (AWD) system.

4.2.2 Differential-Based Torque Distribution Between Left and Right Wheels

As described in Sect. 4.2.1, a conventional differential allows the left and right wheels to rotate at different speeds, which is necessary when the vehicle is turning. The traditional differential is also called open differential.

The conventional differential distributes torque equally between the left and right wheels to which it is attached. If one wheel is off the ground, or on a very slippery surface, they have only a small torque. Since the torque is evenly distributed, another wheel obtains small torque, too. This is the main disadvantage of traditional differentials.

To make up for the disadvantages of conventional differentials, the locking differential is developed. By using locking differential, the driver can lock the left and right wheels together, which ensures that the sum of the torques equals the total torque. If one wheel is on a slippery road, the other wheel still gets enough torque to keep the vehicle driving forward. As a result, the locking differential allows the vehicle to obtain more drive on slippery surfaces than conventional differential, which can be operated by the driver when required.

Another type of differential is the limited-slip differential (LSD). It progressively locks the left and right wheels via a clutch, allowing for some slip initially, which makes the inner and outer wheels to achieve different speeds when cornering. But the wheels will be automatically locked when the speed difference is too large to help the vehicle get enough drive on slippery surfaces.

From the above discussion, it can be seen that the proportion of torque transmitted to the left and right wheels determines the type of differential. In the open differential, the torque transmitted to the left and right wheels is always equal. In the locking differential, the left and right wheels rotate at the same speed and they get the total torque as a whole. Whereas in the limited-slip differential, the wheels with lower speed are able to get more torque.

4.2.3 Active Control of All-Wheel Torque

For the ideal all-wheel drive system, four wheels can be independently controlled. Currently, the automotive industry has developed dual-clutch differentials that transmit torque to the inside or outside wheels in different proportions, as required by active control systems. Torque transfer between the front and rear axles can also be similarly actively controlled using the center differential. By independently controlling the torque delivered to the four wheels, both drive force and yaw stability of the vehicle can be controlled simultaneously. In this way, yaw stability can be controlled simultaneously during acceleration without the need for active differential braking intervention, which would cause the vehicle to decelerate [6].

Figure 4.2 shows three types of yaw stability control systems that can be used during vehicle acceleration, along with their respective characteristics.

The following analysis can be used to model the torque transfer of the differential control with dual-clutch to each wheel.

When the right clutch is engaged with torque T_{clutch}, the driving torque transmitted to the left wheel is

$$T_{dl} = \frac{1}{2}T_d - qT_{clutch} \tag{4.1}$$

And the driving torque transmitted to the right wheel is

$$T_{dr} = \frac{1}{2}T_d + qT_{clutch} \tag{4.2}$$

where q is the ratio determined by the dual-clutch differential system, T_d is the total torque transmitted to the considered shaft.

Likewise, when the left clutch engages with clutch torque, the drive torque transmitted to the left wheel becomes

$$T_{dl} = \frac{1}{2}T_d + qT_{clutch} \tag{4.3}$$

and the driving torque transmitted to the right wheel is expressed as

$$T_{dr} = \frac{1}{2}T_d - qT_{clutch} \tag{4.4}$$

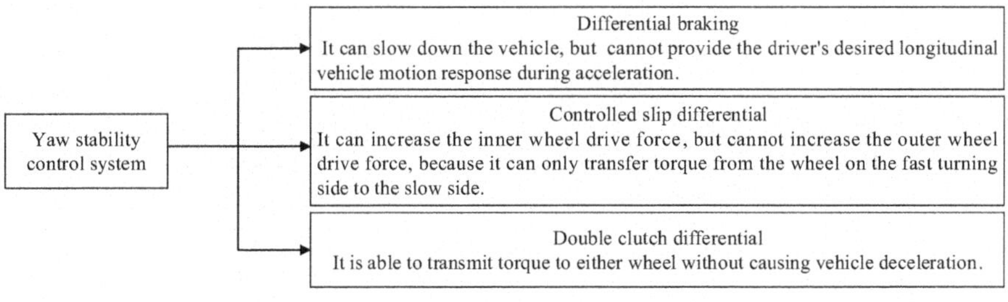

Fig. 4.2 Types of yaw stability control systems and their performance during braking

Therefore, the driving torque transmitted to the left and right wheels can be controlled by changing the clutch torque.

The optimal structural scheme to independently control the torque of each wheel is the application of dual clutches in conjunction with the all-wheel drive transfer case with a central differential, which could transmit torque to the front and rear wheels respectively [7]. However, this is not an attractive scheme due to weight and economic considerations. The alternative is to use a center differential and a dual-clutch to transfer torque.

4.3 Estimation of Tire Effective Cornering Stiffness

4.3.1 Estimation of Tire Effective Cornering Stiffness Based on the Time Rate of Change of Acceleration

1. Vehicle dynamics model of three degrees of freedom

Figure 4.3 shows a 3-degree-of-freedom vehicle model with lateral, yaw, and roll motions [8]. According to D'Alembert's principle, the force balance along the y-axis and the moment balance around the z-axis can be listed as the following differential equations:

$$ma_y - m_s h\ddot{\phi} = 2F_{yf}\cos\delta + 2F_{yr} \tag{4.5}$$

$$I_z\dot{r} + I_{xz}\ddot{\phi} = 2aF_{yf}\cos\delta - 2bF_{yr} \tag{4.6}$$

where, m is the mass of the vehicle, m_s is the sprung mass, a and b are the distances from the center of mass to the front and rear axles, h_c is the roll force arm, I_z is the moment of inertia around the z-axis, I_{xz} is the moment of inertia around the x and

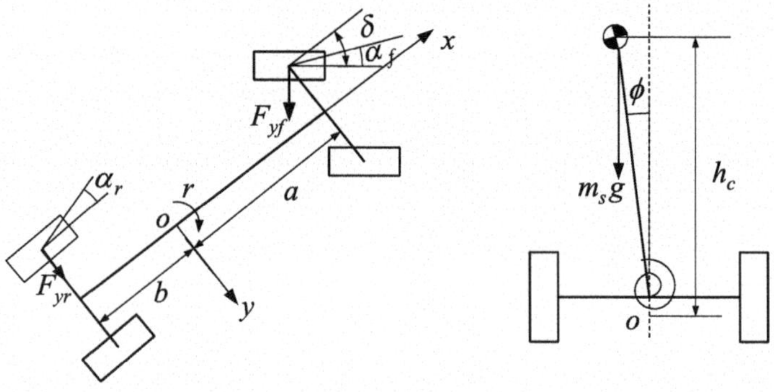

Fig. 4.3 3-degree-of-freedom vehicle model

z axis, ϕ is the roll angle, F_{yf} is the front wheel lateral force, F_{yr} is the rear wheel lateral force, a_y is the lateral acceleration, r is the yaw rate, δ is the front wheel steering angle.

Considering the small δ in the case of large radius steering in this section, Eqs. (4.5) and (4.6) can be written in the state space form:

$$\begin{bmatrix} F_{yf} \\ F_{yr} \end{bmatrix} = \begin{bmatrix} \frac{mb}{2L} & \frac{I_z}{2L} & \frac{I_{xz}-m_s h_c b}{2L} \\ \frac{ma}{2L} & -\frac{I_z}{2L} & -\frac{m_s h_c a+I_{xz}}{2L} \end{bmatrix} \begin{bmatrix} a_y \\ \dot{r} \\ \ddot{\phi} \end{bmatrix} \tag{4.7}$$

Differentiate both sides of Eq. (4.7) to time:

$$\begin{bmatrix} \dot{F}_{yf} \\ \dot{F}_{yr} \end{bmatrix} = \begin{bmatrix} \frac{mb}{2L} & \frac{I_z}{2L} & \frac{I_{xz}-m_s h_c b}{2L} \\ \frac{ma}{2L} & -\frac{I_z}{2L} & -\frac{m_s h_c a+I_{xz}}{2L} \end{bmatrix} \begin{bmatrix} J_y \\ J_r \\ J_\phi \end{bmatrix} \tag{4.8}$$

where, $J_y = \dot{a}_y$ is the rate of change of lateral acceleration, $J_r = \ddot{r}$ denotes the rate of change of yaw acceleration, $J_\phi = \dddot{\phi}$ stands for the rate of change of roll angular acceleration.

2. Estimation of tire effective cornering stiffness [9]

According to the theory of tire mechanical properties, ignoring the camber angle, the lateral force F_y and longitudinal force F_x of the tire can be expressed as functions of the tire slip angle α, the tire slip rate s, and the tire vertical load F_z, namely

$$F_y = F_y(\alpha, s, F_z) \tag{4.9}$$

$$F_x = F_x(\alpha, s, F_z) \tag{4.10}$$

Taking the derivative of Eqs. (4.9) and (4.10) gives

$$\frac{dF_y}{dt} = \frac{\partial F_y}{\partial \alpha}\frac{d\alpha}{dt} + \frac{\partial F_y}{\partial s}\frac{ds}{dt} + \frac{\partial F_y}{\partial F_z}\frac{dF_z}{dt} \tag{4.11}$$

$$\frac{dF_x}{dt} = \frac{\partial F_x}{\partial \alpha}\frac{d\alpha}{dt} + \frac{\partial F_x}{\partial s}\frac{ds}{dt} + \frac{\partial F_x}{\partial F_z}\frac{dF_z}{dt} \tag{4.12}$$

Based on relevant definition, the cornering stiffness refers to the slope of the curve $F_y - \alpha$ at $\alpha = 0$, which means $k = \left(\frac{\partial F_y}{\partial \alpha}\right)_{\alpha=0}$. While the value just reflects lateral force of slip angle when is α small, it refers to the gain of lateral force relative to slip angle when $\alpha = 0°$. As the curve $F_y - \alpha$ enters the nonlinear stage, only the effective cornering stiffness can accurately reflect the relationship between the slip angle and the lateral force at any time [10]:

$$k_e = \frac{\partial F_y}{\partial \alpha} \tag{4.13}$$

It can be seen that replacing the constant cornering stiffness with the instantaneous effective cornering stiffness information is of great significance for the analysis of vehicle handling stability, especially for large slip angles under severe working conditions.

When $\dot{\alpha} \neq 0$, Eq. (4.11) can be written as

$$\frac{\partial F_y}{\partial \alpha} = k_e = \frac{1}{\dot{\alpha}}\left(\frac{\mathrm{d}F_y}{\mathrm{d}t} - \frac{\partial F_y}{\partial F_z}\dot{Z} - \frac{\partial F_y}{\partial s}\dot{s}\right) \tag{4.14}$$

Supposing there are no couple moment of inertia resistance and air lift, the vertical load of the front and rear wheels when the car turns right is calculated as [8]:

$$F_{zfl} = \frac{mgb}{2L} - \frac{ma_x h}{2L} + \Delta F_{zfy} \tag{4.15}$$

$$F_{zfr} = \frac{mgb}{2L} - \frac{ma_x h}{2L} - \Delta F_{zfy} \tag{4.16}$$

$$F_{zrl} = \frac{mga}{2L} + \frac{ma_x h}{2L} + \Delta F_{zry} \tag{4.17}$$

$$F_{zrr} = \frac{mga}{2L} + \frac{ma_x h}{2L} - \Delta F_{zry} \tag{4.18}$$

where, F_{zfl}, F_{zfr}, F_{zrl}, and F_{zrr} are the vertical loads of the left front, right front, left rear, and right rear wheels, respectively. The first term is the static load component; the second term is the load transfer component under the action of the longitudinal force; The third term is the load transfer component under the action of the lateral force; a_x is the longitudinal acceleration; h is the height of the center of mass. It can be seen that the vertical loads of the left and right wheels are different and change continuously when the car is moving in a curve, but as long as the longitudinal acceleration a_x remains unchanged, the total vertical load of the front and rear axles does not change.

The average vertical load of the front and rear wheels is:

$$F_{zf} = \frac{F_{zfl} + F_{zfr}}{2} = \frac{mgb}{2L} - \frac{ma_x h}{2L} \tag{4.19}$$

$$F_{zr} = \frac{F_{zrl} + F_{zrr}}{2} = \frac{mga}{2L} + \frac{ma_x h}{2L} \tag{4.20}$$

Taking the derivative of Eqs. (4.19) and (4.20), we get

$$\frac{dF_{zf}}{dt} = -\frac{mh}{2L}J_x \tag{4.21}$$

$$\frac{dF_{zr}}{dt} = \frac{mh}{2L}J_x \tag{4.22}$$

If the vehicle longitudinal acceleration $a_x = const$ (constant), then $\dot{a}_x = J_x = 0$, $\dot{F}_z = 0$ and $\frac{dF_x}{dt} = mJ_x = 0$.

Substitute $\dot{F}_z = 0$ and $\frac{dF_x}{dt} = 0$ into Eq. (4.12), we have

$$\frac{\partial F_x}{\partial \alpha}\dot{\alpha} + \frac{\partial F_x}{\partial s}\dot{s} = 0$$

There is

$$\dot{s} = -\frac{\frac{\partial F_x}{\partial \alpha}}{\frac{\partial F_x}{\partial s}}\dot{\alpha} \tag{4.23}$$

Substituting Eq. (4.23) and $\dot{F}_z = 0$ into Eq. (4.14) produces

$$k_e = \frac{1}{\dot{\alpha}}\left(\frac{dF_y}{dt} + \frac{\partial F_y}{\partial s}\frac{\frac{\partial F_x}{\partial \alpha}}{\frac{\partial F_x}{\partial s}}\dot{\alpha}\right) = \frac{1}{\dot{\alpha}}\frac{dF_y}{dt} + \frac{\frac{\partial F_y}{\partial s}\cdot\frac{\partial F_x}{\partial \alpha}}{\frac{\partial F_x}{\partial s}} \tag{4.24}$$

According to the theory of tire characteristics, the slope of the curve $F_x - \alpha$ is large (the curve is steep) at small slip rates of $\frac{dF_x}{dt} = 0$, and $\frac{\partial F_x}{\partial s}$ is therefore numerically large. In addition, since the influence of slip rate on lateral force and the influence of slip angle on the longitudinal force are small, both $\frac{\partial F_y}{\partial s}$ and $\frac{\partial F_x}{\partial \alpha}$ are small, which means their product $\frac{\partial F_y}{\partial s}\cdot\frac{\partial F_x}{\partial \alpha}$ is also small. Considering the condition that $\left(\frac{\partial F_y}{\partial s}\cdot\frac{\partial F_x}{\partial \alpha}\right)\Big/\frac{\partial F_x}{\partial_s}$ is small, it can be negligible compared to the magnitude of cornering stiffness.

According to Eq. (4.24), the effective cornering stiffness of the front and rear wheels can be written as

$$k_{ef} = \left(\frac{1}{\dot{\alpha}_r}\frac{dF_{yf}}{dt}\right)_{J_x=0},\quad k_{er} = \left(\frac{1}{\dot{\alpha}_r}\frac{dF_{yr}}{dt}\right)_{J_x=0} \tag{4.25}$$

It can be seen from Eq. (4.25) that the effective cornering stiffness of the tire can be approximated by the ratio of the time change rate of the lateral force to the time rate of the slip angle under the condition $J_x = 0$. The time rate of the lateral force is determined by the acceleration rate of change J_y, J_r and J_ϕ from Eq. (4.8). Since the estimated environment in this chapter is constant radius steering, the lateral forces of the left and right wheels are not equal, where F_{yf} and F_{yr} refer to the average lateral forces of the left and right wheels of the front and rear axles at a certain moment. Therefore, the effective cornering stiffness in the chapter is the effective cornering stiffness under static axle loads.

After obtaining the effective cornering stiffness, we can approximate the tire lateral force of a certain slip angle according to the recursive Eq. (4.26), and then obtain the cornering characteristic curve of tire slip angle and lateral force by sorting the slip angles.

$$F_y^k = F_y^{k-1} + k_e^{k-1}\left(\alpha^k - \alpha^{k-1}\right) \tag{4.26}$$

4.3.2 *Feasibility Analysis of Estimation Method*

While employing Eq. (4.25) to estimate the effective cornering stiffness, the constant radius steering (CRC) condition with constant acceleration and the constant speed single lane shift condition are both ideal estimation environments. Because the use of these working conditions can keep the longitudinal motion steady ($J_x = 0$) at each moment, which is a necessary condition for the estimation method in this chapter. On the other hand, they can keep the lateral acceleration, yaw acceleration, and roll acceleration at a certain rate of change, which means J_y, J_r and J_ϕ are all non-zero quantities. Eventually, the effects of lateral motion, yaw motion, and roll motion can be taken into account.

1. Estimation of acceleration rate

There are two methods to obtain the acceleration rate in Eq. (4.8):

(1) Direct measurement through a jerk transducer/sensor, also known as a differential acceleration sensor, which can sense acceleration and convert it into a usable output signal [11];
(2) Referring to the serpentine test, the lateral acceleration, yaw rate and body roll angle are measured in constant radius cornering test. Conduct classical Savitzky-Golay least-squares smoothing filtering for these discrete high-frequency noise signals [12, 13] and the ideal signal of J_y, J_r and J_ϕ can be obtained. The terms $\mathrm{d}F_{yf}/\mathrm{d}t$ along with $\mathrm{d}F_{yr}/\mathrm{d}t$ in Eq. (4.25) are then estimated according to Eq. (4.8).

The Savitzky-Golay algorithm in the second method is a common filtering algorithm of data smoothing. This algorithm does not use a simple average calculation, but introduces a simplified polynomial least squares fitting convolution in the moving operation of smoothing window. By using this method, the curve can be smoothed and the derivatives of the smoothed curve can be calculated [12]. The algorithm can not only eliminate noise and improve the signal-to-noise ratio of the analyzed signal, but also better maintain the useful information, so it is a relatively good smoothing filtering method. Although it was proposed as early as the 1960s, it is still frequently used in many signals processing fields [13]. The specific equation of the algorithm can be found in the literature [12], which will not be repeated here due to space limitations.

Taking the double lane change maneuvering at the speed of 120 km/h in ADAMS/Car as an example, the effectiveness of Savitzky-Golay (S-G) differential smoothing filter to smooth and differentiate the noise-containing measurement signal to obtain $\mathrm{d}F_{yf}/\mathrm{d}t$ and $\mathrm{d}F_{yr}/\mathrm{d}t$. Figure 4.4 shows the measurement signals of lateral acceleration, yaw rate, and body roll angle with noise interference, which contains uniform white Gaussian noise sequences with a mean value of 0 and a standard deviation of 5% of their respective amplitudes.

The Savitzky-Golay filter parameters are set by trying. For the signal J_y, the width of the smoothing window is chosen as 21 points. The polynomial is three order and the first derivative is chosen; for the signal J_r, the width of the smoothing window is selected as 21 points. The polynomial order is taken as the 7th order with the second-order derivative; for J_ϕ, the smoothing window width is taken as 21 points and the polynomial order is taken as the 5th order, and the third-order derivative is selected. Figure 4.5 is the comparison between the J_y, J_r and J_ϕ obtained by the Savitzky-Golay filtering and the actual ideal signals.

It can be seen from Fig. 4.5 that the curves of signals J_y, J_r and J_ϕ obtained by the Savitzky-Golay smooth differential filter along with the ideal ones are in good agreement. Because of the smooth filtering, the amplitude of the filtered signal at some sharp points is somewhat different from the ideal signal, but it does not have

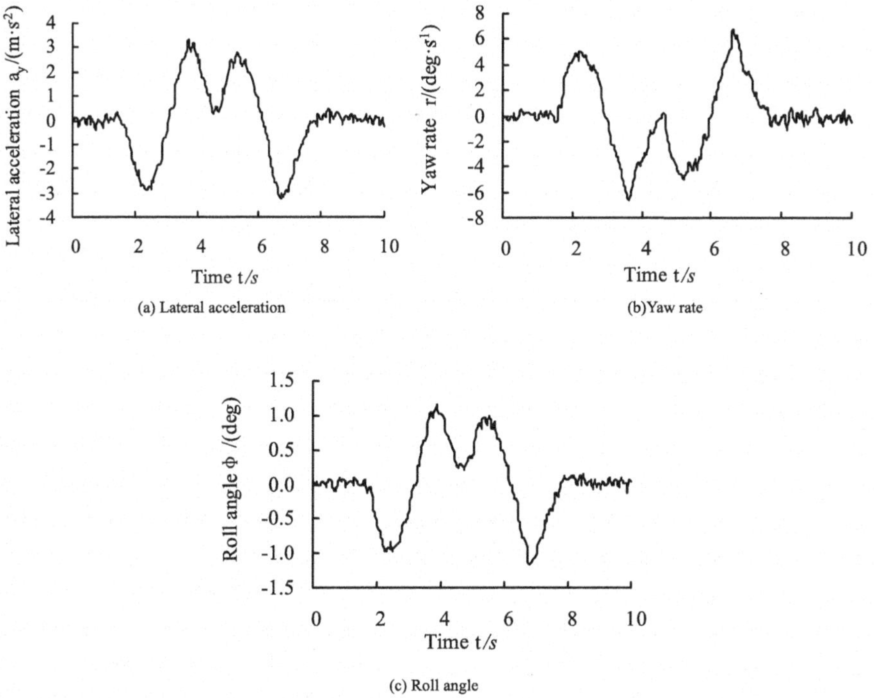

Fig. 4.4 Measured signal with noise

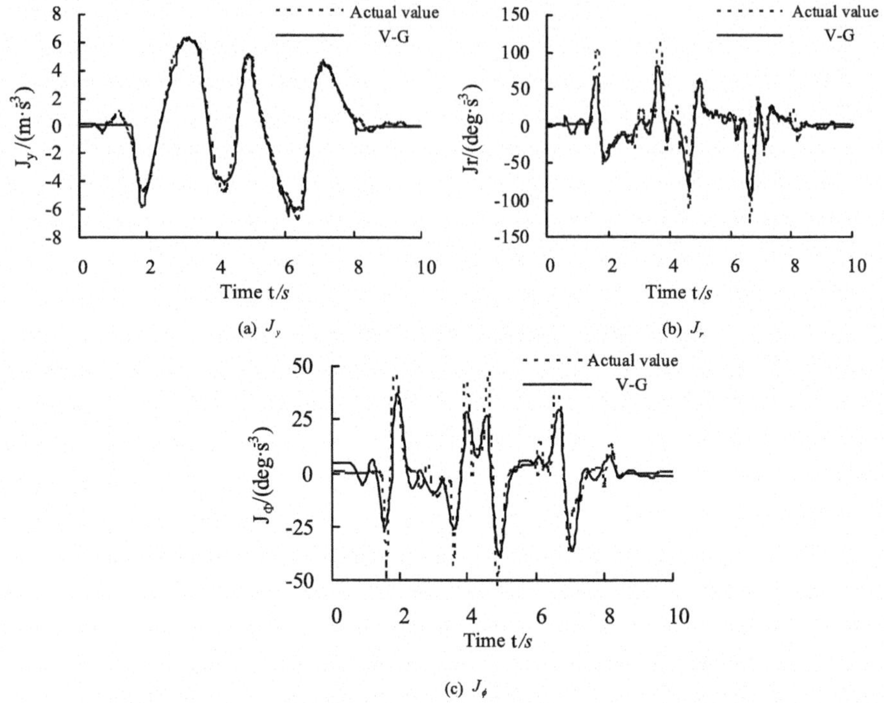

Fig. 4.5 Comparison between signals J_y, J_r, J_ϕ obtained by S-G method and ideal signals

much effect on the estimate of the effective cornering stiffness. With the acceleration rate, the tire lateral force change rate can be calculated by the Eq. (4.8).

2. Estimation of tire slip angle

In the literature [8], the slip angles of the front and rear wheels can be presented as $\alpha_f = \delta - \arctan \frac{v+ar}{u}$ and $\alpha_r = -\arctan \frac{v-br}{u}$, respectively.

In practical applications, longitudinal speed u and lateral speed v can be measured by RTK or non-contact photoelectric fifth wheel system; steering wheel angle is measured by steering angle tester. After being filtered and denoised, the measurement signal could estimate α_f and α_r. In order to simplify the research steps, the measurement signal without noise is directly extracted to calculate the ideal slip angle signal in the following virtual experiment, and $\dot{\alpha}_f$ along with $\dot{\alpha}_r$ is obtained through interpolation-type numerical differentiation.

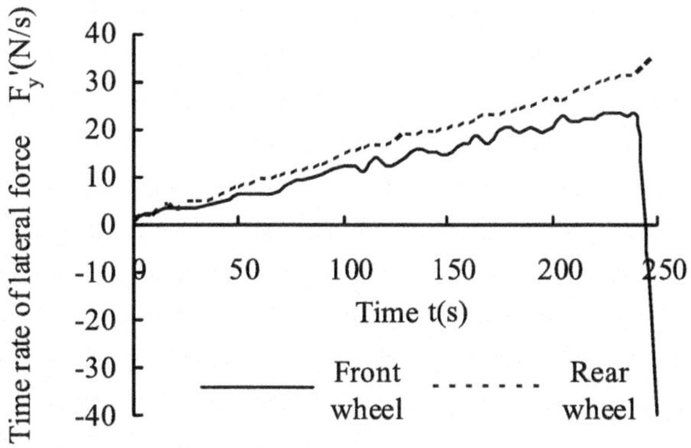

Fig. 4.6 Time rates of lateral force (constant radius steering)

4.3.3 Simulation Test Verification

In order to verify the correctness of the proposed method, a simulation test of a certain type of car was carried out on the virtual prototype software ADAMS. The parameters of the car are set as: $m = 1528\text{kg}$, $m_s = 1312\text{kg}$, $a = 1.38\text{m}$, $b = 1.18\text{m}$, $h = 0.45\text{m}$, $I_z = 2488\text{kg} \cdot \text{m}^2$, $I_{xz} = 0$.

1. Constant radius cornering condition

Firstly, the constant-radius cornering manipulation environment in the vehicle simulation is adopted. In manipulation, the vehicle turning radius is kept constant and the throttle opening is small so that the car could accelerate uniformly along a given radius. Set the simulation time as 250 s. The step size is 0.05 s and the turning radius is set to 40 m. Let the car accelerate until the lateral acceleration reaches 1 g and the simulation ends.

The calculated time rates of the lateral force of the front and rear wheels are shown in Fig. 4.6. The time rates of the slip angle of the front and rear wheels are displayed as Fig. 4.7.

From Fig. 4.6, it can be seen that the time rate of the lateral force of the front wheel is slightly smoother than that of the rear wheel. The engine rear-mounted model is used in this chapter, so the center of mass of the vehicle is close to the rear axle and the vertical distribution on the rear axle is larger, resulting in larger rate of change in the lateral force of the rear wheels. In addition, the time rate of the front wheel lateral force changes greatly in the end, from a positive value to a negative value. With the increase of slip angle, the lateral force begins to decreased, which is consistent with the tire cornering characteristic curve. Figure 4.7 shows that when the lateral acceleration is small, the curves of the front and rear wheel slip angles are relatively flat. In the end of the simulation, the tire slip angle increases at a large rate when there is larger lateral acceleration.

Figure 4.8 is the time-dependent curve of the effective cornering stiffness (absolute value) of the front and rear wheels estimated by the method in this chapter. As can be seen from Fig. 4.8, the effective cornering stiffness of the front and rear wheels

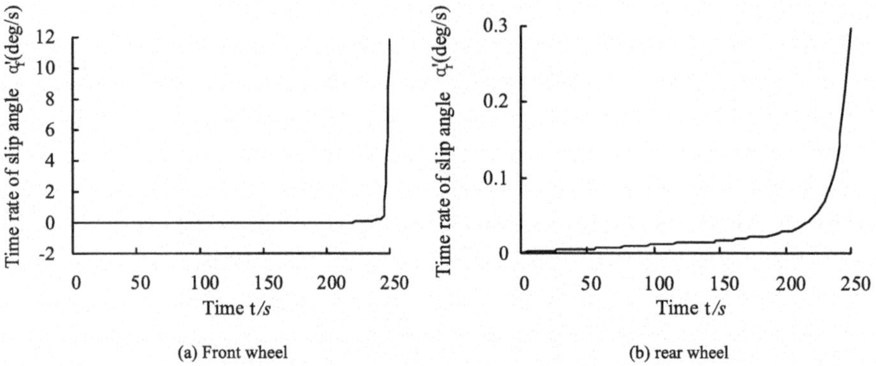

Fig. 4.7 Time rates of slip angle (constant radius steering)

gradually decreases with time, and the change is relatively gentle before about 150 s. At this time, the cornering characteristics are in an approximate linear state; after about 150 s, the cornering stiffness decreases rapidly and the tire enters a nonlinear working condition. The slip angle-lateral force curve becomes flat rapidly.

Sorted according to the slip angle, Fig. 4.9 shows the comparison between the estimated values and the actual ones of the front and rear tire slip characteristic curves.

In this section, the actual values of the cornering characteristic curves of the front and rear wheels are obtained by fitting the parameters of the Magic Fomula tire model based on pac89 in the tire subsystem of ADAMS. The vertical loads of the front and rear wheels are $F_{zf} = 3.458$ kN and $F_{zr} = 4.037$ kN.

It can be seen from Fig. 4.9 that the estimated values are in good correspondence with the actual ones overall. The estimation method can provide information of the tire cornering characteristics without the tire cornering characteristics test bench.

Fig. 4.8 Effective cornering stiffness (constant radius steering)

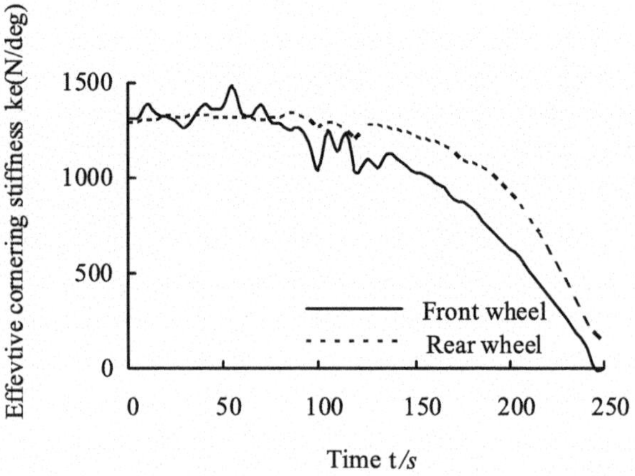

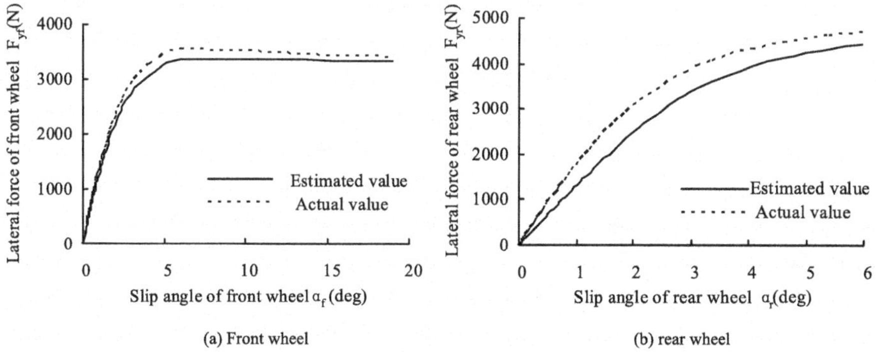

Fig. 4.9 Estimated and actual front and rear wheel cornering characteristic curves (constant radius cornering)

After analysis, the errors between the estimated and the actual values mainly comes from the following aspects:

(1) J_y, J_r and J_ϕ obtained by Savitzky-Golay smoothing filter are different from actual values;
(2) The tire lateral force change rate calculated by Eq. (4.8) is based on a simple three-degree-of-freedom vehicle dynamics model, and there is a certain error compared with the actual value of the complex multi-degree-of-freedom prototype model of the virtual test;
(3) Since Eq. (4.26) is an iterative summation process, the accumulated error cannot be ignored.

2. Single lane change maneuver

As mentioned above, the estimation method in this chapter can also estimate the effective cornering stiffness in the single lane change maneuver with constant velocity. The single lane change maneuver in the open-loop simulation of ADAMS is equivalent to the one-cycle sinusoidal input of the steering wheel angle. The simulation time is selected as 4 s and the step size is set as 0.01 s. The vehicle speed is chosen as 110 km/h. The single lane change maneuver with no straight transitions is constructed with the one-cycle sinusoidal steering wheel angle input with an amplitude of 90°. The steering wheel angle and the lateral displacement of centroid are shown in Figs. 4.10 and 4.11.

The calculated time rate of the lateral force of the front and rear wheels is shown in Fig. 4.12; the time rate of the slip angle of the front and rear wheels is shown in Fig. 4.13.

As can be seen from Fig. 4.12, similar to the constant radius steering condition, the time change rate of the lateral force of the front wheel is slightly smoother than that of the rear wheel, and changes abruptly around 2 s. Due to the positive position transition of steering wheel, the lateral force changes from negative to positive

Fig. 4.10 Steering wheel angle figure (single lane change)

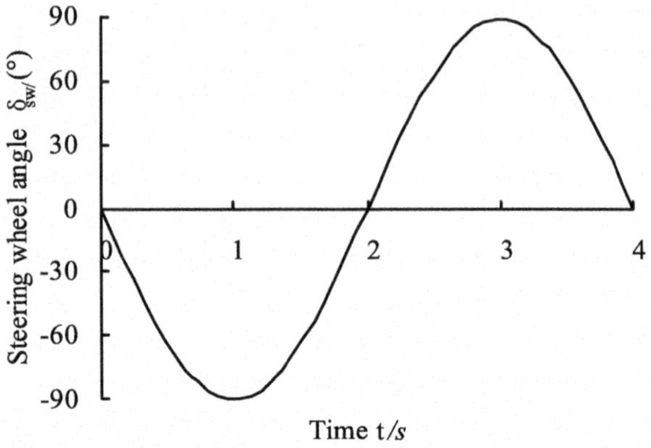

Fig. 4.11 Lateral displacement of centroid (single lane change)

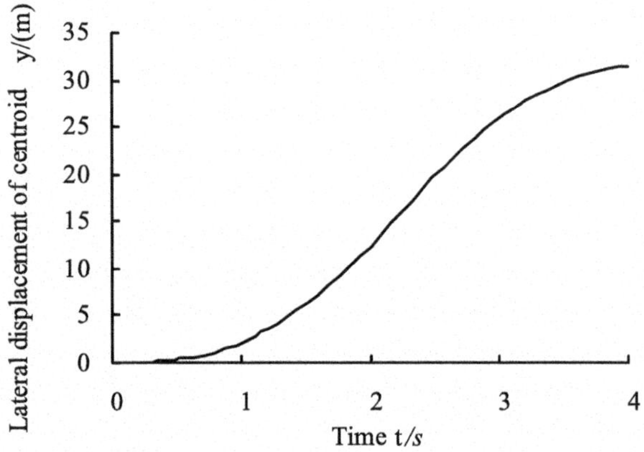

Fig. 4.12 Time change rate of lateral force (single lane change)

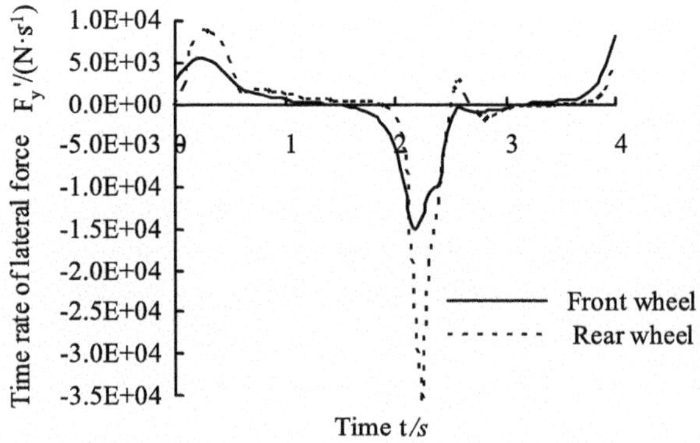

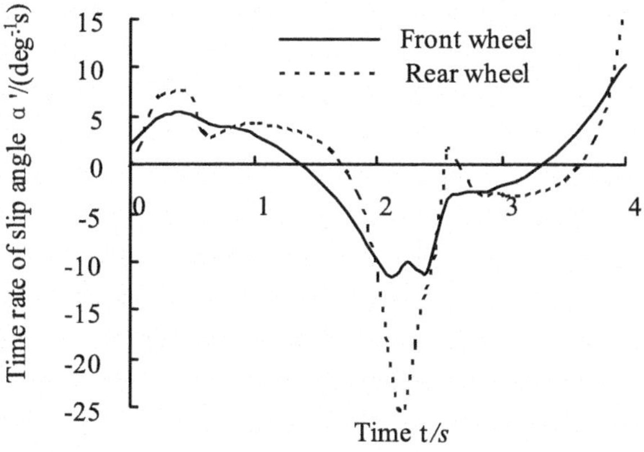

Fig. 4.13 Time change rate of slip angle (single lane change)

severely. Figure 4.13 presents the similar trend for the slip angle and lateral force of the front and rear wheels.

Different from the constant radius steering condition, the time rate curve of the slip angle has two zero-crossing points in the single lane change maneuver (as shown in Fig. 4.13), thereby causing the infinity of the effective cornering stiffness estimated value. To avoid this situation, the rate limiter module in MATLAB/Simulink is added to reduce the influence on the result curve when the change rate is close to zero. The rate-of-change controller block constrains the first derivative of its signal (the rate of change), which is calculated by the following equation:

$$\text{rate} = \frac{u(i) - y(i-1)}{t(i) - t(i-1)} \tag{4.27}$$

where $u(i)$ and $t(i)$ are the input and time of the current module. $y(i-1)$ and $t(i-1)$ are the output and time of the previous step. The output of the module depends on the comparison with the rising slope threshold R_s and the falling slope threshold F_s:

If the rate of change is greater than the rising slope R_s, the output is calculated by:

$$y(i) = \Delta t \cdot R_s + y(i-1) \tag{4.28}$$

If the rate of change is less than the falling slope F_s, the output is calculated by:

$$y(i) = \Delta t \cdot F_s + y(i-1) \tag{4.29}$$

If the rate of change is between R_s and F_s, the input is equal to the output.

$$y(i) = u(i) \tag{4.30}$$

In this section, R_s and F_s are chosen as 10^4 and -10^4, respectively. Then, the sudden change of the estimated curve of the effective cornering stiffness is smoothed around 1.5 and 3.5 s.

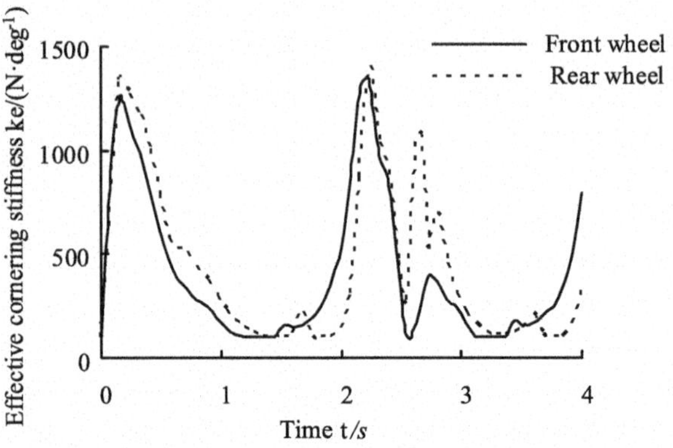

Fig. 4.14 Estimated value of effective cornering stiffness of front and rear wheels (single lane change)

Figure 4.14 shows the time-dependent curve of the effective cornering stiffness (absolute value) of the front and rear wheels estimated by the method in this chapter.

From Fig. 4.14, it can be seen that the variation trend is similar to the W shape. When the steering wheel angle reaches the amplitude, the effective cornering stiffness of the tire is the smallest because the nonlinear trend reaches the maximum and the slope of the lateral force-slip angle curve is the smallest. In the same way, when the steering wheel returns to center, the effective cornering stiffness of the tire is the largest. Meanwhile, the tire nonlinear trend reaches the minimum and the slope of curve becomes maximum.

In the same order according to the tire side slip angle from small to large, Fig. 4.15 shows the comparison between the estimated values and the actual ones of the cornering characteristic curve of the front and rear wheels under the single lane change maneuver.

It can be observed from Fig. 4.15 that the alignment between the estimated and actual values of the front wheel is higher than that of the rear wheel, mainly because the rear wheel has more severe changes in the slip angle and lateral force than the

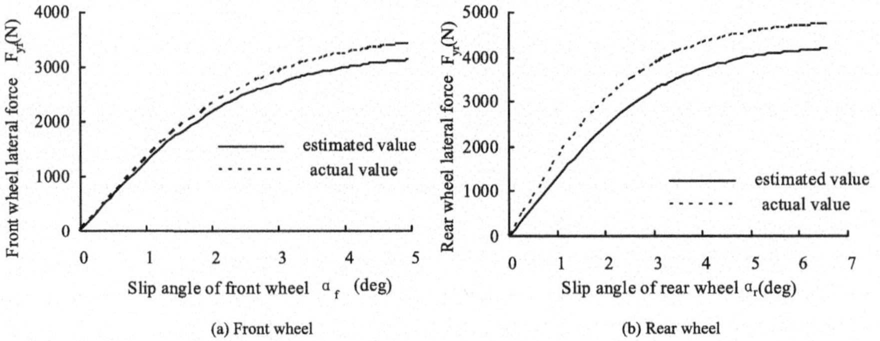

Fig. 4.15 Estimated and actual front and rear wheel cornering characteristic curves (single lane change)

front wheel. With the same step size, the error accumulation generated by Eq. (4.26) will also be slightly larger. Therefore, if the step size is controlled relatively smaller, the result will be more accurate.

References

1. Rajamani R. Vehicle dynamics and control. Minneapocis: Springer; 2011.
2. Yu F, Lin Y. Automotive system dynamics. Beijing: Machinery Industry Press; 2005.
3. Irie N, Kuroki J. 4WS technology and the prospects for improvement of vehicle dynamics. In: Vehicle electronics in the 90s: international congress on transportation electronics; 1990.
4. Ackermann J. Robust control prevents car skidding. IEEE Control Syst Mag. 1997;17(3):23–31.
5. Bevly DM, Ryu J, Gerdes JC. Integrating INS sensors with GPS measurements for continuous estimation of vehicle sideslip, roll, and tire cornering stiffness. IEEE Trans Intell Transp Syst. 2006;7(4):483–93.
6. Drakunov SV, Ashrafi B, Rosiglioni A. Yaw control algorithm via sliding mode control. In: American Control Conference, Chicago; 2000.
7. Rajesh R. Lateral vehicle dynamics. Boston: Springer; 2012. p. 15–46.
8. Guo K. Vehicle handling dynamics. Chang Chun: Jilin Science and Technology Press; 1991.
9. Lin F, Zhao Y. Estimation of tire effective cornering stiffness based on the time rate of change of acceleration. Mech Sci Technol Aerosp Eng. 2008;27(8):1092–6.
10. Zhao Y, Yin H, Zhang L, et al. Nonlinear motion stability analysis of vehicle by two kinds of steering input. Mech Sci Technol Aerosp Eng. 2005;24(10):1138–40.
11. National Standard of the People's Republic of China. General terminology for sensors. GB/T 7665-2005.
12. Savitzky A, Golay M. Smoothing and differentiation of data by simplified least squares procedures. Anal Chem. 1964;36(8):1627–39.
13. Nakazawa S, Ishihara T, Inooka H. Real-time algorithms for estimating jerk signals from noisy acceleration data. Int J Appl Electromagn Mech. 2003;18:149–63.

Chapter 5
Evaluation of Active Safety Based on Driver-Vehicle Closed-Loop Control System Dynamics

5.1 Theoretical Basis

Introduce the basic knowledge of deterministic driver-vehicle–road closed-loop system model, evolutionary stochastic process, Kronecker algebra and second-order moment techniques for vector values and matrix functions used in this chapter.

5.1.1 Driver-Vehicle–Road Closed-Loop System Model

1. Driver model

According to the optimal curvature control theory [1], the block diagram of the driver-vehicle–road closed-loop system can be obtained as shown in Fig. 5.1. This figure is the PO preview model in the optimal preview acceleration model, that is, the position plus square preview model.

Define $f(s)$ as road input (m), λ is the azimuth action coefficient, $V(s)$ is the transfer function of vehicle lateral acceleration a_y to steering wheel angle δ_{sw}, T_d is the delay time of driver's neural response (s), T_h is the delay time of driver's manipulation response (s), T is the preview time of the driver (s), $T_c(a, T)$ is the driver's correction time (s), a is the following order of the driver, C_o is the driver correction parameter, $\ddot{y}^*$ is the ideal lateral acceleration (m/s^2), δ^*_{sw} is the ideal steering wheel angle (°), δ_{sw} is the actual steering wheel angle (°).

$$V(S) = \frac{a_y}{\delta_{sw}}(S) = G_{ay}\frac{1 + T_{y1}S + T_{y2}S^2 + \cdots}{1 + T_1S + T_2S^2 + \cdots} \tag{5.1}$$

where, G_{ay} is the steady-state gain of lateral acceleration, T_{y1}, T_{y2}, T_1, T_2 etc. are constants.

Y. Zhao and F. Lin, *Vehicle System Dynamics*,
https://doi.org/10.1007/978-981-97-2019-4_5

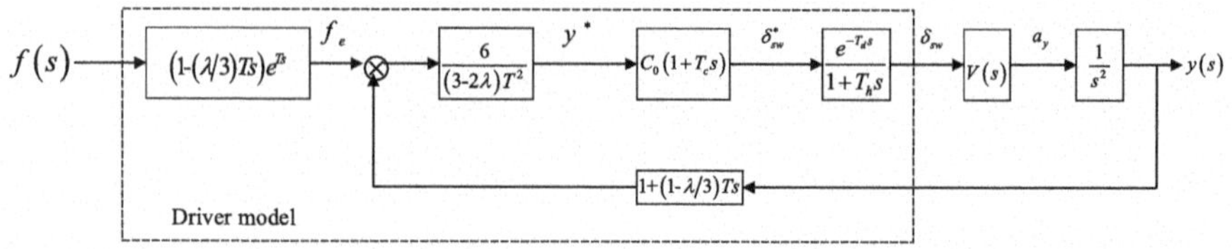

Fig. 5.1 Block diagram of driver-vehicle–road closed loop system

Therefore, the state equation of the driver model can be derived as (the specific derivation process is omitted):

$$\begin{cases} \dot{x}_1 = x_2 \\ \dot{x}_2 = -\dfrac{2}{T_h t_d}x_1 - \dfrac{2T_h + t_d}{T_h t_d}x_2 \\ +\dfrac{12C_o}{T^2 T_h t_d(3-2\lambda)}\left(f(t+T) + \left(T_c - \dfrac{\lambda T}{3}\right)\dot{f}(t+T) - \dfrac{\lambda}{3}TT_c\ddot{f}(t+T) \right. \\ \left. -\left(y + \left(T - \dfrac{\lambda}{3}T + T_c\right)\dot{y} + \left(1 - \dfrac{\lambda}{3}\right)TT_c\ddot{y}\right)\right) \end{cases} \tag{5.2}$$

where, $f(t+T)$ is the road input at time $(t+T)$ (m), y, $\dot{y}$ and $\ddot{y}$ are the lateral displacement, lateral velocity and lateral acceleration of the vehicle at time t respectively.

The output of the driver link is the steering wheel angle, and its expression is

$$\delta_{sw} = x_1 - \frac{t_d}{2}x_2 \tag{5.3}$$

where, x_1 and x_2 are the state variables of the driver link.

2. Vehicle model

This chapter will study the lateral motion and yaw motion of the car moving at a uniform speed, and only do small angle motion, so the two degree of freedom linear car model can be used, as shown in Fig. 5.2. The linear tire model used is:

$$\begin{cases} Y_1 = k_1\beta_1 \\ Y_2 = k_2\beta_2 \end{cases} \tag{5.4}$$

The state equation of the two degree of freedom linear vehicle model is:

$$\begin{bmatrix} \dot{r} \\ \dot{\beta} \end{bmatrix} = \begin{bmatrix} a_{11} & a_{12} \\ a_{21} & a_{22} \end{bmatrix}\begin{bmatrix} r \\ \beta \end{bmatrix} + \begin{bmatrix} b_1 \\ b_2 \end{bmatrix}\delta_{sw} \tag{5.5}$$

where, Y_1 and Y_2 are the ground lateral force on the front wheel and the rear wheel (N), k_1 and k_2 are the sum of the combined lateral stiffness of front and rear axle tires

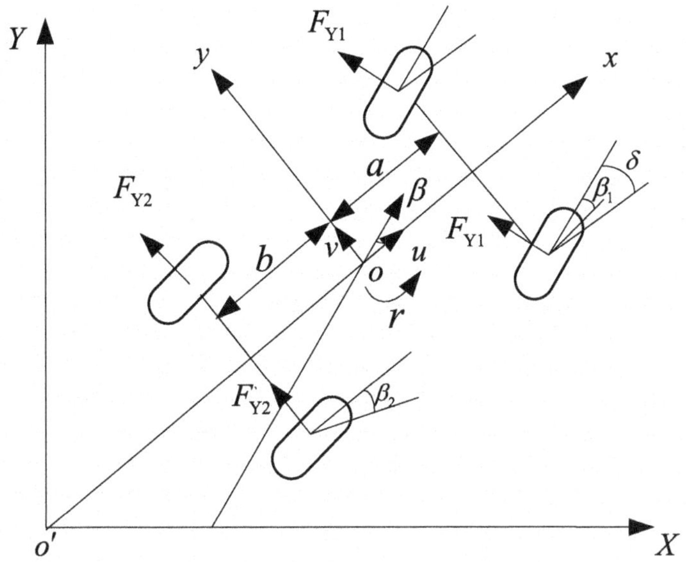

Fig. 5.2 Two degree of freedom vehicle handling dynamics model

(N/rad), β_1 and β_2 are front and rear axle side angles respectively (°), r is the yaw rate of the vehicle (°/s), β is the side slip angle at the center of mass of the vehicle (°), δ_{sw} is the steering wheel angle, $\delta_{sw} = i\delta$ (°), δ is the front wheel angle (°), i is the total transmission ratio of the steering system.

3. Mathematical model of road input

The road in the figure refers to the center line of the actual road, as shown in Fig. 5.3. The abscissa in the figure represents the forward direction of the car, and the ordinate represents the left-hand direction of the driver. The mathematical model of the actual road centerline is

$$f(x) = \begin{cases} 0 & x \in [0 \ a_0) \\ \frac{b}{a_1 - a_0}(x - a_0) & x \in [a_0 \ a_1) \\ b & x \in [a_1 \ a_2) \\ \frac{b}{a_2 - a_3}(x - a_3) & x \in [a_2 \ a_3) \\ 0 & x \in [a_3 \ \infty) \end{cases} \quad (5.6)$$

When doing small angle simulation calculation, requirements $a_0 \geq d = uT, \alpha' \leq 10°, \alpha \leq 10°$, here $\alpha' = \arctan \frac{b}{a_2 - a_3}, \alpha = \arctan \frac{b}{a_1 - a_0}$. Coordinate transformation is needed in large angle simulation calculation. In the actual driving process, the ideal target trajectory in the driver's mind should be a low-order smooth curve closest to the broken line in Fig. 5.3. After cubic spline fitting, a third-order curve with continuous first-order derivatives can be obtained:

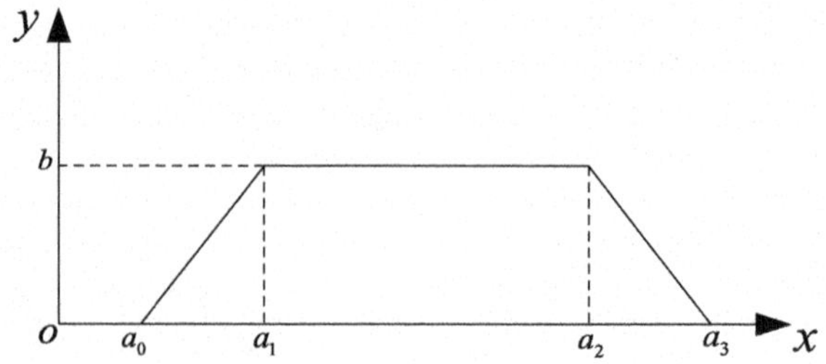

Fig. 5.3 Schematic diagram of secondary lane change

$$f(x) = \begin{cases} 0 & x \in [0\ a_0) \\ e_0 + e_1 x + e_2 x^2 + e_3 x^3 & x \in [a_0\ a_1) \\ b & x \in [a_1\ a_2) \\ e_0' + e_1' x + e_2' x^2 + e_3' x^3 & x \in [a_2\ a_3) \\ 0 & x \in [a_3\ \infty) \end{cases} \tag{5.7}$$

where, $e_0 = a_0^2(3a_1 - a_0)b/d^3$, $e_1 = -6a_0a_1b/d^3$, $e_2 = 3(a_0 + a_1)b/d^3$, $e_3 = -2b/d^3$, $d = a_1 - a_0$, $e_0' = a_3^2(3a_2 - a_3)b/d^3$, $e_1' = -6a_3a_2b/d^3$, $e_2' = 3(a_3 + a_2)b/d^3$, $e_3' = -2b/d^3$, $d' = a_2 - a_3$.

When the car starts from the origin and travels along the x-axis at a constant speed u, $f(x)$ can be transformed into a function of time t, and $x = ut$ can be substituted into Eq. (5.7).

4. Mathematical model of driver-vehicle–road closed-loop control system dynamics

Under the condition of a small yaw angle, for the two degrees of freedom linear vehicle model, there are

$$\begin{cases} v \approx u\beta \\ a_y \approx u(r + \dot{\beta}) \end{cases} \tag{5.8}$$

order

$$v_a = \dot{y} \tag{5.9}$$

where y is the horizontal coordinate of the vehicle's center of mass in the horizontal plane.

By combining Eqs. (5.2), (5.5), (5.8), (5.9) and (5.3), the following equation is obtained:

$$\begin{bmatrix} \dot{r} \\ \dot{\beta} \\ \dot{y} \\ \dot{v}_a \\ \dot{x}_1 \\ \dot{x}_2 \end{bmatrix} = \begin{bmatrix} a_{11} & a_{12} & 0 & 0 & a_{15} & a_{16} \\ a_{21} & a_{22} & 0 & 0 & a_{25} & a_{26} \\ 0 & 0 & 0 & 0 & 0 & 0 \\ a_{41} & a_{42} & 0 & 0 & a_{45} & a_{46} \\ 0 & 0 & 0 & 0 & 0 & 0 \\ a_{61} & a_{62} & a_{63} & a_{64} & a_{65} & a_{66} \end{bmatrix} \begin{bmatrix} r \\ \beta \\ y \\ v_a \\ x_1 \\ x_2 \end{bmatrix} + \begin{bmatrix} 0 \\ 0 \\ 0 \\ 0 \\ 0 \\ b_6 \end{bmatrix} f_r \tag{5.10}$$

The output variable is

$$\begin{bmatrix} y \\ a_y \\ \delta_{sw} \\ \dot{\delta}_{sw} \\ s_1 \\ s_2 \end{bmatrix} = \begin{bmatrix} 0 & 0 & c_{13} & 0 & 0 & 0 \\ c_{21} & c_{22} & 0 & 0 & c_{25} & c_{26} \\ 0 & 0 & 0 & 0 & c_{35} & c_{36} \\ c_{41} & c_{42} & c_{43} & c_{44} & c_{45} & c_{46} \\ c_{51} & c_{52} & 0 & 0 & c_{55} & c_{56} \\ c_{61} & c_{62} & 0 & 0 & 0 & 0 \end{bmatrix} \begin{bmatrix} r \\ \beta \\ y \\ v_a \\ x_1 \\ x_2 \end{bmatrix} + \begin{bmatrix} 0 \\ 0 \\ 0 \\ d_4 \\ 0 \\ 0 \end{bmatrix} f_r \tag{5.11}$$

where, f_r is the effective road input function of the system (m), v_a is the lateral speed of the vehicle (m/s), s_1 and s_2 are the lateral force coefficients of the front and rear tires respectively.

Elements a_{ij}, b_i, c_{ij} and d_i in the system matrix are functions of vehicle structural parameters and driver model parameters.

In the above derivation process, if the azimuth action coefficient $\lambda = 0.0$, the driver's single point preview optimal acceleration model commonly used in the simulation can be obtained. The derivation process is omitted.

Unless otherwise specified below, the driver's single point preview optimal acceleration model is used.

5.1.2 Evolutionary Random Processes

1. Stationary random process

Stationary random processes are random processes in which the statistical properties do not change with time or with the selection of time origins.

One of the most important properties of stationary random processes is that they can be approximated arbitrarily by a linear combination of simple harmonic oscillations with random amplitudes using the overlay principle. So far, the research and application of stationary random processes are mature.

2. Spectral density

The spectral density of random process $x(t)$ is the average power of $x(t)$ with respect to the distribution of frequency ω.

$$p = \int_{-\frac{T}{2}}^{\frac{T}{2}} |x(t)|^2 dt \tag{5.12}$$

$$\begin{aligned} E[P] &= E\left(\int_{-\frac{T}{2}}^{\frac{T}{2}} (x(t))^2 dt\right) \\ &= \int_{-\frac{T}{2}}^{\frac{T}{2}} E\left((x(t))^2\right) dt = \int_{-\frac{T}{2}}^{\frac{T}{2}} R(0) dt \\ &= \left(\int_{-\infty}^{\infty} s(\omega) d\omega\right) \cdot T \end{aligned} \tag{5.13}$$

3. Evolutionary random processes

Evolutionary random processes can be defined by Stieltjes integrals for stationary random processes.

$$y(t) = \int_{-\infty}^{\infty} A(t, \omega) e^{i\omega t} ds(\omega) \tag{5.14}$$

where, $A(t, \omega)$ is a modulation function and it is a deterministic function of t and ω.

The autocorrelation function of A is given by Eq. (5.15):

$$\begin{aligned} R_y(\tau) &= \int_{-\infty}^{\infty} \int_{-\infty}^{\infty} A(t+\tau, \omega_1) \overline{A}(t, \omega_2) e^{j\omega_1 (t+\tau) - j\omega_2 t} E(dz(\omega_1) d\overline{z}(\omega_2)) \\ &= \int_{-\infty}^{\infty} A(t+\tau, \omega) \overline{A}(t, \omega) e^{j\omega t} s_x(\omega) d\omega \end{aligned} \tag{5.15}$$

Let $\tau = 0$ then get the mean square of $y(t)$.

$$\begin{aligned} E\left(y^2(t)\right) &= \int_{-\infty}^{\infty} (A(t, \omega))^2 s_x(\omega) d\omega \\ &= \int_{-\infty}^{\infty} s_y(t, \omega) d\omega \end{aligned} \tag{5.16}$$

where

$$s_y(t, \omega) = (A(t, \omega))^2 s_x(\omega) \tag{5.17}$$

Since $s_y(t, w)$ is a function of time, it is called an evolutionary spectral density function.

5.1.3 Kronecker Algebra Basis

1. Element Matrix

I_n: unit matrix of $n \times n$; $\underset{(q)}{e_k}$: at point "$k - k$" is the "1" and elsewhere is the zero q-dimensional matrix; $E_{ik}^{(p\times q)} = \underset{(p)}{e_i}\, \underset{(q)}{e_k^{\mathrm{T}}}$: at point "$i - k$" is the "1" and elsewhere is the zero $(p \times q)$-dimensional matrix.

2. *Kronecker* product

The *Kronecker* product is defined as $A \otimes B$:

$$A \otimes B = \begin{bmatrix} a_{11}B & \cdots & a_{1q}B \\ a_{21}B & \cdots & a_{2q}B \\ \vdots & & \vdots \\ a_{p1}B & \cdots & a_{pq}B \end{bmatrix} \tag{5.18}$$

where, $A(p \times q)$, $B(s \times t)$, then the dimension of $A \otimes B$ is $(ps \times qt)$.

3. Transformation Matrix

The transformation matrix is defined as

$$U_{p\times q} = \sum_i^p k \sum^q E_{ik}^{(p\times q)} \otimes E_{ki}^{(q\times p)} \tag{5.19}$$

where, its dimension is $(pq \times pq)$.

4. Column and row vector functions of matrices

The column vector function of matrix $A_{p\times q}$ is

$$\underset{pq\times 1}{cs\ (A)} = \sum_{j=1}^{q} \left(\underset{(q)}{e^{\mathrm{j}}} \otimes I_p \right) A \underset{(q)}{e^{\mathrm{j}}} \tag{5.20}$$

The row vector function of matrix $A_{p\times q}$ is

$$\underset{1\times pq}{rs\ (A)} = \sum_{j=1}^{p} \underset{(p)}{e^{\mathrm{jT}}} A \left(\underset{(p)}{e^{\mathrm{jT}}} \otimes I_q \right) \tag{5.21}$$

5. Properties of *Kronecker* Product

$$k(A \otimes B) = (kA) \otimes B = A \otimes (kB) \tag{5.22}$$

$$A \otimes (B + C) = A \otimes B + A \otimes C \tag{5.23}$$

$$(A + B) \otimes (C + D) = A \otimes C + A \otimes D + B \otimes C + B \otimes D \tag{5.24}$$

$$(A \otimes B)^{\mathrm{T}} = A^{\mathrm{T}} \otimes B^{\mathrm{T}} \tag{5.25}$$

$$A \otimes (B \otimes C) = (A \otimes B) \otimes C = A \otimes B \otimes C \tag{5.26}$$

$$(A \otimes B)(C \otimes D) = AC \otimes BD \tag{5.27}$$

$$(A \otimes B)^{-1} = A^{-1} \otimes B^{-1} \tag{5.28}$$

$$B \otimes A = U_{s\times p}(A \otimes B)U_{q\times t} \tag{5.29}$$

6. *Kronecker* Power

$$A^{[0]} = 1 \tag{5.30}$$

$$A^{[1]} = A \tag{5.31}$$

$$A^{[k]} = A \otimes A^{[k-1]} \tag{5.32}$$

$$(AB)^{[k]} = A^{[k]} \otimes B^{[k]} \tag{5.33}$$

where, k is any positive integer.

7. Matrix Differentiation

$$\frac{\partial A}{\partial B} = \left[\frac{\partial A}{\partial b_{ij}}\right]_{ps\times qt} \tag{5.34}$$

$$\left(\frac{\partial^n A}{\partial B^n}\right)^T = \frac{\partial^n A^T}{\partial B^{Tn}} \tag{5.35}$$

$$\frac{\partial (AD)}{\partial B} = \frac{\partial A}{\partial B}(I_t \otimes D) + (I_s \otimes A)\frac{\partial D}{\partial B} \tag{5.36}$$

$$\frac{\partial (A \otimes C)}{\partial B} = \frac{\partial A}{\partial B} \otimes C + (I_s \otimes U_{\mathrm{p}\times r})\left(\frac{\partial C}{\partial B} \otimes A\right)(I_t \otimes U_{l\times q}) \tag{5.37}$$

where, $A(p \times q)$, $B(s \times t)$, $C(r \times l)$, $D(q \times u)$.

8. Taylor Series Expansion of Vector Value and Matrix Value Functions

If matrix $A(p \times q)$ is a function of matrix $B(s \times t)$, then the Taylor series expansion of A at the nominal value $\overline{B}$ is

$$A(B) = A(\overline{B}) + \sum_{k=1}^{n} \frac{1}{k!} A_k \left(\left(cs\left(B - \overline{B}\right) \right)^{[k]} \otimes I_q \right) + R_{n+1}(\overline{B}, B) \tag{5.38}$$

$$A_k = \frac{\partial^k A}{\partial (cs(B))^{Tk}} \Big|_{B=\overline{B}} \tag{5.39}$$

$$R_{n+1}\left(\overline{B}, B\right) = o\left(\left(\mathrm{cs}\left(B - \overline{B}\right) \right)^{[n+1]} \right) \tag{5.40}$$

where, $(cs(B - \overline{B}))^{[k]}$ is the k-order *Kronecker* power of $(cs(B - \overline{B}))$.

5.1.4 Second Moment Technique

1. Second moment technique

Assume that random vector $Z = (Z_i)_{r \times 1}$ is a function vector of random parameter vector $b = (b_i)_{s \times 1}$. The mean value is $E(Z) = \overline{Z} = Z(\overline{b})$ and the covariance is $Cov(Z_i, Z_j) = E((Z_i - \overline{Z}_i)(Z_j - \overline{Z}_j))$. $Z = Z(b)$ is expanded to a second order Taylor series at $E(b) = \overline{b}$.

$$Z = Z\left(\overline{b}\right) + \sum_{i=1}^{s} \frac{\partial \overline{Z}}{\partial b_i}\left(b_i - \overline{b}_i\right) + \frac{1}{2} \sum_{i=1}^{s} \sum_{j=1}^{s} \frac{\partial^2 \overline{Z}}{\partial b_i \partial b_j}\left(b_i - \overline{b}_i\right)\left(b_j - \overline{b}_j\right) \tag{5.41}$$

All partial derivatives here take their values at the random parameter vector $\overline{b} = E(b)$.The expressions of second-order approximate mean and first-order approximate second-order moment (covariance) are:

$$E(Z_m) = Z_m\left(\overline{b}\right) + \frac{1}{2} \sum_{i=1}^{s} \sum_{j=1}^{s} \frac{\partial^2 \overline{Z}_m}{\partial b_i \partial b_j} \mathrm{Cov}\left(b_i, b_j\right) \tag{5.42}$$

$$\mathrm{Cov}(Z_m, Z_n) = \sum_{i=1}^{s} \sum_{j=1}^{s} \left(\frac{\partial \overline{Z}_m}{\partial b_i} \right) \left(\frac{\partial \overline{Z}_n}{\partial b_j} \right) \mathrm{Cov}\left(b_i, b_j\right) \tag{5.43}$$

If each random parameter is independent of each other, then Eqs. (5.42) and (5.43) become:

$$\mu = E(Z_m) = Z_m\left(\overline{b}\right) + \frac{1}{2}\sum_{i=1}^{s}\frac{\partial^2 \overline{Z}_m}{\partial b_i^2}\mathrm{Var}(b_i) \tag{5.44}$$

$$\sigma^2 = \mathrm{Var}(Z_m) = \sum_{i=1}^{s}\left(\frac{\partial \overline{Z}_m}{\partial b_i}\right)^2 \mathrm{Var}(b_i) \tag{5.45}$$

2. Second moment technique for vector valued and matrix valued functions

Assume that random vector $Z = (Z_i)_{r\times 1}$ is a function vector of random parameter vector $B = (b_{ij})_{s\times t}$. $E(Z) = \overline{Z} = Z(\overline{B})$ represents the mean vector, and $\mathrm{Var}(Z) = E\left((Z - E(Z))^{[2]}\right)$ represents the variance and covariance matrix. Expand $Z = Z(B)$ into a second-order Taylor series at $E(B) = \overline{B}$.

$$Z = Z\left(\overline{B}\right) + \frac{\partial Z\left(\overline{B}\right)}{\partial(\mathrm{cs}B)^T}\left(cs\left(B - \overline{B}\right)\right) + \frac{1}{2}\frac{\partial^2 Z\left(\overline{B}\right)}{\partial(csB)^{T2}}\left(cs\left(B - \overline{B}\right)\right)^{[2]} \tag{5.46}$$

All partial derivatives here are taken at random parameter matrix $\overline{B} = E(B)$. Take the expression of the second-order approximate mean, the first-order approximate second-order moment (variance and covariance) as:

$$E(Z) = Z\left(\overline{B}\right) + \frac{1}{2}\frac{\partial^2 Z\left(\overline{B}\right)}{\partial(csB)^{T2}}\mathrm{Var}(csB) \tag{5.47}$$

$$\mathrm{Var}(Z) = \left(\frac{\partial Z\left(\overline{B}\right)}{\partial(csB)^T}\right)^{[2]}\mathrm{Var}(csB) \tag{5.48}$$

5.2 Response Analysis Method of Driver-Vehicle–Road Closed Loop Control System

When the vehicle is nonlinear, such as violent movement with large lateral acceleration or movement on the road with low friction coefficient, it must be evaluated on the basis of analyzing the time-domain response of the driver-vehicle–road closed loop system [2]. In addition, if you want to know the transient history of system response, you must do the simulation calculation of time-domain response.

5.2.1 Existing Precise Integration Algorithm

Now, the state Eq. (5.10) of the driver-vehicle–road closed loop system is expressed as the following equation.

$$\dot{x} = Hx + \overline{r} \tag{5.49}$$

where, $x = \left[r \ \beta \ y \ v_a \ x_1 \ x_2 \right]^{\mathrm{T}}, \overline{r} = \left[0\,0\,0\,0\,0\, b_6 f_r \right]^{\mathrm{T}}$

$$H = \begin{bmatrix} a_{11} & a_{12} & 0 & 0 & a_{15} & a_{16} \\ a_{21} & a_{22} & 0 & 0 & a_{25} & a_{26} \\ 0 & 0 & 0 & 0 & 0 & 0 \\ a_{41} & a_{42} & 0 & 0 & a_{45} & a_{46} \\ 0 & 0 & 0 & 0 & 0 & 0 \\ a_{61} & a_{62} & a_{63} & a_{64} & a_{65} & a_{66} \end{bmatrix}$$

the elements in x, $\overline{r}$ and H are described in Chap. 2. The homogeneous solution of Eq. (5.49) is

$$x_h(t) = T(\tau)C \tag{5.50}$$

where

$$T(\tau) = \exp(H \times \tau) \tag{5.51}$$

within the integration step $t \in [t_k, t_{k+1}]$, $\tau = t - t_k$. Temporarily assume that the special solution $x_p(t)$ of Eq. (5.49) has been obtained. Then $t = t_k$ at the initial time of this step, that is, $\tau = 0$, $T(\tau) = [I]$ (unit matrix). Thus, the integral constant vector C can be determined, and then the general solution of Eq. (5.49) is:

$$x(t) = T(\tau)\big(x(t_k) - x_p(t_k)\big) + x_p(t) \tag{5.52}$$

The problem boils down to solving the special solution $x_p(t)$ and the precise calculation $T(\tau)$.

1. Calculation of matrix $T(\tau)$

Integral step τ is internally subdivided into $m = 2^N$ equal divisions

$$\Delta t = \tau/m = 2^{-N}\tau \tag{5.53}$$

generally, $N = 20$ is taken, so $\Delta t \approx 10^{-6}\tau$.$\Delta t$ is generally much smaller than all natural vibration periods of the system. Substitute Eq. (5.53) into Eq. (5.51) to obtain:

$$
\begin{aligned}
T(\tau) &= (\exp(H \times \Delta t))^m \\
&\approx \left(I + H \times \Delta t + (H \times \Delta t)^2 / 2! + (H \times \Delta t)^3 / 3! + (H \times \Delta t)^4 / 4!\right)^m \\
&\equiv (I + T_{a0})^m
\end{aligned}
\tag{5.54}
$$

order

$$
\begin{aligned}
I + T_{ai} &\equiv \left(I + T_{a,i-1}\right)^2 \\
&= \left(I + 2 \times T_{a,i-1} + T_{a,i-1} \times T_{a,i-1}\right) \\
i &= 1,\ 2,\ \cdots,\ N
\end{aligned}
\tag{5.55}
$$

then

$$
\begin{aligned}
I + T_{aN} &= \left(I + T_{a,N-1}\right)^2 \\
&= \left(I + T_{a,N-2}\right)^4 = \cdots = (I + T_{a0})^m \\
&= T(\tau)
\end{aligned}
\tag{5.56}
$$

The above two equations show that T_{a1}, T_{a2}, ..., T_{aN} can be calculated in the following way to obtain $T(\tau)$.

$$
T_{ai} = 2 \times T_{a,i-1} + T_{a,i-1} \times T_{a,i-1} \tag{5.57}
$$

The recursive calculation according to Eq. (5.57) excludes the participation of I in addition operation, so it avoids the serious loss of significant digits of elements in T_{ai}, thus ensuring the extremely high calculation accuracy of T_{ai} and $T(\tau)$. The only approximation introduced in the above derivation is that Δt^5 and higher order terms are omitted in Taylor expansion of Eq. (5.54). Its order is $O(\Delta t)^5 = 10^{-30} O(\tau^5)$, which is the rounding error range of general computers. Therefore, the $T(\tau)$ calculated according to the above steps can be regarded as accurate within the range of computer accuracy. For linear equal step time domain integration, $T(\tau)$ only needs to be calculated once and the amount of calculation is very small [3].

2. *HPD-L* Precise Integration Format

It is assumed that $\overline{r}$ varies linearly over each time step in $t \in [t_k, t_{k+1}]$

$$
\overline{r} = r_0 + r_1 \times (t - t_k) \tag{5.58}
$$

where, r_0 and r_1 are time-invariant vectors, then the special solution of Eq. (5.49) is:

$$
x_p(t) = \left(H^{-1} + I \times t\right)\left(-H^{-1} r_1\right) - H^{-1}(r_0 - r_1 \times t_k) \tag{5.59}
$$

The *HPD-L* format of precise integration by substituting Eq. (5.59) for Eq. (5.52) is:

$$x(t_{k+1}) = T(\tau) \times \left(x(t_k) + H^{-1}\left(r_0 + H^{-1}r_1\right)\right) - H^{-1}\left(r_0 + H^{-1}r_1 + r_1 \times \tau\right) \tag{5.60}$$

where, $\tau = t_{k+1} - t_k$ is the integral step. If the load varies linearly in strict accordance with Eq. (5.58) at each step, the integral scheme (5.59) gives an accurate numerical solution on the computer.

5.2.2 *Extension of Precise Integration Algorithms*

When the system is nonlinear, Eq. (5.49) can be written as follows:

$$\dot{x} = H_0 x + H_1 x + \overline{r} \tag{5.61}$$

where, H_0 is the linear part of the system matrix, H_1 is the non-linear part of the system. The second and third terms of the right end of the Eq. (5.61) can be combined into input parts of the system.

$$\dot{x} = H_0 x + \tilde{r} \tag{5.62}$$

where

$$\tilde{r} = H_1 x + \overline{r} \tag{5.63}$$

Assuming that the response of known $t = t_k$ time is $x_{(k)}$, an iteration method can be used to obtain the response $x_{(k+1)}$ of $t = t_{k+1} = t_k + \tau$ time.

The response of Eq. (5.62) is continuous and the time interval is usually small. Therefore, it is most appropriate to take $x_{(k)}$ as the initial value of $t = t_{k+1}$ response at all times. order:

$$x_{(k+1)}^{(0)} = x_{(k)} \tag{5.64}$$

Substitute Eq. (5.64) into Eq. (5.63) and solve Eq. (5.62) to obtain $x_{(k+1)}^{(1)}$. Compare $x_{(k+1)}^{(0)}$ with $x_{(k+1)}^{(1)}$, such as $\left(x_{(k+1)}^{(1)} - x_{(k+1)}^{(0)}\right) \le \varepsilon$ (ε is the accuracy requirement of response and is a small positive number), therefore

$$x_{(k+1)} = x_{(k+1)}^{(1)} \tag{5.65}$$

Otherwise, substitute Eq. (5.64) into Eq. (5.63) and solve Eq. (5.62) to obtain $x_{(k+1)}^{(2)}$. Compare $x_{(k+1)}^{(1)}$ with $x_{(k+1)}^{(2)}$, such as $\left(x_{(k+1)}^{(2)} - x_{(k+1)}^{(1)}\right) \le \varepsilon$ (ε is the accuracy requirement of response and is a small positive number), therefore

$$x_{(k+1)} = x_{(k+1)}^{(2)} \tag{5.66}$$

after such repetition, $x_{(k+1)} = x_{(k+1)}^{(m)}$ can be found to meet the accuracy requirements.

Using the above iterative method, the response value at each moment can be obtained.

5.2.3 Numerical Simulation Example

For CA141 (CA1091) vehicle, the step input of front wheel angle is simulated as an example. The tire model is the *E*-index model proposed by Professor Guo Konghui. The friction coefficients of front and rear tires are 0.6251 and 0.6138 respectively (general friction coefficient). The friction coefficients of front and rear tires in the high friction coefficient model are 0.9062 and 0.8898 respectively. The friction coefficients of front and rear tires in the low friction coefficient model are 0.2035 and 0.1998 respectively.

1. Steering wheel angle ramp step input

The steering wheel speed is 400°/s during ramp step input (the steering wheel speed should reach 200–500°/s as far as possible according to ISO/7401 standard). In Figs. 5.4 and 5.5, R curve represents yaw rate r, B curve represents vehicle centroid side slip angle β, BD curve represents vehicle centroid side slip angular velocity $\dot{\beta}$, and AY curve represents vehicle centroid lateral acceleration a_y. In Figs. 5.4 and 5.5, the step value of the front wheel angle is 1°, that is, $\delta_0 = 1° \times i$, and i is the total transmission ratio of the steering system.

It can be seen from Figs. 5.4 and 5.5 that the response trend of nonlinear vehicle model and linear vehicle model with time is basically the same under different vehicle speeds. However, the overshoot of the response at high speed is significantly higher than that at low speed, that is, both nonlinear vehicle models and linear vehicle models show higher response errors at high speed than at low speed. In addition, the transition time of high-speed vehicles is significantly longer than that of low-speed vehicles when turning the same steering wheel angle. The main differences between the responses of nonlinear vehicle models and linear vehicle models are: (1) The steady-state value of the response of the nonlinear vehicle model (except for the side slip angular velocity of the mass center) is smaller than the steady-state value of the response of the corresponding linear vehicle model; (2) The yaw rate and the lateral acceleration of the mass center of the whole vehicle in the initial period of the response of the nonlinear vehicle model in the transient process are slightly higher than the response value of the corresponding linear vehicle model.

2. Discussion on side slip characteristics of vehicle with step input of steering wheel angle

The lateral force coefficient of the tire, that is, the ratio of the lateral force acting on the tire by the road surface to the vertical load on the tire, can reveal the sideslip

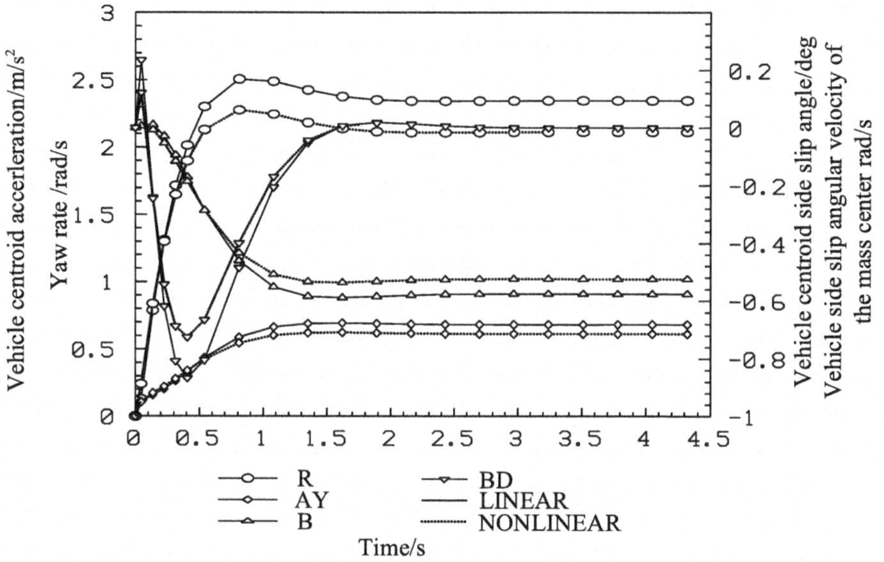

Fig. 5.4 Response curve of steering wheel angle step input when the vehicle speed is 60 km/h

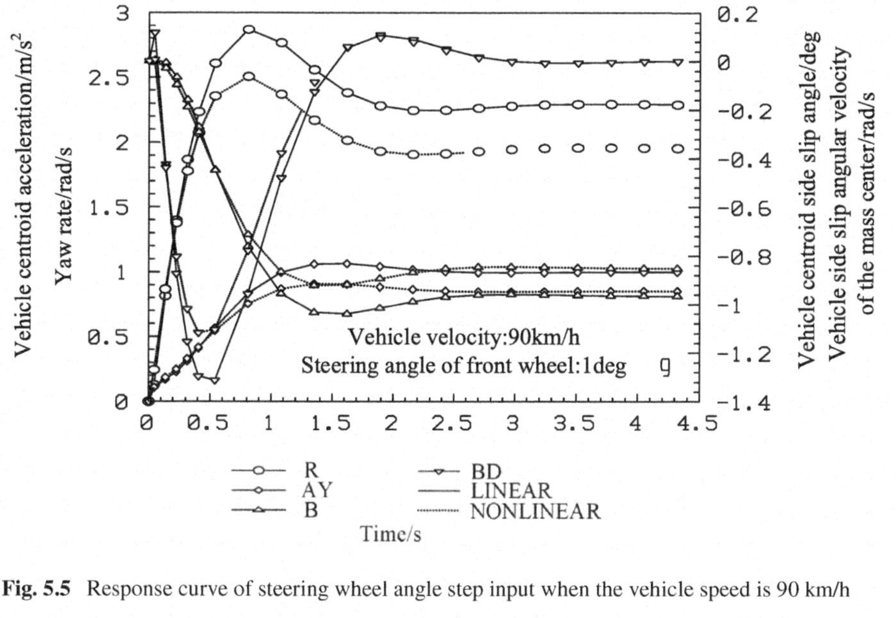

Fig. 5.5 Response curve of steering wheel angle step input when the vehicle speed is 90 km/h

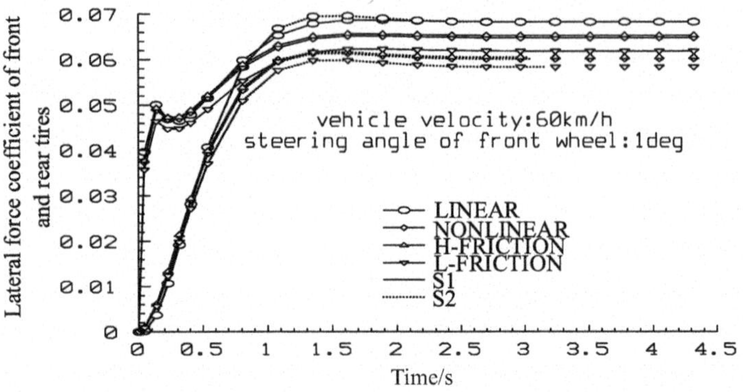

Fig. 5.6 Tire lateral force coefficient curve at 60 km/h

characteristics of the tire. When the lateral force coefficient of the tire is greater than the road friction coefficient (or adhesion coefficient) that can be provided by the road surface, the vehicle will have the phenomenon of front axle sliding or tail flicking.

In Figs. 5.6 and 5.7, S1 curve represents the lateral force coefficient curve of front axle tire, and S2 curve represents the lateral force coefficient curve of rear axle tire. It can be seen from the figure that the S1 curve reaches a maximum value almost linearly and quickly with the steering wheel angle from zero to the step value according to the slope law; When the steering wheel reaches the step value and remains unchanged, the S1 curve reaches a steady-state value according to the nonlinear law. All S2 curves do not respond quickly and have a maximum value as S1 curve when the steering wheel reaches the step value, and S2 curve slowly approaches the steady-state value relative to S1 curve. Nevertheless, it cannot be considered that the first thing that occurs when the driver turns the rudder sharply in a straight line is the front wheel sideslip. When the vehicle speed is low or the steering is small, the simulation results of linear model and nonlinear model are not much different. However, when the steering is large, even if the vehicle speed is low, the simulation results of linear model and nonlinear model are quite different. In addition, the low friction nonlinear model simulation can predict whether the front axle slip measurement or the rear axle slip measurement occurs first in real time.

The nonlinear system simulation based on this method can more realistically predict the dynamic motion process of the vehicle in real time.

5.3 Evaluation Index of Vehicle Active Safety

Recently, many chassis control systems, such as active suspension, four-wheel steering, traction control and antilock brake system, have been continuously developed and widely put into the market. However, the dynamic characteristics of these systems have not been fully evaluated objectively and their respective advantages

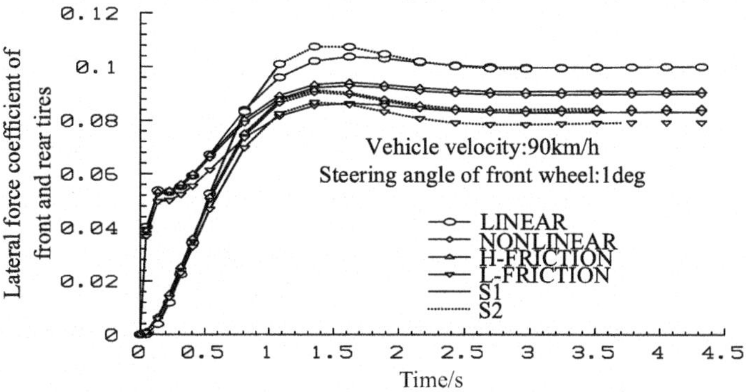

Fig. 5.7 Tire lateral force coefficient curve at 90 km/h

have not been quantitatively analyzed. This makes it difficult to optimize the design of vehicle dynamic characteristics in theory.

In the 1980s, Japanese scholar Ankawa proposed a theoretical method for predicting maneuverability based on the preview model [4]. Scholars believe that when driving a car, one does not fully aim at following performance and adjusts his own dynamic characteristics parameters, taking into account the safety of roll and his own physical factors. The manoeuvrability of the vehicle and the control burden required by the dynamic parameters of the driver selected therefrom depend mainly on the pre-aiming distance (pre-aiming time) and the performance index achieved by the entire closed-loop system. The following can be used to forecast:

$$E = L^{*^2} \times J^* \tag{5.67}$$

where, J^* is the best performance indicator value; L^* is the pre-aiming distance to achieve the best performance. The performance indexes used are the sum of weighted square integrals for vehicle lateral position deviation, vehicle lateral acceleration and steering wheel angle.

Professor Guo Konghui has studied the driver-vehicle closed-loop control system model and considered many factors affecting the vehicle handling, and put forward each individual total variance evaluation index (track, lateral acceleration, steering wheel angular speed, front and rear wheel lateral force coefficient) with clear physical meaning. A comprehensive evaluation and optimization design method for active safety of closed-loop system is proposed by applying frequency-domain statistical analysis method, which is convenient for engineering application.

5.3.1 Supplementary to the Closed-Loop Individual Evaluation Indexes

Various evaluation methods have been proposed for maneuverability and tracking accuracy. For example, the track total variance index J_E represents tracking accuracy, which is one of the most important indexes to ensure driving safety. The total variance of lateral acceleration J_{ay} is related to whether it is easy to overturn. The total variance J_{s1} and J_{s2} of the front and rear wheel lateral force coefficients are related to whether it is easy to sideslip. The total variance $J_{\dot{\delta}}$ of steering wheel angular speed can reflect the driver's busy degree (physical burden). Therefore, J_{ay}, J_s and $J_{\dot{\delta}}$ represent the maneuverability of the vehicle. Among them, J_{ay} and J_s are important indicators to ensure driving safety respectively, and $J_{\dot{\delta}}$ also indirectly affects driving safety.

As we all know, when the side slip angle β is small, the side slip angle β, the yaw rate r and the lateral acceleration $\ddot{y}$ have the following relationship:

$$\ddot{y} = V\left(r + \dot{\beta}\right) \tag{5.68}$$

Equation (5.68) shows that $\ddot{y}$ is affected by r and $\dot{\beta}$ at the same time. Experience has proved that cars with good handling always hope that Vr, the effective part of $\ddot{y}$, plays a leading role, while $V\dot{\beta}$, the part that interferes with the stability of $\ddot{y}$, the smaller the better. That is, in order to make $\ddot{y}$ stable quickly ($\ddot{y}$ stability will be very easy to manipulate), always hope that $\dot{\beta}$ tends to zero. Therefore, the total variance $J_{\dot{\beta}}$ of the sideslip angular velocity $\dot{\beta}$ of the center of mass can be used as a single evaluation index.

Zhu Hongwei pointed out: "the information of the vehicle's lateral acceleration may not always be obtained in time through the steering wheel angle, but the phase difference between the steering torque applied by the driver on the steering wheel and the vehicle's lateral acceleration is always small. This consistency between the steering torque and the lateral acceleration is an important factor for the driver to obtain a good road feel through the steering wheel." Therefore, the total variance $J_{T_{sw}}$ of the steering torque T_{sw} applied by the driver on the steering wheel is taken as a single evaluation index.

The individual indicators proposed by Professor Guo Konghui are shown as follows:

$$J_E = \int_0^{T_n} \left(\frac{f - y}{\widehat{E}}\right)^2 \mathrm{d}t \; or \; \frac{1}{T_n}\int_0^{T_n} \left(\frac{f - y}{\widehat{E}}\right)^2 \mathrm{d}t \tag{5.69}$$

$$J_{\dot{\delta}} = \int_0^{T_n} \left(\frac{\dot{\delta}}{\widehat{\dot{\delta}}}\right)^2 \mathrm{d}t \text{ or } \frac{1}{T_n}\int_0^{T_n} \left(\frac{\dot{\delta}}{\widehat{\dot{\delta}}}\right)^2 \mathrm{d}t \tag{5.70}$$

$$J_{ay} = \int_0^{T_n} \left(\frac{\ddot{y}}{\overset{\frown}{a}_y} \right)^2 \mathrm{d}t \text{ or } \frac{1}{T_n} \int_0^{T_n} \left(\frac{\ddot{y}}{\overset{\frown}{a}_y} \right)^2 \mathrm{d}t \tag{5.71}$$

$$J_{s1} = \int_0^{T_n} \left(\frac{s_i}{\overset{\frown}{s}_i} \right)^2 \mathrm{d}t \text{ or } \frac{1}{T_n} \int_0^{T_n} \left(\frac{s_i}{\overset{\frown}{s}_i} \right)^2 \mathrm{d}t \tag{5.72}$$

where, T_n is a long enough time (s), $\overset{\frown}{E}$ is the standard threshold value of track error, usually 0.3 m; $\overset{\frown}{\dot{\delta}}$ is the standard threshold value of steering wheel angular speed $\dot{\delta}$, usually 1.0 rad/s, $\overset{\frown}{a}_y$ is the standard threshold value of lateral acceleration $\ddot{y}$, usually 0.3 g, $\overset{\frown}{s}_1$ and $\overset{\frown}{s}_2$ are the standard threshold values of the lateral force coefficients s_1 and s_2 of the front and rear wheels, usually taken as (0.2–0.4) g (0.2 g on the sliding road). If f is limited input, the formula before "or" shall prevail; If f is a fairly long input or random input, the formula after "or" is taken as the index.

Now the added indicators are expressed as follows:

$$J_{\dot{\beta}} = \int_0^{T_n} \left(\frac{\dot{\beta}}{\overset{\frown}{\dot{\beta}}} \right)^2 \mathrm{d}t \ or \ \frac{1}{T_n} \int_0^{T_n} \left(\frac{\dot{\beta}}{\overset{\frown}{\dot{\beta}}} \right)^2 \mathrm{d}t \tag{5.73}$$

$$J_{T_{sw}} = \int_0^{T_n} \left(\frac{T_{sw}}{\hat{T}_{sw}} \right)^2 \mathrm{d}t \ or \ \frac{1}{T_n} \int_0^{T_n} \left(\frac{T_{sw}}{\hat{T}_{sw}} \right)^2 \mathrm{d}t \tag{5.74}$$

where, $\overset{\frown}{\dot{\beta}}$ is the standard threshold value of the side deflection angular speed $\dot{\beta}$ of the vehicle mass center, usually $0.1g/V$, and V is the vehicle speed, $\hat{T}_{sw}$ is the standard threshold value of the steering torque T_{sw} applied by the driver on the steering wheel, usually $10\text{N} \cdot m$.

5.3.2 Supplement to the Open-Loop Individual Evaluation Indexes

In the open-loop evaluation index, the total variance index of lateral acceleration and the total variance index of yaw rate have been proposed. Now, two new single evaluation indicators are proposed. From the analysis in Sect. 5.3.1, it can be seen that the yaw rate r of the car plays a positive role in the handling of the car, and the drastic change of the side deflection angle of the car's mass center will make the car tend to be unstable. Therefore, the total variance of vehicle centroid sideslip angular velocity is taken as an evaluation index. For the steering movement of the car at constant speed, the car must have enough lateral acceleration to accurately respond to the driver's steering command, and the smaller the rotation of the driver's steering wheel, the less busy it is. Now the rotation of steering wheel with unit lateral acceleration is taken as a single evaluation index.

The single index given by Professor Guo Konghui is shown as follows:

$$E_r = \int_0^{T_n} \left(1 - \frac{r}{\delta_{sw} \cdot G_r}\right)^2 \mathrm{d}t \; or \; \frac{1}{T_n} \int_0^{T_n} \left(1 - \frac{r}{\delta_{sw} \cdot G_r}\right)^2 \mathrm{d}t \tag{5.75}$$

$$E_{\ddot{y}} = \int_0^{T_n} \left(1 - \frac{\ddot{y}}{\delta_{sw} G_{\ddot{y}}}\right)^2 \mathrm{d}t \; or \; \frac{1}{T_n} \int_0^{T_n} \left(1 - \frac{\ddot{y}}{\delta_{sw} G_{\ddot{y}}}\right)^2 \mathrm{d}t \tag{5.76}$$

Now the added indicators are expressed as follows:

$$E_{\dot{\beta}} = \int_0^{T_n} \left(\frac{\dot{\beta}}{\delta_{sw} G_r}\right)^2 \mathrm{d}t \; or \; \frac{1}{T_n} \int_0^{T_n} \left(\frac{\dot{\beta}}{\delta_{sw} G_r}\right)^2 \mathrm{d}t \tag{5.77}$$

$$L_{G_{\ddot{y}}^{-1}} = \frac{\delta_{sws}}{\ddot{y}_s} = \frac{1}{G_{\ddot{y}}} \tag{5.78}$$

where r is the yaw rate, G_r is the steady-state gain of yaw rate, δ_{sws} is the steady-state value of steering wheel angle, $\ddot{y}_s$ is the steady-state value of lateral acceleration, $G_{\ddot{y}}$ is the steady-state gain of lateral acceleration.

5.3.3 Selection of Comprehensive Evaluation Index and Weighting Coefficient

First, the comprehensive evaluation index of active safety of driver vehicle closed-loop system in medium and high speed is given:

$$J_T = \left(1 + \left(\frac{T_c}{\underset{c}{\widehat{T}}}\right)^2 + \left(\frac{TV}{\widehat{D}}\right)^4\right) \cdot \sqrt{\frac{w_1 J_{\dot{E}}^2 + w_2 J_{\dot{\delta}}^2 + w_3 J_{\dot{ay}}^2 + w_4 J_{\dot{s}}^2 + w_5 J_{\dot{\beta}}^2 + w_6 J_{\dot{T}_{sw}}^2}{w_1 + w_2 + w_3 + w_4 + w_5 + w_6}} \tag{5.79}$$

where, $\underset{c}{\widehat{T}}$ is the standard threshold value of driver correction parameter $T_c(a, T)$, a is the following order, T is the preview time, usually $\hat{T}_c = 0.3$ s, $\widehat{D}$ is the standard threshold of effective visual range, usually $\widehat{D} = 50 \sim 1000$ m, when visibility is very poor, $\widehat{D} = 50$ m, $w_1 \sim w_6$ is the weight coefficient, six kinds of experiments were done in the simulation study: $w_5 = w_6 = 0$, $w_1 = w_2 = w_3 = w_4 = 1$, its comprehensive evaluation index is expressed by J_{T1},$w_4 = w_6 = 0$, $w_1 = w_2 = w_3 = w_5 = 1$, its comprehensive evaluation index is expressed by J_{T2}, $w_6 = 0$, $w_1 = w_2 = w_3 = w_4 = w_5 = 1$, its comprehensive evaluation index is expressed by J_{T3}, $w_4 = 0$, $w_1 = w_2 = w_3 = w_5 = w_6 = 1$, its comprehensive evaluation index is expressed by J_{T4}, $w_5 = 0$, $w_1 = w_2 = w_3 = w_4 = w_6 = 1$, its comprehensive evaluation index is expressed by J_{T5}. When the six weight coefficients are all 1, its comprehensive evaluation index is expressed by J_{T6}.

Now the comprehensive evaluation index of the open-loop system is given:

$$E_T = \sqrt{\frac{\omega_1 \cdot E_r^2 + \omega_2 \cdot E_\beta^2 + \omega_3 \cdot E_{\ddot{y}}^2 + \omega_4 \cdot L_{G_{\ddot{y}}^{-1}}}{\omega_1 + \omega_2 + \omega_3 + \omega_4}} \tag{5.80}$$

where, $\omega_1 \sim \omega_4$ is the weight coefficient, and the optimal value is taken in the following way in the simulation study.

$$\begin{cases} \min & E_T(\omega_1, \omega_2, \omega_3, \omega_4) \\ \text{s.t.} & 0 \le \omega_i \le \hat{\omega}_i \ i = 1, \ldots, 4 \end{cases} \tag{5.81}$$

5.3.4 Correlation Between Closed-Loop Comprehensive Evaluation Index and Open-Loop Comprehensive Evaluation Index

Here, the correlation between the closed-loop comprehensive evaluation index and the open-loop comprehensive evaluation index adopts the formal prediction method, that is, the obtained information of the closed-loop comprehensive evaluation index and the open-loop comprehensive evaluation index is formally calculated according to a certain mathematical model, to obtain the prediction results.

The correlation between the two information Y and X can be expressed by the following correlation coefficient R:

$$R = \frac{\sum_{i=1}^{n} (x_i - \overline{x})(y_i - \overline{y})}{\sqrt{\sum_{i=1}^{n} (x_i - \overline{x})^2 \sum_{i=1}^{n} (y_i - \overline{y})^2}} \tag{5.82}$$

where, $\overline{x}$ and $\overline{y}$ represent the average values of x and y respectively, $R \in \left[-1\ 1\right]$.

5.3.5 Numerical Simulation Example

Taking the parameters of the Audi100 car as an example, the driver-vehicle closed-loop simulation is a double-lane road input experiment with time as the standard, and the vehicle open-loop simulation is a steering wheel angle step experiment. To study the correlation between open-loop comprehensive evaluation and closed-loop comprehensive evaluation, the author squared E_T in the correlation analysis.

Under the condition that the relative comprehensive cornering stiffness K_2 / F_{Z2} of the rear axle two tires is 18.125 and remains unchanged, the influence of the relative comprehensive cornering stiffness K_2 / F_{Z2} of the front axle two tires on the open-loop comprehensive evaluation index $(E_T)^2$ and the closed-loop comprehensive evaluation index J_{T1} and J_{T2} are shown in Figs. 5.8, 5.9 and 5.10 respectively. From Figs. 5.8, 5.9 and 5.10, it can be seen that the open-loop comprehensive evaluation index $(E_T)^2$ is relatively small when $K_1 / F_{Z1} = 12–16$, the closed-loop comprehensive evaluation index J_{T1} is relatively small when $K_1 / F_{Z1} = 8–12$, and J_{T2} is relatively small when $K_1 / F_{Z1} = 8–12$. At different speeds, the open-loop comprehensive evaluation index $(E_T)^2$ and the closed-loop comprehensive evaluation index J_{T1} and J_{T2} have the same trend with K_1 / F_{Z1}. Moreover, when the vehicle speed is high, the larger understeer compared with the larger oversteer can make E_T relatively large and J_{T1} and J_{T2} relatively small, indicating that the driver-vehicle closed-loop system formed after the driver is embedded is easier to operate the understeer vehicle than the vehicle open-loop system. However, excessive understeer (K_1 value is too small) and oversteer (K_1 value is too large) will make the handling worse.

Figures 5.11 and 5.12 respectively show the relationship between the open-loop comprehensive evaluation index $(E_T)^2$ and the closed-loop comprehensive evaluation index J_{T1} and J_{T2}. Figure 5.13 shows the correlation between the open-loop comprehensive evaluation index $(E_T)^2$ and the closed-loop comprehensive evaluation indexes J_{T1} and J_{T2}. From Figs. 5.11, 5.12 and 5.13, it can be seen that the

Fig. 5.8 Influence of K_1 / F_{Z1} on open-loop comprehensive evaluation index $(E_T)^2$

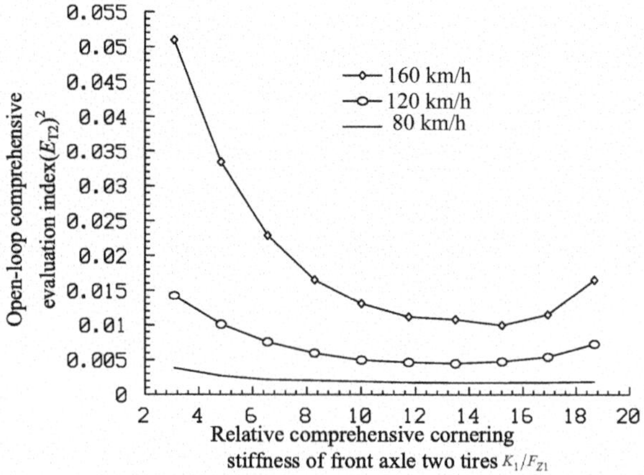

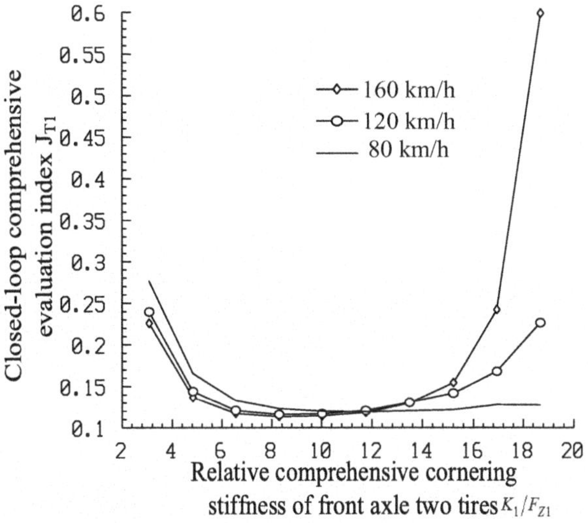

Fig. 5.9 Influence of K_1 / F_{Z1} on closed-loop comprehensive evaluation index J_{T1}

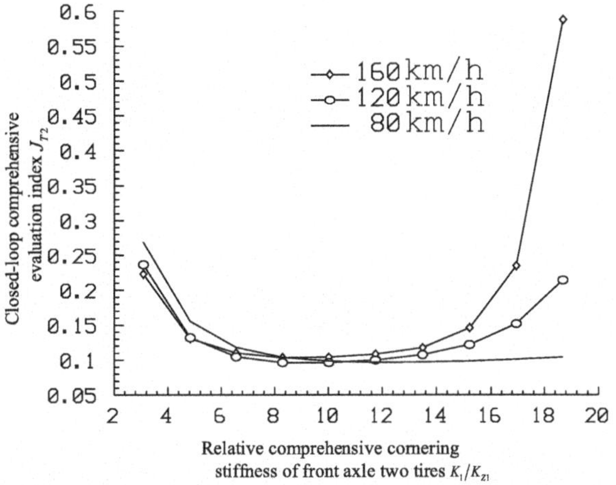

Fig. 5.10 Influence of K_1 / F_{Z1} on closed-loop comprehensive evaluation index J_{T2}

open-loop comprehensive evaluation index $(E_T)^2$ and the closed-loop comprehensive evaluation index J_{T1} and J_{T2} are basically mapped one by one at low speed, and the correlation is good; With the increase of vehicle speed, there is a one-to-two relationship between the open-loop comprehensive evaluation index $(E_T)^2$ and the closed-loop comprehensive evaluation indexes J_{T1} and J_{T2}, and the correlation is poor when the vehicle speed is high.

Under the condition that the relative comprehensive cornering stiffness K_2 / F_{Z2} of the rear axle two tires is 18.125 and remains unchanged, the influence of the relative comprehensive cornering stiffness K_1 / F_{Z1} of the front axle two tires on the closed-loop comprehensive evaluation indicators J_{T1}, J_{T2}, J_{T3}, J_{T4}, J_{T5} and J_{T6} are

Fig. 5.11 Relationship between open-loop evaluation index $(E_T)^2$ and closed-loop evaluation index J_{T1}

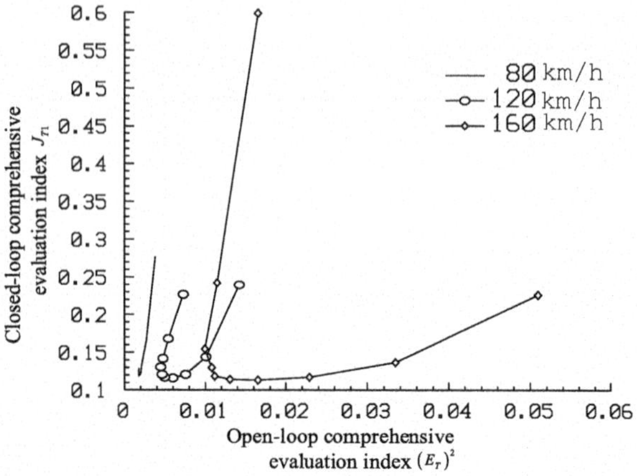

Fig. 5.12 Relationship between open-loop index $(E_T)^2$ and closed-loop index J_{T2}

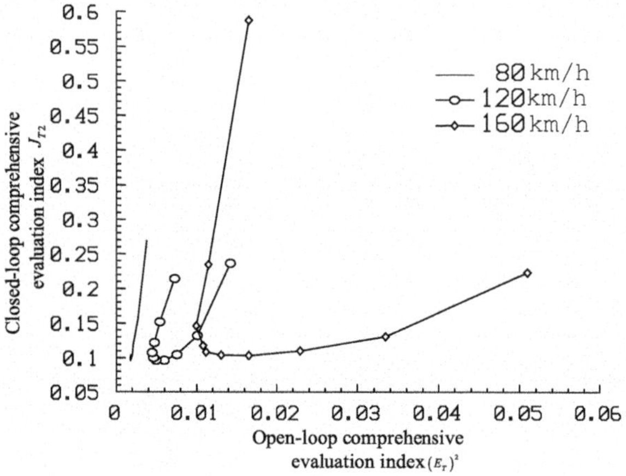

Fig. 5.13 Correlation between open-loop indicator $(E_T)^2$ and closed-loop indicators J_{T1} and J_{T2}

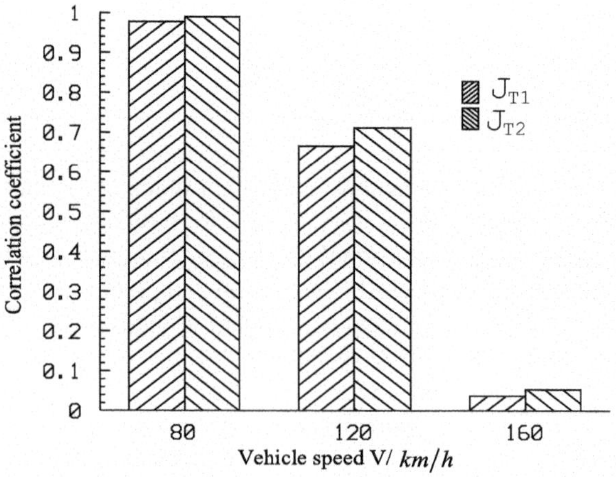

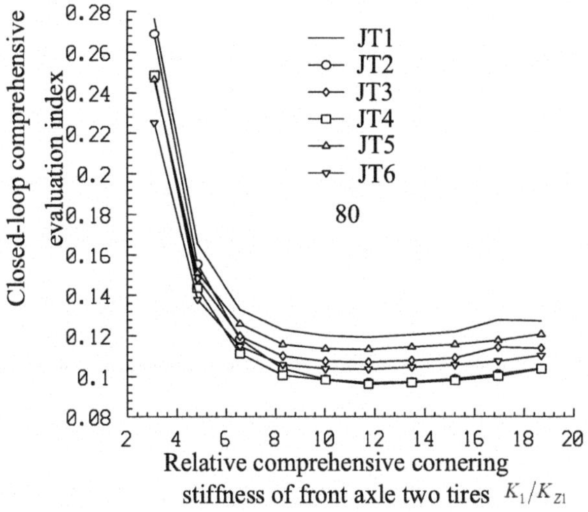

Fig. 5.14 Influence of K_1/F_{Z1} on closed-loop comprehensive evaluation index (80 km/h)

shown in Figs. 5.14, 5.15 and 5.16 respectively. It can be seen from Figs. 5.14, 5.15 and 5.16 that under the same speed, the change trend of closed-loop comprehensive evaluation indicators J_{T1}, J_{T2}, J_{T3}, J_{T4}, J_{T5} and J_{T6} with K_1/F_{Z1} is basically the same. Moreover, the J_{T2} value with J_{T2} as the comprehensive evaluation index is the most sensitive to the change of K_1/F_{Z1}, and the J_{T6} value is the least sensitive to the change of K_1/F_{Z1}. The optimal value of K_1/F_{Z1} under various closed-loop comprehensive evaluation indicators at different vehicle speeds is shown in Table 5.1. According to the existing experience, the ratio of K_1/F_{Z1} to K_2/F_{Z2} is roughly 1:2, which is more reasonable. Therefore, $K_1/F_{Z1} = 9.063$ is in good agreement with the results in Table 5.1, while the $K_1/F_{Z1} = 6.456$ of the original vehicle is smaller. The comprehensive evaluation indicators closest to $K_1/F_{Z1} = 9.063$ in Table 5.1 are J_{T3} and J_{T6}.

5.4 Summary

From the perspective of active safety of driver-vehicle closed-loop system, the closed-loop evaluation index given in this chapter includes the driver's adaptation process to vehicle dynamic characteristics, and is directly related to the path following error of driver-vehicle closed-loop system, the driver's handling burden and the statistical characteristics of vehicle driving safety; For the driving conditions with low speed, the corresponding supplement is made in the closed-loop evaluation; In order to correspond to the closed-loop evaluation index, the open-loop evaluation index considers the steering wheel rotation index. Each single evaluation index added has concise and clear physical significance, which is convenient for engineering design and application.

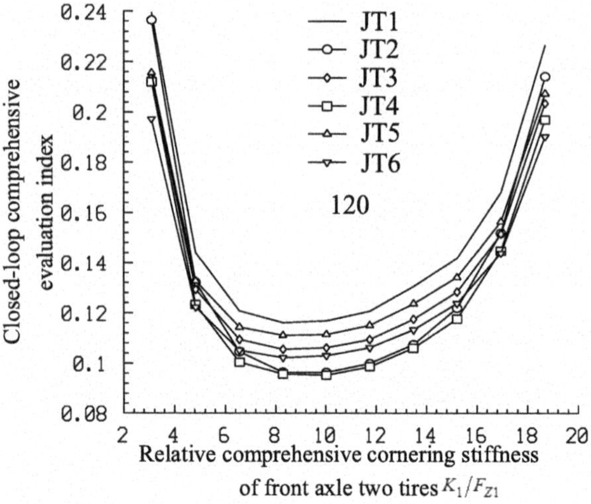

Fig. 5.15 Influence of K_1/F_{Z1} on closed-loop comprehensive evaluation index (120 km/h)

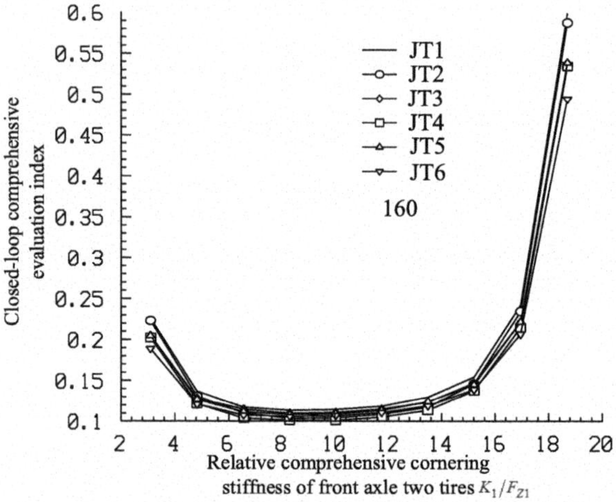

Fig. 5.16 Influence of K_1/F_{Z1} on closed-loop comprehensive evaluation index (160 km/h)

Table 5.1 K_1/F_{Z1} optimal value under various closed-loop comprehensive evaluation indexes at different vehicle speeds

V (km/h)	J_{T1}	J_{T2}	J_{T3}	J_{T4}	J_{T5}	J_{T6}
80	11.5	12.0	11.0	11.4	12.0	10.8
120	8.5	9.0	8.5	8.7	9.8	8.4
160	8.6	9.0	8.8	8.8	9.0	8.6

The simulation research and correlation analysis of closed-loop evaluation and open-loop evaluation show that: (1) When the structural parameters of the vehicle change in a considerable range (the larger the range of parameter changes, the more sufficient and complete the correlation analysis is), at medium and low speeds, the correlation between the closed-loop comprehensive evaluation index and the open-loop comprehensive evaluation index is good, and the faster the speed is, the worse the correlation is. (2) The active safety evaluation index of driver-vehicle closed-loop system can be used as the main basis for the objective evaluation of vehicle controllability, that is, to analyze the advantages and disadvantages of vehicle controllability from a quantitative point of view, which lays a foundation for the optimal design of vehicle controllability.

References

1. Guo K. Drivers-vehicle close-loop simulation of handling by "Preselect optimal curvature method." Automot Eng. 1984;6(3):3–18.
2. Zhao Y. Discussion on active safety evaluation index of man-vehicle-road closed loop operating system. Autom Res Dev 1997;11(3):25–29, 42
3. Zhao Y. Operating safety evaluation of driver-vehicle closed-loop system with random road input. China Mech Eng. 2000;11(4):465–8.
4. Zhao Y. A research on active safety evaluation of vehicle. Automot Eng. 2001;23(1):1–4.

Chapter 6
A New Evaluation Method for Driver-Vehicle Closed-Loop Handling System Dynamics

6.1 Maneuverability Evaluation of the Driver-Vehicle Closed-Loop System with Random Road Inputs

The study of vehicle handling stability has been around for at least fifty years. Currently, the vehicle itself has been studied in considerable depth. The in-depth study of tire characteristics has made it possible to make a fairly comprehensive and realistic simulation of the dynamic response of the vehicle. A great deal of research and discussion has also been done on the driver model theory, which has led to breakthroughs in the research of driver-vehicle–road closed-loop system dynamics simulation [1–3], vehicle handling stability, and vehicle active safety research [4–6]. However, the research topic of the dynamic characteristics of the driver-vehicle–road closed-loop control system still needs a lot of in-depth and detailed work [7].

So far, the mathematical model of road input in the simulation of the driver-vehicle–road closed-loop system is mainly deterministic. There are various road alignments in the objective world, i.e., the roads encountered by different drivers and the roads encountered by the same driver at different times are random. Therefore, the simulation of the closed-loop driver-vehicle–road system based on a skilled driver driving the same type of vehicle can predict the possible results of a real-world test more realistically, treating the road input as a stochastic process is reasonable and necessary. The road input can be regarded as a smooth random process, then the simulation of the driver-vehicle–road closed-loop system is the simulation of a skilled driver doing many real-world tests with different alignment routes, or the simulation of a skilled driver going through a long route (including many kinds of alignment). A skilled driver here means that the optimal values are taken as the model parameters [8].

If computer prediction of the active safety of a closed-loop driver-vehicle–road system with random road inputs is feasible, then it will solve the limitations of the danger, economic burden and evaluation results due to the real-world tests [9].

Y. Zhao and F. Lin, *Vehicle System Dynamics*,
https://doi.org/10.1007/978-981-97-2019-4_6

6.1.1 Driver-Vehicle–Road Closed-Loop System Model with Random Road Input

Generally, the driver-vehicle–road closed-loop system includes the randomness of road input and the randomness of driver model parameters [10]. This chapter is limited to the drivers whose model parameters take the optimal value (i.e., skilled drivers) and stochastic road inputs. The stochastic road inputs include both stationary and evolving random road inputs.

According to the optimal curvature control theory [11], the driver-vehicle–road system diagram shown in Fig. 6.1 can be obtained.

In this section, we will study the lateral motion and yaw motion of the vehicle moving at a constant speed. The two-degree-of-freedom vehicle model is used [12]. Therefore, from Fig. 6.1, the state equation of the system can be derived as

$$\dot{x} = Hx + bf_r \tag{6.1}$$

The output variables are

$$z = Gx + df_r \tag{6.2}$$

where

$$x = \left[r \; \beta \; y \; v_a \; q_1 \; q_2 \right]^{\mathrm{T}}, b = \left[0\,0\,0\,0\,0\; b_6 \right]^{\mathrm{T}}$$

$$z = \left[y \; a_y \; \delta_{sw} \; \dot{\delta}_{sw} \; s_1 \; s_2 \right], d = \left[0\,0\,0\; d_4\; 0\,0 \right]^{\mathrm{T}}$$

where r is the yaw rate, β is the sideslip angle, v_a is the vehicle lateral speed, q_1 and q_2 are the two state variables in the driver model, s_2 is the lateral acceleration, s_2 and s_2 are the lateral force coefficient of front and rear tires, respectively, H, G and b, d are the system matrix and vector, functions of vehicle parameters and driver model parameters, $f_r(t)$ is the effective road input function of the system, which can be expressed as

$$f_r(t) = f(t+T) + T_c\dot{f}(t+T) \tag{6.3}$$

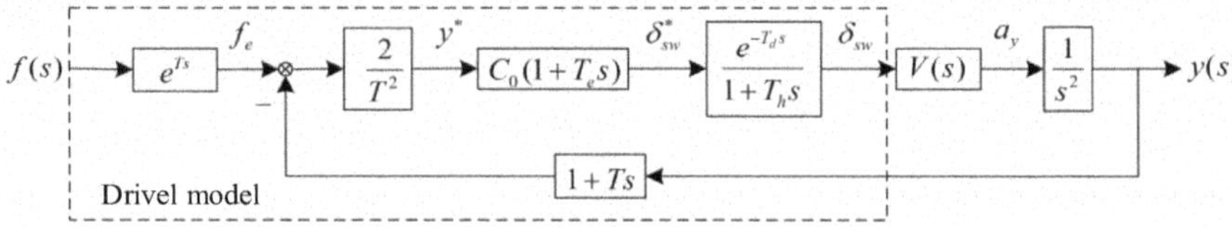

Fig. 6.1 Driver-vehicle–road closed-loop system

where $f(t+T)$ is the road input at $s_\rho(\Omega) = \begin{cases} const\ \Omega_L \leq \Omega \leq \Omega_U \\ 0 \qquad \text{others} \end{cases}$.

1. Stationary Stochastic Road Input Model

The road input model to be discussed first is a stationary stochastic process, which is suitable for the simulation of the driver-vehicle–road closed-loop system with constant speed. Therefore, it can be assumed that the curvature ρ of the road input is generated by a pseudo-random generator and is a band-limited white noise random process. Its self-spectral density in the spatial domain is

$$s_\rho(\Omega) = \begin{cases} \text{const}, & \Omega_L \leq \Omega \leq \Omega_U \\ 0, & \text{others} \end{cases} \tag{6.4}$$

where Ω is the spatial frequency. It can be deduced that the relationship between the self-spectral density $s_\rho(\Omega)$ in the spatial domain and the self-spectral density $s_\rho(\omega)$ in the time domain ($s_\rho(\omega) = \frac{1}{U}s_\rho(\Omega)$ is the time frequency) is

$$s_\rho(\omega) = \frac{1}{U}s_\rho(\Omega) \tag{6.5}$$

where

$$\omega = U\Omega \tag{6.6}$$

where U is the vehicle speed.

It can be deduced that the self-spectral density of the road input $f(t)$ and the effective road input $f_r(t)$ in the time domain are respectively

$$s_f(\omega) = U^4 s_\rho(\omega)/\omega^4 \tag{6.7}$$

$$s_{f_r}(\omega) = \left(1 + T_c^2\omega^2\right)^2 s_f(\omega) \tag{6.8}$$

Now construct a virtual road input

$$f_r(t) = \sqrt{s_{f_r}(\omega)}\exp(i\omega t) \tag{6.9}$$

In this way, the random road input is transformed into deterministic input.

2. Evolving Stochastic Road Input Model

To make the model of road input more truly reflect the actual driving process, the road input should be a non-stationary random process. The response analysis of the system subjected to non-stationary random excitation has received high attention in the past decade. At present, the evolutionary random excitation model is widely used in engineering [13].

Let the input of the virtual road be

$$f(t, \Omega, U) = e^{\mathrm{i}U\Omega t} \tag{6.10}$$

The vehicle speed is a slowly changing process, thus

$$U = U_0(1 + \varepsilon(t)) \tag{6.11}$$

Therefore

$$f(t, \Omega, U) = e^{\mathrm{i}U_0(1+\varepsilon(t))\Omega t} = p(t, \Omega)e^{\mathrm{i}U_0\Omega t} \tag{6.12}$$

where

$$p(t, \Omega) = e^{\mathrm{i}U_0\varepsilon(t)\Omega t} \tag{6.13}$$

Road input is a slow-changing process. Therefore, it can be assumed that the evolving random road input model is as follows

$$F(t) = g(t)f(t) \tag{6.14}$$

where $g(t)$ is a slowly varying modulation function, $f(t)$ is a stationary random process.

Now construct a virtual road input

$$f_r(t) = \sqrt{s_{f_r}(\omega)}g(t)\exp(\mathrm{i}\omega t) \tag{6.15}$$

In this way, the random road input is transformed into deterministic input.

6.1.2 Self-Spectral Density of Effective Road Input

For a stationary random road input $f(t)$, the spectrum of the first realization of its effective road input f_r is

$$\begin{aligned} f_{r_{f_r}}(\omega) &= \int_{-\infty}^{\infty} f_r(t)e^{-\mathrm{i}\omega t}dt \\ &= \int_{-\infty}^{\infty} f(t)e^{-\mathrm{i}\omega t}dt + \int_{-\infty}^{\infty} T_c\dot{f}(t)e^{-\mathrm{i}\omega t}dt \\ &= \left(1 + T_c^2\omega^2\right)f_f(\omega) \end{aligned} \tag{6.16}$$

and the self-spectral density of the effective road input $f_r(t)$ is

$$s_{f_r}(\omega) = \lim_{T\to\infty} \frac{1}{2T} E\left(\left|f_{r_{fr}}(T,\omega)\right|^2\right) \tag{6.17}$$

and

$$\left|f_{r_{fr}}(T,\omega)\right|^2 = \left(1 + T_c^2\omega^2\right)^2 \left|f_f(T,\omega)\right|^2 \tag{6.18}$$

Thus

$$s_{f_r}(\omega) = \left(1 + T_c^2\omega^2\right)^2 s_f(\omega) \tag{6.19}$$

In addition

$$\ddot{f} = U^2 \rho \tag{6.20}$$

where $\ddot{f}$ is the road input acceleration, thus, the self-spectral density of $\ddot{f}, \dot{f}, f$ are

$$s_{\ddot{f}}(\omega) = U^4 s_\rho(\omega) \tag{6.21}$$

$$s_{\dot{f}}(\omega) = U^4 s_\rho(\omega)/\omega^2 \tag{6.22}$$

$$s_f(\omega) = U^4 s_\rho(\omega)/\omega^4 \tag{6.23}$$

Therefore, the self-spectral density of effective road input is

$$s_{f_r}(\omega) = U^4\left(1 + T_c^2\omega^2\right)^2 s_\rho(\omega)/\omega^4 \tag{6.24}$$

6.1.3 Time Domain Analysis of Driver-Vehicle–Road Stochastic Closed-Loop System Response

Equation (6.1) is expressed as the following form

$$\dot{x} = Hx + \overline{r} \tag{6.25}$$

where $\overline{r} = bf_r$. The homogeneous solution of Eq. (6.25) is

$$x_h(t) = T(\tau)C \tag{6.26}$$

where

$$T(\tau) = \exp(H \times \tau) \tag{6.27}$$

In the integration step $t \in [t_k, t_{k+1}]$, $\tau = t - t_k$. Assuming that the special solution of Eq. (6.25) $x_p(t)$ has been obtained. Then at the initial time of this step $[T(\tau)] = [I]$, i.e., $[T(\tau)] = [I]$, $T(\tau) = I$ (unit array). Thus, the integral constant vector C can be determined, and the general solution of Eq. (6.25) can be obtained as follows

$$x(t) = T(\tau)\big(x(t_k) - x_p(t_k)\big) + x_p(t) \tag{6.28}$$

The problem boils down to solving the special solution $x_p(t)$ and fine calculation $T(\tau)$.

1. Calculation of Matrix $T(\tau)$

The integration step τ is internally subdivided into $m = 2^N$ equal parts

$$\Delta t = \tau / m = 2^{-N}\tau \tag{6.29}$$

Generally, take $N = 20$, then $\Delta t \approx 10^{-6}\tau$. It is far less than all the natural period of the system. The substitution of the Eq. (6.29) into the Eq. (6.27) is obtained

$$\begin{aligned} T(\tau) &= (\exp(H \times \Delta t))^m \\ &\approx \left(I + H \times \Delta t + (H \times \Delta t)^2 / 2! + (H \times \Delta t)^3 / 3! + (H \times \Delta t)^4 / 4!\right)^m \\ &\equiv (I + T_{a0})^m \end{aligned} \tag{6.30}$$

Let

$$\begin{aligned} I + T_{ai} &\equiv \left(I + T_{a,i-1}\right)^2 \\ &= \left(I + 2 \times T_{a,i-1} + T_{a,i-1} \times T_{a,i-1}\right) \\ &(i = 1, 2, \cdots, N) \end{aligned} \tag{6.31}$$

then

$$I + T_{aN} = \left(I + T_{a,N-1}\right)^2 = \cdots = \left(I + T_{a,0}\right)^m = T(\tau) \tag{6.32}$$

The above two equations show that $(i = 1, 2, \cdots, N)$ can be calculated in the following way to obtain $T(\tau)$

$$T_{ai} = 2 \times T_{a,i-1} + T_{a,i-1} \times T_{a,i-1} \tag{6.33}$$

The recursive calculation according to Eq. (6.33) excludes I from the addition operation, thus retaining the significant digits in T_{ai}, so that the calculation accuracy of T_{ai} and $T(\tau)$ is extremely high. The only approximation introduced in the above derivation is that $O(\Delta t^5) = 10^{-30}O(\tau^5)$ and higher-order terms are omitted from the Taylor series expansion of Eq. (6.30). Its magnitude order is $O(\Delta t^5) = 10^{-30}O(\tau^5)$, which is already the rounding error range of general computers. $T(\tau)$ calculated in the above steps can be considered to be accurate within the computer accuracy range.

For the linear equal step time domain integral, the computation of $T(\tau)$ is very small and only needs to be calculated once.

2. HPD-S Precise Integration Scheme

According to the characteristics of Eq. (6.9), it is assumed that $\overline{r}$ changes according to the simple harmonic law within each integration step $t \in \left[t_k, t_{k+1}\right]$

$$\overline{r}(t) = r_1 \sin(\omega t) + r_2 \cos(\omega t) \tag{6.34}$$

where r_1, r_2 are time-invariant vectors.

Substituting Eq. (6.34) into the right end of Eq. (6.25), the special solution of the equation can be obtained as follows

$$x_p(t) = a \sin(\omega t) + b \cos(\omega t) \tag{6.35}$$

where

$$\begin{cases} a = \left(\omega I + H^2/\omega\right)^{-1}(r_2 - Hr_1/w) \\ b = \left(\omega I + H^2/\omega\right)^{-1}(r_1 - Hr_2/w) \end{cases} \tag{6.36}$$

Therefore, the format of the $HPD-S$ $HDP-S$((High Precision Direct Integration Method—Sinusoidal) is

$$x(t_{k+1}) = T(\tau)(x(t_k) - a \sin(\omega t_k) + b \cos(\omega t_k)) + a \sin(\omega t_{k+1}) + b \cos(\omega t_{k+1}) \tag{6.37}$$

where $\tau = t_{k+1} - t_k$.

As long as $\overline{r}$ changes harmonically within the time step $\overline{r}$, Eq. (6.37) always gives accurate results, even if $\overline{r}$ is large enough to include a plurality of $\overline{r}$ change periods.

For the case of Eq. (6.15), in time step $\overline{\tau}$, $\overline{r}$ is generally still a simple harmonic function, or can be approximately regarded as a simple harmonic function. $g(t)$ is taken as the average value in this step, which is regarded as a constant.

6.1.4 Numerical Simulation Example

Taking the parameters of a domestic vehicle as an example, considering the random road input generated by the pseudo-random generator (6.4), if the slow rate of speed is 1/10 [14] of the acceleration capacity of the vehicle, then there is $\varepsilon(t) = 0.694t/U$. The active safety evaluation index of the driver-vehicle–road closed-loop system given in the previous chapter and the calculation method given in this chapter are used to evaluate the closed-loop system in time domain.

The comprehensive evaluation indexes of the active safety of the driver-vehicle–road closed-loop system in Figs. 6.2, 6.3, 6.4 and 6.5 are

$$J_T = \left(1 + (T_c/\hat{T}_c)^2 + (TU/\hat{D})^4\right)\sqrt{\frac{w_1 J_E^2 + w_2 J_{\dot{\delta}}^2 + w_3 J_{a_y}^2 + w_4 J_s^2}{w_1 + w_2 + w_3 + w_4}} \tag{6.38}$$

where J_E, $J_{\dot{\delta}}$, J_{a_y}, J_s are the evaluation index of track error, steering wheel angle velocity, lateral acceleration and lateral force coefficient, $w_i (i = 1, 2, \cdots, 4)$ is the weighting coefficient, $\hat{T}_c$ is the "threshold" of T_c, $\hat{D}$ is the "threshold" of the effective visual range.

The solid line in the figure indicates the result when the road is input smoothly, and the broken line indicates the result when the road is input randomly.

It can be seen from Fig. 6.2 that the influence of evolving random road input and stationary random road input on the optimal preview time is roughly the same. With the increase of vehicle speed (U) (100 km/h, 140 km/h, 180 km/h), the optimal

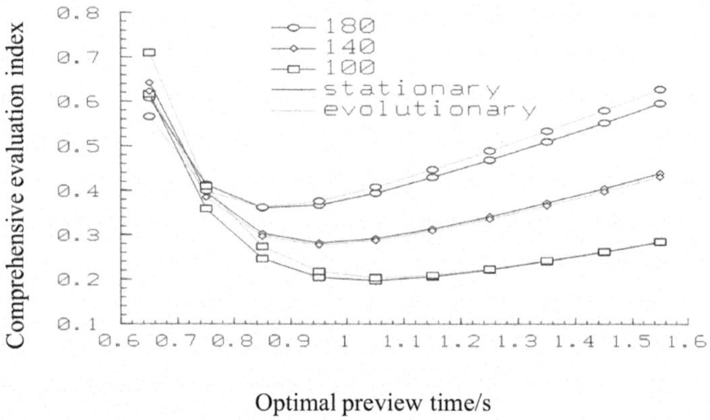

Fig. 6.2 Influence of optimal preview time

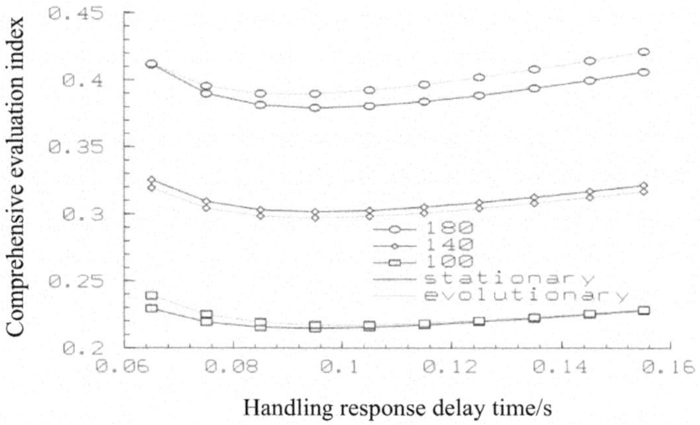

Fig. 6.3 Influence of handling response delay time

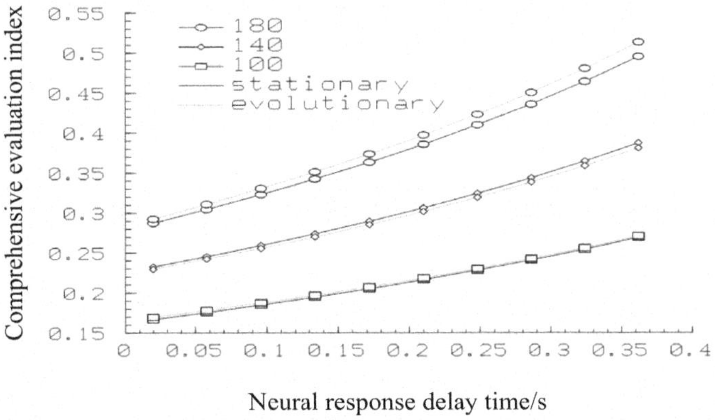

Fig. 6.4 Influence of neural response delay time

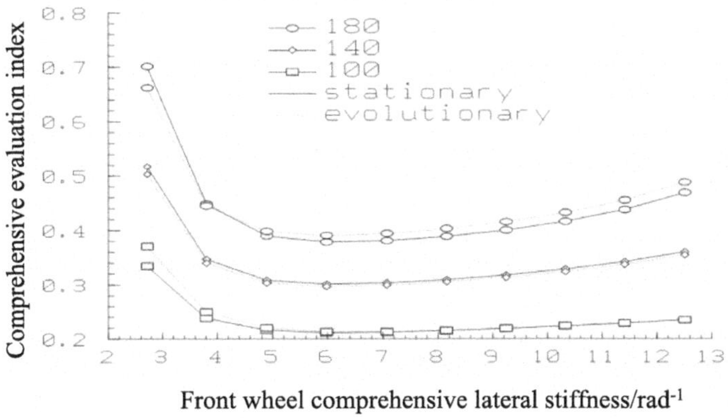

Fig. 6.5 Influence of front wheel comprehensive lateral stiffness

preview time (T) will decrease (1.05 s, 0.95 s and 0.85 s, respectively). Moreover, with the increase of vehicle speed, the influence of the change of T around the optimal value on the active safety becomes significant. The difference between the influence of evolutionary random road input and stationary random road input on the optimal preview time is that compared with stationary random road input, evolutionary random road input makes the optimal preview time area (each point near the optimal preview time) move left when the speed is high and right when the speed is low. This phenomenon can be explained as that the slow change of speed has higher requirements on the driver's technical level, and the driver's visual busyness has increased [15].

It can be seen from Fig. 6.3 that under the condition of stable random road input, the driver's handling response delay time (T_h) decreases with the change of vehicle speed. When the vehicle speed is 100 km/h, 140 km/h and 180 km/h, the optimal value of the driver's handling response delay time is around 0.95 s, while under the condition of evolving random road input, the driver's handling response delay time is 0.105 s, 0.095 s and 0.90 s at the vehicle speeds of 100 km/h, 140 km/h and 180 km/h, respectively. It indicates that the driver's handling response should be accelerated as the vehicle speed increases. According to human physiological characteristics and existing experience, when simulating the driver-vehicle–road closed-loop system, generally, T_h is taken as 0.1 s [16].

It can be seen from Fig. 6.4 that the influence of evolving random road input and stationary random road input on the driver's neural response delay time is roughly the same. The smaller the driver's neural response delay time is, the better. Moreover, at high speed ($U \geq 180$ km/h), when T_d exceeds 0.15 s, the deterioration of active safety is aggravated. At medium and high speed ($U = 140$ km/h), when T_d exceeds 0.2 s, the deterioration of active safety is aggravated. According to human physiological characteristics and existing experience, when simulating the driver-vehicle–road closed-loop system, generally, T_d is taken as 0.2 s.

It can be seen from Fig. 6.5 that under the condition that the relative comprehensive lateral stiffness of the rear wheel (K_2 / F_{z2}) remains unchanged, the influence of evolutionary random road input and stationary random road input on the optimal relative comprehensive lateral stiffness of the front wheel is roughly the same. When the vehicle speed (U) is 100 km/h, 140 km/h and 180 km/h respectively, the optimal relative comprehensive lateral stiffness of the front wheel (K_1 / F_{z1}) is 7.18 rad^{-1}, 6.45 rad^{-1} and 6.24 rad^{-1} respectively, which basically matches the parameters of the original vehicle. Moreover, when the comprehensive cornering stiffness of the front wheel changes within a certain range near the optimal value, the impact on the active safety is not too great, especially when the vehicle speed is low. In addition, it can also be seen that excessive under-steer (the value of K_1 is too small) and over-steer (the value of K_1 is too large) will deteriorate the active safety.

6.1.5 Virtual Input Algorithm

If the system has a stationary random process input $r(t)$ from the initial state, its self-spectral density function $s_r(\omega)$ is known. Any response of the system $x(t)$ can be expressed by Duhamel integral as

$$x(t) = \int_0^t h(t - \tau) r(\tau) \mathrm{d}\tau \tag{6.39}$$

Its autocorrelation function is

$$R_x(t_1, t_2) = E(x(t_1)x(t_2))$$

$$= E\left(\int_0^{t_1}\int_0^{t_2} h(t_1-\tau_1)h(t_2-\tau_2)r(\tau_1)r(\tau_2)\mathrm{d}\tau_1\mathrm{d}\tau_2\right)$$

$$= \int_0^{t_1}\int_0^{t_2} h(t_1-\tau_1)h(t_2-\tau_2)R_r(\tau_1,\tau_2)\mathrm{d}\tau_1\mathrm{d}\tau_2 \tag{6.40}$$

where

$$R_r(\tau_1,\tau_2) = E(r(\tau_1)r(\tau_2)) \tag{6.41}$$

Since $r(t)$ is a stationary random process, the Wiener-Sinchin relationship is applied

$$R_r(\tau_1,\tau_2) = R_r(\tau) = \int_{-\infty}^{\infty} s_r(\omega)e^{i\omega\tau}\mathrm{d}\omega \tag{6.42}$$

where $\tau = \tau_2 - \tau_1$.

Substituting Eq. (6.42) into Eq. (6.40)

$$R_x(t_1,t_2) = \int_{-\infty}^{\infty} I^*(\omega,t_1)I(\omega,t_2)s_r(\omega)\mathrm{d}\omega \tag{6.43}$$

where

$$I(\omega,t) = \int_0^t h(t-\tau)e^{\mathrm{i}\omega t}\mathrm{d}\tau \tag{6.44}$$

In Eq. (6.43), let $t_1 = t_2 = t$, and get the variance of $x(t)$

$$\sigma_x^2(t) = R_x(t,t)$$

$$= \int_{-\infty}^{\infty} I^*(\omega,t)I(\omega,t)s_r(\omega)\mathrm{d}\omega \tag{6.45}$$

The integrand function at the right end is the self-spectral density of $x(t)$

$$s_x(\omega,t) = I^*(\omega,t)I(\omega,t)s_r(\omega) \tag{6.46}$$

It is known from Eq. (6.44) that $f(t) = \sqrt{s_r(\omega)}e^{\mathrm{i}\omega t}$ is the response of the system generated by the deterministic input $e^{\mathrm{i}\omega t}$ in the initial state.

Therefore, if a virtual road input is constructed

$$f(t) = \sqrt{s_r(\omega)}e^{\mathrm{i}\omega t} \tag{6.47}$$

The response generated by this virtual road input is

$$x(\omega, t) = \sqrt{s_r(\omega)}I(\omega, t) \tag{6.48}$$

Obviously

$$x^*(\omega, t)x(\omega, t) = I^*(\omega, t)I(\omega, t)s_r(\omega) \tag{6.49}$$

Comparing Eq. (6.46) with Eq. (6.49)

$$\begin{aligned} s_x(\omega, t) &= x^*(\omega, t)x(\omega, t) \\ &= |x(\omega, t)|^2 \end{aligned} \tag{6.50}$$

And its variance is calculated by Eq. (6.51)

$$\sigma_x^2(t) = 2\int_0^{\infty} s_x(\omega, t)\mathrm{d}\omega \tag{6.51}$$

If the system has an evolving random process input $g(t)r(t)$ from the initial state, where $g(t)$ is a slowly varying modulation function, $r(t)$ is a stationary random process, and its self-spectral density function $s_r(\omega)$ is known. Any response of the system $x(t)$ can be expressed by Duhamel integral as

$$x(t) = \int_0^t h(t-\tau)g(\tau)r(\tau)\mathrm{d}\tau \tag{6.52}$$

Its autocorrelation function is

$$\begin{aligned} R_x(t_1, t_2) &= E(x(t_1)x(t_2)) \\ &= E\left(\int_0^{t_1}\int_0^{t_2} h(t_1-\tau_1)h(t_2-\tau_2)g(\tau_1)g(\tau_2)r(\tau_1)r(\tau_2)\mathrm{d}\tau_1\mathrm{d}\tau_2\right) \\ &= \int_0^{t_1}\int_0^{t_2} h(t_1-\tau_1)h(t_2-\tau_2)g(\tau_1)g(\tau_2)R_r(\tau_1, \tau_2)\mathrm{d}\tau_1\mathrm{d}\tau_2 \end{aligned} \tag{6.53}$$

where

$$R_r(\tau_1, \tau_2) = E(r(\tau_1)r(\tau_2)) \tag{6.54}$$

Since $r(t)$ is a stationary random process, the Wiener-Sinchin relationship is applied

$$R_r(\tau_1, \tau_2) = R_r(\tau) = \int_{-\infty}^{\infty} s_r(\omega)e^{\mathrm{i}\omega\tau}\mathrm{d}\omega \tag{6.55}$$

where $R_x(t_1, t_2) = \int_{-\infty}^{\infty} I^*(\omega, t_1)I(\omega, t_2)s_r(\omega)d\omega$.

Substituting Eq. (6.55) into Eq. (6.53)

$$R_x(t_1, t_2) = \int_{-\infty}^{\infty} I^*(\omega, t_1)I(\omega, t_2)s_r(\omega)\mathrm{d}\omega \tag{6.56}$$

where

$$I(\omega, t) = \int_{0}^{t} h(t-\tau)g(\tau)e^{\mathrm{i}\omega t}\mathrm{d}\tau \tag{6.57}$$

In Eq. (6.56), let $x(t)$, and get the variance of $x(t)$

$$\begin{aligned} \sigma_x^2(t) &= R_x(t, t) \\ &= \int_{-\infty}^{\infty} I^*(\omega, t)I(\omega, t)s_r(\omega)\mathrm{d}\omega \end{aligned} \tag{6.58}$$

The integrand function at the right end is the self-spectral density of $x(t)$

$$s_x(\omega, t) = I^*(\omega, t)I(\omega, t)s_r(\omega) \tag{6.59}$$

It is known from Eq. (6.57) that $f(t) = \sqrt{s_r(\omega)}e^{i\omega t}$ is the response of the system generated by the deterministic input $g(t)e^{\mathrm{i}\omega t}$ in the initial state.

Therefore, if a virtual road input is constructed

$$f(t) = \sqrt{s_r(\omega)}g(t)e^{\mathrm{i}\omega t} \tag{6.60}$$

The response generated by this virtual road input is

$$x(\omega, t) = \sqrt{s_r(\omega)}I(\omega, t) \tag{6.61}$$

Obviously

$$x^*(\omega, t)x(\omega, t) = I^*(\omega, t)I(\omega, t)s_r(\omega) \tag{6.62}$$

Comparing Eq. (6.59) with Eq. (6.62)

$$\begin{aligned} s_x(\omega, t) &= x^*(\omega, t)x(\omega, t) \\ &= |x(\omega, t)|^2 \end{aligned} \tag{6.63}$$

and its variance is calculated by Eq. (6.64)

$$\sigma_x^2(t) = 2\int_0^\infty s_x(\omega, t)\mathrm{d}\omega \tag{6.64}$$

6.1.6 Conclusion

When the driver, the vehicle and the road are deterministic, the simulation and evaluation of the driver-vehicle–road closed-loop system corresponds to a driving test in which the same driver drives the same model of vehicle through the determined route. The driver-vehicle–road closed-loop system simulation based on random road input can comprehensively reflect the active safety of vehicles, since its simulation and evaluation correspond to many driving tests of the same driver driving the same model of vehicle through random routes. Therefore, it is of great practical value to carry out the simulation and active safety evaluation of the driver-vehicle–road closed-loop system from the perspective of random road input.

It is proved that the computer prediction of the active safety of the driver-vehicle–road closed-loop system with random road input is feasible, a new way is found to solve the limitations of danger, economic burden and evaluation results caused by real vehicle tests, and a new method is found for the joint design of the driver-vehicle–road closed-loop system.

6.2 Influence of Driver's Dynamic Characteristics on Vehicle Handling Safety

Driver-vehicle–road closed-loop system is the basic basis of vehicle handling safety evaluation. The research results of McRuer and others also show that there is observation noise in the driver's observation link. In fact, the driver's judgment and operation are also uncertain. Therefore, it is very necessary and valuable to establish a closed-loop evaluation method of vehicle handling considering the uncertainty of driver link [17] for the prediction of driver handling feeling and the optimization design of vehicle structural parameters based on handling safety. At present, this work is one of the main problems to be solved in the automotive engineering field at home and abroad.

Up to now, the evaluation of vehicle dynamic characteristics is mainly carried out on the road. Although the development type driving simulator adopts advanced real-time simulation, digital imaging, hydraulic control and other technologies, combines the real driver with the modeled vehicle, and replaces the field test with indoor computer simulation, which can ease the contradiction between the development level of theoretical research and the closed-loop design requirements of vehicle handling safety [18]. It is one of the irreplaceable technical tools to shorten the automobile development cycle and improve the competitiveness of automobile products. However, the driving simulator is very expensive, it is not available to ordinary automobile manufacturers.

6.2.1 *Driver-Vehicle–Road Stochastic Closed-Loop System Model*

1. Selection of Driver Direction Control Model

The comparative study of existing driver direction control models shows that the optimal preview acceleration model not only has good tracking performance and a wide application range [19], but also has a simple structure, clear physical meaning of model parameters, and describes various characteristics of the driver.

According to the optimal curvature control theory, the PO preview model [20] in the optimal acceleration model shown in Fig. 6.6 can be obtained, i.e., the block diagram of the position and azimuth preview model. In Fig. 6.6, λ is the azimuth action coefficient.

2. Statistical Characteristics of Random Driver Model Parameters

Mathematically, the uncertainty of a random system can be described by a deterministic model structure [21], plus randomly varying model parameters. That is, after

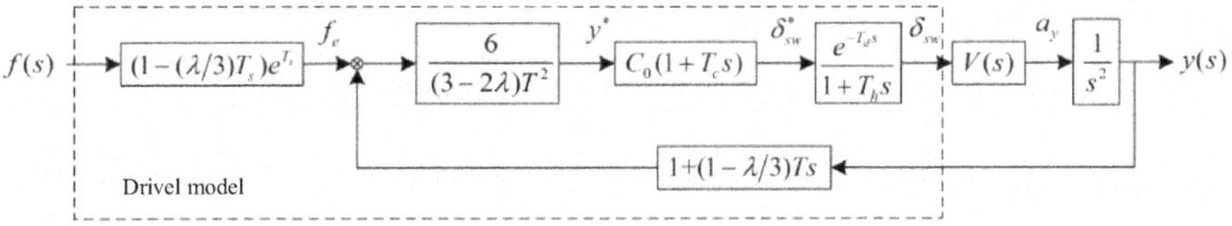

Fig. 6.6 Block diagram of driver-vehicle–road closed-loop system

the model structure is determined, appropriate model parameters should be selected and randomized.

According to the preview optimal curvature model, the dynamic characteristics of the driver are divided into preview decision and correction decision. The preview decision represents the driver's ability to deal with the road environment, while the correction decision describes the driver's mastery of the dynamic characteristics of the vehicle. The preview decision model parameters include preview and follow-up parameters. The preview parameters include preview time T and azimuth action coefficient λ. The following parameter is mainly the third-order term action coefficient a. While the correction decision parameters only include the correction gain C_0 and the differential correction time constant T_c. Therefore, the random parameters of the model shown in Fig. 6.6 only include preview time T, azimuth action coefficient λ, third-order term action coefficient a, correction gain C_0 and differential correction time constant T_c. The test investigation results show that the dynamic delay parameters T_d and T_h of the actual driver are also different from person to person. Therefore, the random variables in the driver model shown in Fig. 6.6 are T, λ, a, C_0, T_c, T_d, T_h. C_0 and a can be determined by the structural parameters of the vehicle and other driver parameters. In addition, if it is noted that the above random parameters represent the most basic preview characteristics of the driver, it can be assumed that the random variables are independent of each other. Therefore, the independent random variables are T, λ, T_c, T_d, T_h. If the driver model is randomized, the physical process described by the mathematical model of the whole closed-loop system is the sum of all driving processes of the same model of the vehicle after it is put into actual operation.

3. Statistical Properties of Parameters

After considering the randomness of drivers, the driver-vehicle–road closed-loop system shown in Fig. 6.6 is a random closed-loop system. Therefore, to solve the statistical characteristics of the system response, we must first determine the statistical characteristics of each random variable and their relationship (i.e., the probability distribution characteristics, mean and variance of each random variable, and the joint probability density between them). Among these statistical characteristics, the mean and variance can be accurately estimated through small batch experiments. However, a large number of experiments are needed to obtain the probability distribution density. Under the current situation, it is difficult to use real vehicle experiments. Therefore, reasonable selection can only be made according to the results of theoretical analysis and simulation calculation.

Experience has shown that there is always an optimal value for the driver dynamic characteristic parameters in the preview optimal curvature model, which makes the closed-loop system achieve the best tracking performance. If people's learning and adaptability are fully considered, it can be considered that most of the parameters adopted by actual drivers are close to the optimal values of the corresponding parameters. Therefore, if it is impossible to obtain a large number of sufficient test data, it can be assumed that the probability distribution of the above independent random parameters is approximately normal, and their joint probability distribution density is the product of their respective probability densities.

4. Driver-Vehicle-Road Stochastic Closed-Loop System Model

Using the two-degree-of-freedom linear vehicle model under the condition of a small heading angle, according to the driver structural model in Fig. 6.6, the state equation of the driver-vehicle–road closed-loop system can be expressed as follows

$$\dot{x} = Hx + bf_r \tag{6.65}$$

The output variables are

$$z = Gx + df_r \tag{6.66}$$

where

$x = \left[r \; \beta \; y \; v_a \; q_1 \; q_2 \right]^{\mathrm{T}}, b = \left[0\,0\,0\,0\,0\; b_6 \right]^{\mathrm{T}}.$

$z = \left[y \; a_y \; \delta_{sw} \; \dot{\delta}_{sw} \; s_1 \; s_2 \right], d = \left[0\,0\,0\; d_4 \; 0\,0 \right]^{\mathrm{T}}$

$$f_r(t) = f(t+T) + \left(T_c - \frac{\lambda}{3} T \right) \dot{f}(t+T) - \frac{\lambda}{3} T T_c \ddot{f}(t+T)$$

Now Eq. (6.65) is expressed as the following form

$$\dot{x} = Hx + \overline{r} \tag{6.67}$$

where

$$\overline{r} = bf_r \tag{6.68}$$

When considering the randomness of the driver, the parameters T, λ, T_c, T_d, T_h in the system matrix H should be randomized (i.e., H should be the function matrix of the above random parameter $R_i (i = 1, 2, \cdots, 5)$). To study the influence of random parameters on the system response characteristics, the random parameter R, the system matrix H, the response quantities x and $\dot{x}$, and the system input $\overline{r}$ should be transformed as follows

$$R = R_d + \varepsilon R_r \tag{6.69}$$

$$H = H_d + \varepsilon H_r \tag{6.70}$$

$$x = x_d + \varepsilon x_r \tag{6.71}$$

$$\dot{x} = \dot{x}_d + \varepsilon \dot{x}_r \tag{6.72}$$

$$\overline{r} = \overline{r}_d + \varepsilon \overline{r}_r \tag{6.73}$$

where ε is a small parameter, subscript d represents the deterministic part of the random variable, that is, the mathematical model of the optimal driver model corresponding to the deterministic, subscript r represents the random part of the random variable, and the mean value is zero. Obviously, the random part required here is much smaller than the deterministic part. Take mathematical expectation for Eq. (6.69), Eqs. (6.71)–(6.73), there are

$$E(R) = E(R_d) + \varepsilon E(R_r) = R_d \tag{6.74}$$

$$E(x) = E(x_d) + \varepsilon E(x_r) = x_d \tag{6.75}$$

$$E(\dot{x}) = E(\dot{x}_d) + \varepsilon E(\dot{x}_r) = \dot{x}_d \tag{6.76}$$

$$E(\overline{r}) = E(\overline{r}_d) + \varepsilon E(\overline{r}_r) = \overline{r}_d \tag{6.77}$$

Similarly, take the variance and covariance of Eq. (6.69), Eqs. (6.71)–(6.73). According to Kronecker algebra and corresponding random analysis theory, there are

$$\mathrm{Var}(R) = E\left((R - E(R))^{[2]}\right) = \varepsilon^2 E\left(R_r^{[2]}\right) \tag{6.78}$$

$$\mathrm{Var}(x) = E\left((x - E(x))^{[2]}\right) = \varepsilon^2 E\left(x_r^{[2]}\right) \tag{6.79}$$

$$\mathrm{Var}(\dot{x}) = E\left((\dot{x} - E(\dot{x}))^{[2]}\right) = \varepsilon^2 E\left(\dot{x}_r^{[2]}\right) \tag{6.80}$$

$$\mathrm{Var}(\overline{r}) = E\left((\overline{r} - E(\overline{r}))^{[2]}\right) = \varepsilon^2 E\left(\overline{r}_r^{[2]}\right) \tag{6.81}$$

where $(\cdot)^{[2]} = (\cdot) \otimes (\cdot)$ is the Kronecker power, the symbol "$\otimes$" represents the Kronecker product, and $\mathrm{Var}(\cdot)$ contains all variances and covariance, which is called the variance vector.

6.2.2 General Stochastic Perturbation Method for Response Analysis of Stochastic Closed-Loop Systems

Since it is difficult to obtain the probability density and joint probability density of random variables in engineering practice, this paper will use the general random perturbation method to calculate the response of the system. The random perturbation method only requires that the first two moments of the random parameters of the system be known, and the first two moments of the response of the random system can be determined. The general stochastic perturbation method not only has standardized mathematical expressions and clear physical meanings, but also makes it possible to solve the calculation problems of the differential equations of general stochastic systems with random parameters.

Substitute Eqs. (6.70)–(6.73) into Eq. (6.67), and

$$\dot{x}_d + \varepsilon \dot{x}_r = (H_d + \varepsilon H_r)(x_d + \varepsilon x_r) + \overline{r}_d + \varepsilon \overline{r}_r \tag{6.82}$$

Expand this equation, compare the same power coefficients of ε, omit the higher-order terms above $O(\varepsilon^2)$, then

$$\varepsilon^0 : \dot{x}_d = H_d x_d + \overline{r}_d \tag{6.83}$$

$$\varepsilon^1 : \dot{x}_r = H_d x_r + H_r x_d + \overline{r}_r \tag{6.84}$$

From Eq. (6.83), the determined parts of the system response $x_d, \dot{x}_d$ can be solved (i.e., the mean value of the response). Obviously, when only the first two moments of the random parameters are given, it is impossible to determine the random part of the response from Eq. (6.84), that is, the part related to variance and covariance, which must be solved after transforming the form.

According to the Taylor expansion of the vector-valued function, when the random part of the random parameter vector is much smaller than the determined part, H_r, x_r, $\dot{x}_r$, $\overline{r}_r$ can be expanded to the first-order term near $E(R) = R_d$, so that

$$H_r = \frac{\partial H_d}{\partial R^{\mathrm{T}}} R_r \tag{6.85}$$

$$x_r = \frac{\partial x_d}{\partial R^{\mathrm{T}}} R_r \tag{6.86}$$

$$\dot{x}_r = \frac{\partial \dot{x}_d}{\partial R^{\mathrm{T}}} R_r \tag{6.87}$$

$$\overline{r}_r = \frac{\partial \overline{r}_d}{\partial R^{\mathrm{T}}} R_r \tag{6.88}$$

Replace Eqs. (6.85)–(6.88) into Eq. (6.84) and expand it, there are

$$\frac{\partial \dot{x}_d}{\partial R_i} = H_d \frac{\partial x_d}{\partial R_i} + \frac{\partial H_d}{\partial R_i} x_d + \frac{\partial \bar{r}_d}{\partial R_i} \quad (i = 1,\ 2,\ \cdots,\ 5) \tag{6.89}$$

The sensitivity of the system response $\partial x_d / \partial R_i$, $\partial \dot{x}_d / \partial R_i$ can be solved from Eq. (6.89), so that

$$\frac{\partial x_d}{\partial R^{\mathrm{T}}} = \begin{bmatrix} \frac{\partial x_d}{\partial R_1} & \cdots & \frac{\partial x_d}{\partial R_m} \end{bmatrix} \tag{6.90}$$

$$\frac{\partial \dot{x}_d}{\partial R^{\mathrm{T}}} = \begin{bmatrix} \frac{\partial \dot{x}_d}{\partial R_1} & \cdots & \frac{\partial \dot{x}_d}{\partial R_m} \end{bmatrix} \tag{6.91}$$

Substituting Eqs. (6.86) and (6.87) into Eqs. (6.79) and (6.80) respectively, and using Eq. (6.78), the variance vector of the system response can be obtained

$$\mathrm{Var}(x) = \varepsilon^2 E\big(x_r^{[2]}\big) = \varepsilon^2 E\left(\left(\frac{\partial x_d}{\partial R^{\mathrm{T}}} R_r\right)^{[2]}\right) = \left(\frac{\partial x_d}{\partial R^{\mathrm{T}}}\right)^{[2]} \mathrm{Var}(R) \tag{6.92}$$

$$\mathrm{Var}(\dot{x}) = \varepsilon^2 E\big(\dot{x}_r^{[2]}\big) = \varepsilon^2 E\left(\left(\frac{\partial \dot{x}_d}{\partial R^{\mathrm{T}}} R_r\right)^{[2]}\right) = \left(\frac{\partial \dot{x}_d}{\partial R^{\mathrm{T}}}\right)^{[2]} \mathrm{Var}(R) \tag{6.93}$$

Substituting Eqs. (6.90) and (6.91) into Eqs. (6.92) and (6.93) respectively, the variance vector of the system response is determined. In this way, the mean and variance vectors of the system response are completely determined.

For static equations such as Eq. (6.66), the mean and variance vectors of the system output z can also be obtained by the above random perturbation method. The specific steps are as follows

(1) $\frac{\partial z_d}{\partial R_i} = \mathrm{OH}_d \frac{\partial x_d}{\partial R_i} + \frac{\partial \mathrm{OH}_d}{\partial R_i} x_d + \frac{\partial \mathrm{OR}_d}{\partial R_i}$, then $\frac{\partial z_d}{\partial R_i}$;

(2) $\mathrm{Var}(z) = \left(\frac{\partial z_d}{\partial R^{\mathrm{T}}}\right)^{[2]} \mathrm{Var}(R)$.

6.2.3 *The Influence of Driver's Dynamic Characteristics on Vehicle Active Safety*

The closed-loop system considered is linear, and it is assumed that the changes of the dynamic characteristic parameters of the driver and the road input conform to the normal random distribution. Therefore, the response of the system must also be normally distributed, and the corresponding standard deviation can be calculated by Eq. (6.94) through the obtained variance

$$\sigma_i = \sqrt{\mathrm{Var}(x_i)},\ i = 1, 2, \ldots, N \tag{6.94}$$

where x_i is an element in the system response vector, σ_i is the standard deviation corresponding to x_i, N is the dimension of the system.

According to the principle of mathematical statistics, for any random variable x_i with a normal distribution mean of x_{id}, as long as its standard deviation σ_i is known, the probability density $P(x_i)$ of the random variable can be determined by Eq. (6.95)

$$P(x_i) = \frac{1}{\sqrt{2\pi}\sigma_i} e^{-\frac{(x_i - x_{id})^2}{2\sigma_i^2}} \tag{6.95}$$

The probability $N(z_i)$ of the absolute value $|x_i|$ of the random variable within a certain threshold value z_i can be obtained by integrating x_i with Eq. (6.95)

$$N(z_i) = \int_{-z}^{z} P(x_i)\mathrm{d}x_i \tag{6.96}$$

Using the above method, the safety information of the vehicle in a single evaluation can be easily obtained (i.e., the probability of no traffic accident). The probability of traffic accident is determined by the following equation

$$\overline{N}(z_i) = 1 - \int_{-z}^{z} P(x_i)\mathrm{d}x_i \tag{6.97}$$

6.2.4 Numerical Simulation Example

The active safety evaluation index of the driver-vehicle–road closed-loop system given in Sect. 6.2.3 and the calculation method given in this paper are used to evaluate the random closed-loop system in the time domain.

1. Numerical Simulation of Audi100 Vehicle

Audi100 vehicle is subject to double lane changing test at speeds of 120 and 160 km/h. The mean and variance of random variables in the driver model are shown in Tables 6.1 and 6.2, respectively.

Table 6.1 Random parameters and statistical values of driver model ($V = 120$ km/h)

Random parameter	T/s	λ	T_c	T_d/s	T_h/s
Mean value	1.290	5.175×10^{-2}	1.457×10^{-3}	0.2	0.1
Variance	1.60×10^{-2}	2.01×10^{-3}	1.00×10^{-3}	1.50×10^{-2}	1.20×10^{-2}
Variance/Mean value%	1.240	3.884	68.627	7.500	12.000

Table 6.2 Random parameters and statistical values of driver model ($V = 160$ km/h)

Random parameter	T/s	λ	T_c	T_d/s	T_h/s
Mean value	1.290	5.330E-02	6.053E-03	0.2	0.1
Variance	1.60×10^{-2}	2.01×10^{-3}	1.00×10^{-3}	1.50×10^{-2}	1.20×10^{-2}
Variance/Mean value%	1.240	3.771	16.522	7.500	12.000

The response of the Audi100 vehicle to the lateral position, lateral acceleration and steering wheel angular velocity at speeds of 120 and 160 km/h and the results of its active safety prediction are shown in Figs. 6.7, 6.8, 6.9 and Figs. 6.10, 6.11, 6.12 respectively. Wherein, curve *B* represents the standard deviation and curve *P* represents the safety probability value of a single response.

In Fig. 6.7, the *Y* curve represents the average value of the driving trajectory, and the *F* curve represents the expected path. It can be seen from Fig. 6.7 that the safety probability of the lateral position is small at the entrance after lane change, and the safety probability at the entrance of the second lane change is smaller than that at the entrance of the first lane change. When the threshold value of lateral position deviation is 0.4 m, the safety probability at the entrance of the first lane change is 0.88, and the safety probability at the entrance of the second lane change is 0.82.

In Fig. 6.8, the AY curve represents the mean value of the lateral acceleration of the vehicle. It can be seen from Fig. 6.8 that the safety probability of lateral acceleration near the exit after the first lane change is small. When the threshold value of the lateral acceleration deviation is 2.0 m/s^2, the probability that the lateral acceleration does not change abnormally is 0.53, which occurs about 0.2 s from the exit after the first lane change, as shown in the figure.

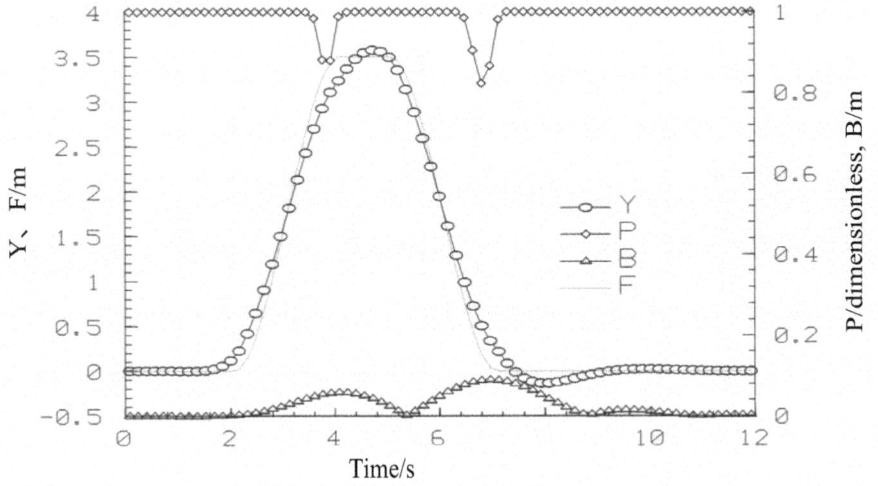

Fig. 6.7 Single prediction results of lateral position response and active safety ($V = 120$ km/h, threshold $= 0.4$ m)

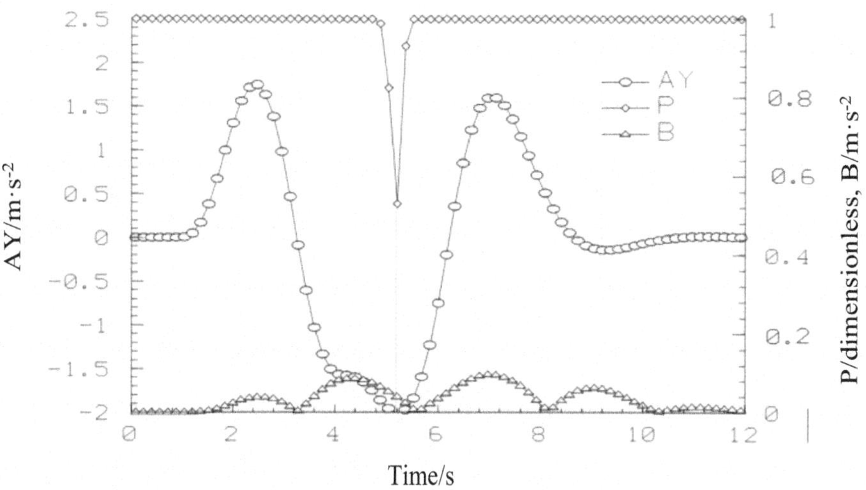

Fig. 6.8 Single prediction results of lateral acceleration response and active safety (V = 120 km/h, threshold = 2.0 m/s^2)

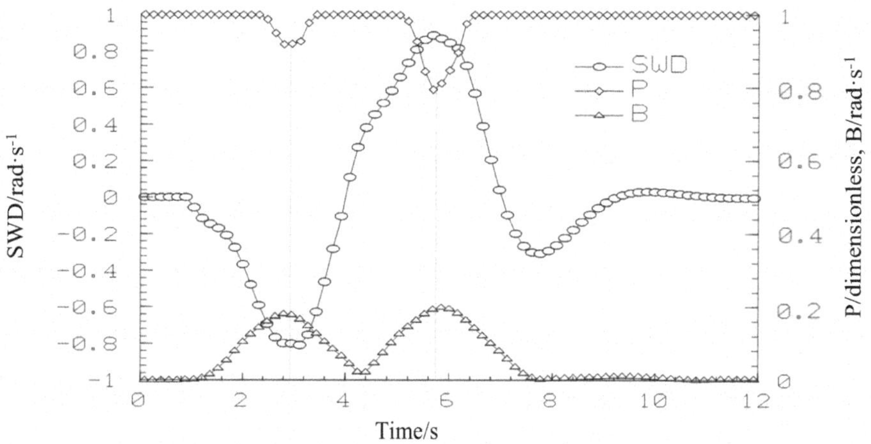

Fig. 6.9 Single prediction results of steering wheel angle velocity response and active safety (V = 120 km/h, threshold = 1.047 rad/s)

In Fig. 6.9, the SWD curve represents the mean angle velocity of the steering wheel. It can be seen from Fig. 6.9 that the response of the steering wheel angle velocity at the position indicated by the straight line in the figure is relatively large (i.e., the busy degree of the driver is relatively high). In addition, in the transition process of each lane change, not only the driver is busy, but also the probability of getting busy is high. The probability of getting busy in the second lane change transition process is slightly smaller than the probability of getting busy in the first

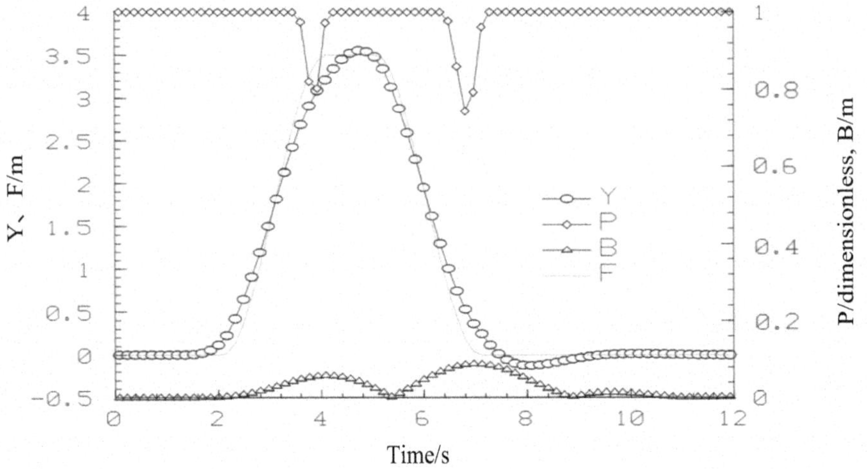

Fig. 6.10 Single prediction results of lateral position response and active safety (V = 160 km/h, threshold = 0.4 m)

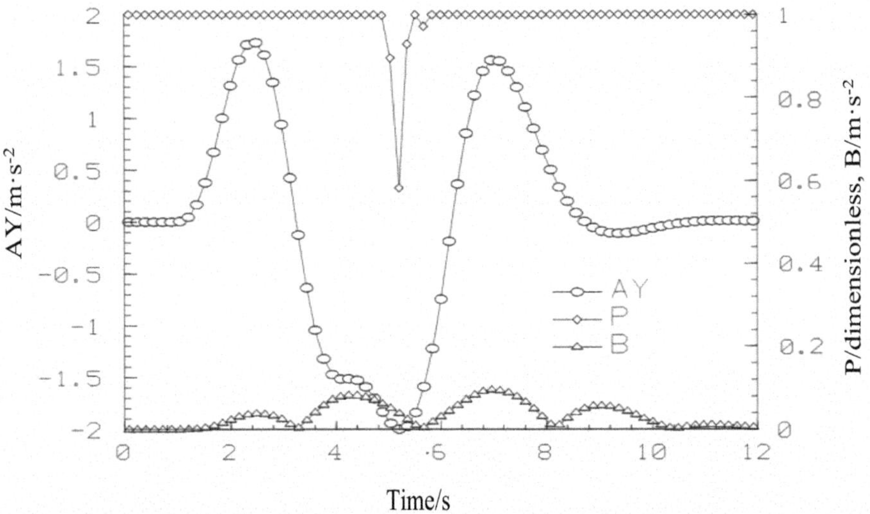

Fig. 6.11 Single prediction results of lateral acceleration response and active safety (V = 160 km/h, threshold = 2.0 m/s^2)

lane change. When the threshold value of the angle velocity deviation of the steering wheel is 1.047 rad/s (i.e., 60°/s), the minimum probability that the driver can easily manipulate the steering wheel in the first lane change transition is 0.92, and the minimum probability that the driver can easily manipulate the steering wheel in the second lane change transition is 0.81.

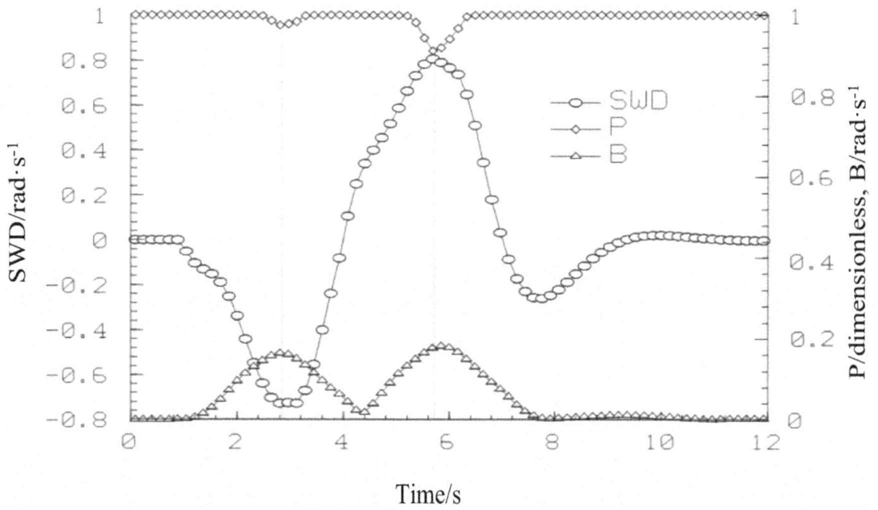

Fig. 6.12 Single prediction results of steering wheel angle velocity response and active safety (V = 160 km/h, threshold = 1.047 rad/s)

In Fig. 6.10, the Y curve represents the average value of the driving trajectory, and the F curve represents the expected path. It can be seen from Fig. 6.10 that the safety probability of the lateral position is small at the entrance after lane change, and the safety probability at the entrance of the second lane change is smaller than that at the entrance of the first lane change. When the threshold value of lateral position deviation is 0.4 m, the safety probability at the entrance of the first lane change is 0.80, and the safety probability at the entrance of the second lane change is 0.74.

In Fig. 6.11, the AY curve represents the mean value of lateral acceleration of the vehicle. It can be seen from Fig. 6.11 that the safety probability of lateral acceleration near the exit after the first lane change is small. When the threshold value of the lateral acceleration deviation is 2.0 m/s^2, the probability that the lateral acceleration does not change abnormally is 0.58, which occurs about 0.2 s from the exit after the first lane change.

In Fig. 6.12, the SWD curve represents the average angle velocity of the steering wheel. It can be seen from Fig. 6.12 that the response of the steering wheel angle velocity at the position indicated by the dotted line in the figure is relatively large (i.e., the driver's busy degree is relatively high). And in the transition process of each lane change, not only the driver is busy, but also the probability of getting busy is high. The probability of getting busy in the transition process of the second lane change is slightly smaller than the probability of getting busy in the first lane change. When the threshold value of the angle velocity deviation of the steering wheel is 1.047 rad/s (i.e., 60°/s), the minimum probability that the driver can easily manipulate the steering wheel in the first lane change transition is 0.97, and the minimum probability that the driver can easily manipulate the steering wheel in the second lane change transition is 0.91.

2. Numerical Simulation of CA141 Truck

CA141 truck is subject to single lane changing test with distance as the standard at speeds of 60 km/h and 90 km/h. The mean and variance of random variables in the driver model are shown in Tables 6.3 and 6.4, respectively.

The responses of the CA141 truck to the lateral position, lateral acceleration and steering wheel angle velocity at speeds of 60 km/h and 90 km/h and the results of its active safety prediction are shown in Figs, 6.13, 6.14, 6.15 and Figs. 6.16, 6.17, 6.18 respectively. Wherein, curve *B* represents the standard deviation and curve *P* represents the safety probability value of a single response.

In Fig. 6.13, the *Y* curve represents the average value of the driving trajectory, and the *F* curve represents the expected path. It can be seen from Fig. 6.13 that the safety probability of lateral position is small at the entrance of lane change. When the threshold value of lateral position deviation is 0.5 m, the safety probability is 0.95. See Table 6.5 for a comparison between simulation value and real vehicle test value. The test values are obtained from 24 real vehicle tests by six drivers of different levels.

In Fig. 6.14, the AY curve represents the mean value of the lateral acceleration of the vehicle. It can be seen from Fig. 6.14 that the safety probability of lateral acceleration at about 0.3 s after the first turning point and about 0.2 s after the second turning point (lane change entrance) is small. When the threshold value of lateral acceleration deviation is taken as 2.0 m/s^2, the probability that the lateral acceleration does not change abnormally is 0.013 and 0.605.

In Fig. 6.15, the SWD curve represents the average angle velocity of the steering wheel. It can be seen from Fig. 6.15 that during the transition process of lane change, not only the driver is busy, but also the probability of getting busy is high. When the

Table 6.3 Random parameters and statistical values of driver model ($V = 60$ km/h)

Random parameter	T/s	λ	T_c	T_d/s	T_h/s
Mean value	1.188	2.320×10^{-2}	3.239×10^{-2}	0.2	0.1
Variance		2.01×10^{-3}	4.00×10^{-3}	1.50×10^{-2}	1.20×10^{-2}
Variance/mean value%	1.347	8.664	12.346	7.500	12.000

Table 6.4 Random parameters and statistical values of driver model ($V = 90$ km/h)

Random parameter	T/s	λ	T_c	T_d/s	T_h/s
Mean value	1.186	1.163×10^{-2}	4.714×10^{-2}	0.2	0.1
Variance	1.60×10^{-2}	2.01×10^{-3}	4.00×10^{-3}	1.50×10^{-2}	1.20×10^{-2}
Variance/Mean value%	1.349	17.283	8.486	7.500	12.000

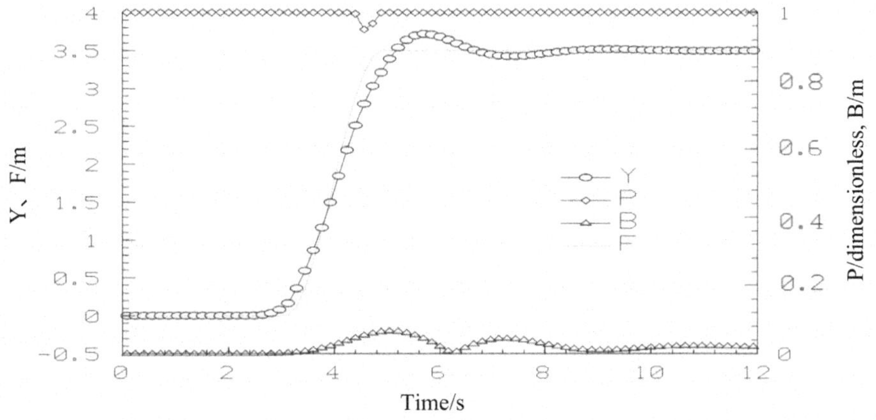

Fig. 6.13 Single prediction results of lateral position response and active safety (V = 60k m/h, threshold = 0.5 m)

Table 6.5 Comparison of safety results of CA141 truck in single lane changing test at the speed of 60 km/h

Position	Entrance	First turning point	Second turning point	Exit
Simulation value test	1.0000	0.9997	0.9502	0.9997
Value	24/24	24/24	21/24	23/24
Error/%	0.00	0.03	8.59	4.33

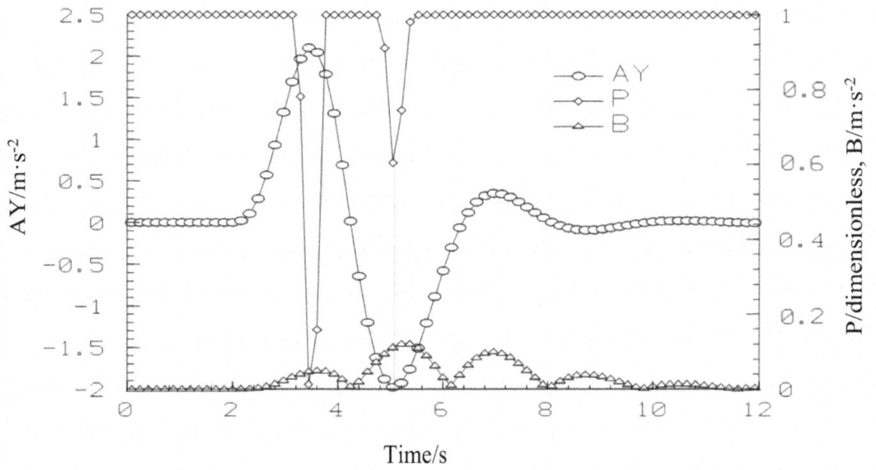

Fig. 6.14 Single prediction results of lateral acceleration response and active safety (V = 60 km/h, threshold = 2.0 m/s^2)

threshold value of steering wheel angle velocity deviation is 3.14 rad/s (i.e., 180°/s), the minimum probability that the driver can easily manipulate the steering wheel during lane change transition is 0.22.

In Fig. 6.16, the *Y* curve represents the average value of the driving trajectory, and the *F* curve represents the expected path. It can be seen from Fig. 6.16 that the safety probability of lateral position is extremely small at the entrance of lane change. When the threshold value of lateral position deviation is taken as 0.5 m, the

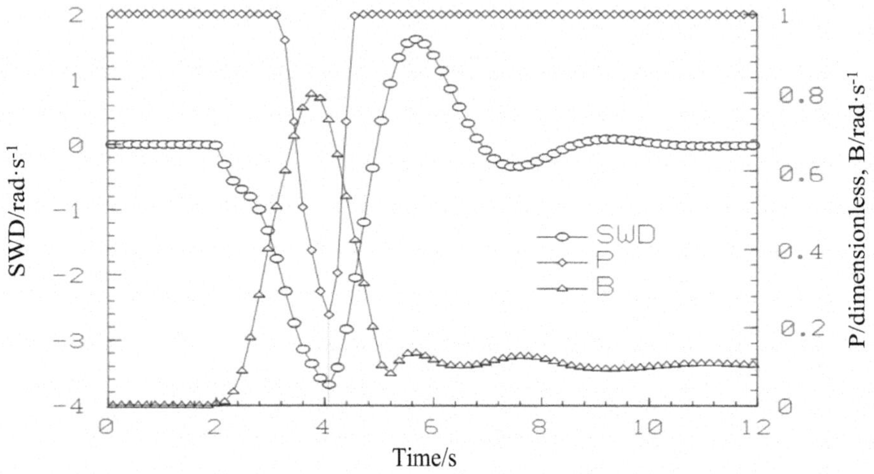

Fig. 6.15 Single prediction results of steering wheel angle velocity response and active safety ($V = 60$ km/h, threshold = 3.14 rad/s)

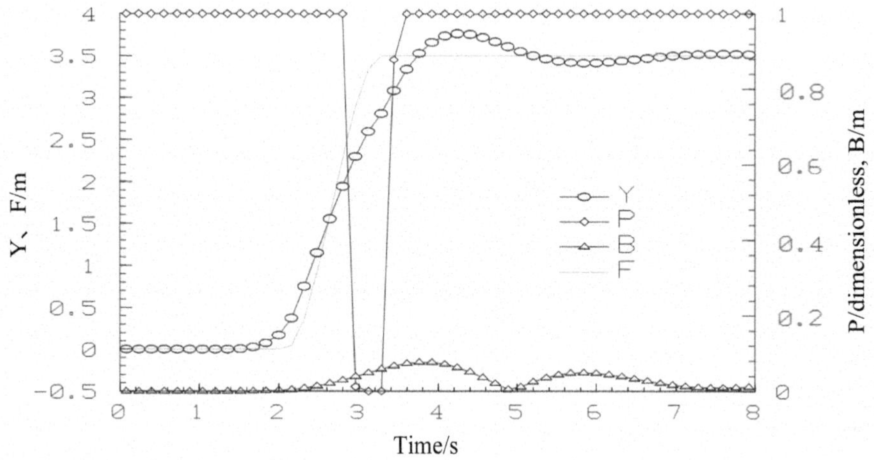

Fig. 6.16 Single prediction results of lateral position response and active safety ($V = 90$ km/h, threshold = 0.5 m)

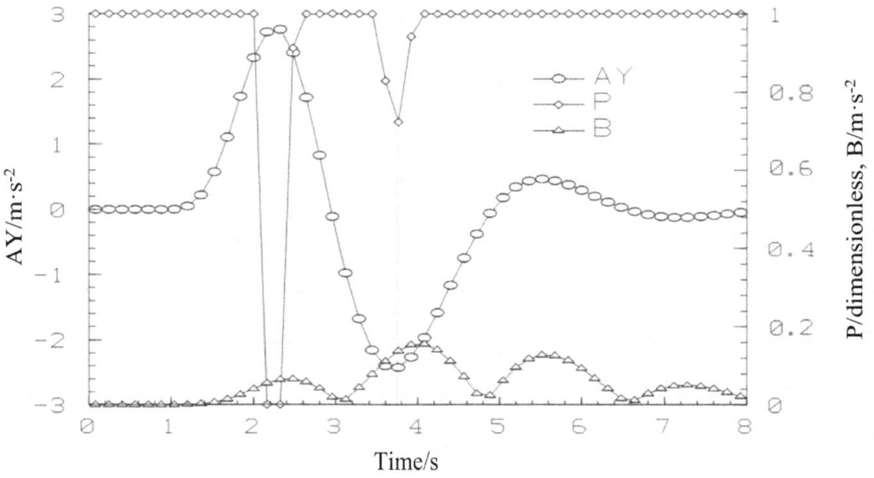

Fig. 6.17 Single prediction results of lateral acceleration response and active safety (V = 90 km/h, threshold = 2.5 m/s^2)

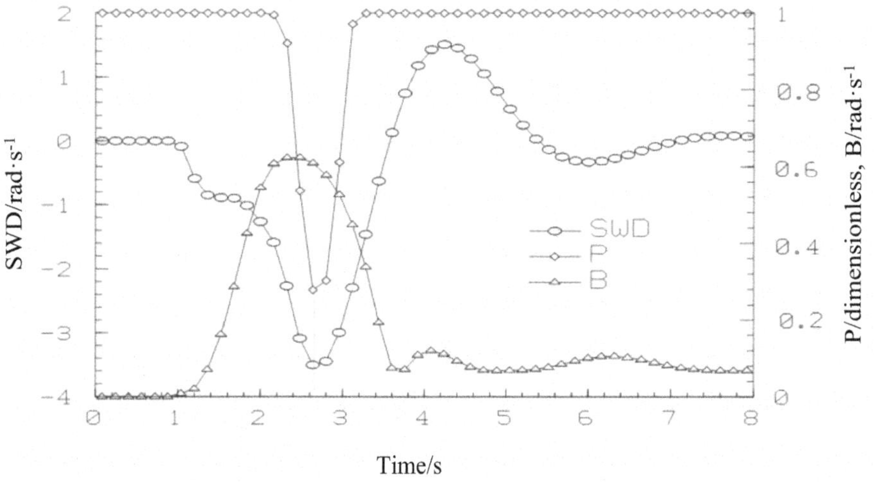

Fig. 6.18 Single prediction results of steering wheel angle velocity response and active safety (V = 90 km/h, threshold = 3.14 rad/s)

safety probability is 0.00. Therefore, it can be predicted that the success rate of the real vehicle test at the speed of 90 km/h will be zero (i.e., the benchmark will be hit every time).

In Fig. 6.17, the AY curve represents the mean value of the lateral acceleration of the vehicle. It can be seen from Fig. 6.17 that the safety probability of lateral acceleration at about 2 s after the first turning point and about 3 s after the second turning point (lane change entrance) is small. When the threshold value of lateral acceleration deviation is taken as 2.5 m/s^2, the probability that the lateral acceleration does not change abnormally is 0.000 and 0.723.

In Fig. 6.18, the SWD curve represents the average angle velocity of the steering wheel. It can be seen from Fig. 6.18 that during the transition process of lane change, not only the driver is busy, but also the probability of getting busy is high. When the threshold value of steering wheel angle velocity deviation is 3.14 rad/s (i.e., 180°/s), the minimum probability that the driver can easily manipulate the steering wheel during lane change transition is 0.279.

6.2.5 Conclusion

In this section, the response calculation of driver-vehicle–road closed-loop system and the computer prediction of active safety are discussed. The results show that after the driver model parameter database is established, the active safety of the vehicle can be predicted and optimized more realistically on the computer. According to the existing experience of real vehicle test and the results of this chapter, setting the line moving benchmark based on time can not only improve the test speed, but also facilitate the comparison and analysis of the correlation under different vehicle speeds. Setting the line moving benchmark based on the distance is convenient for testing the vehicle's handling performance, although the improvement of vehicle speed is limited. Therefore, both of them have their own advantages, which can be regarded as a good way to study the closed-loop controllability of vehicles.

6.3 Motion Stability Analysis of Four-Wheel Steering Vehicle

Four-wheel steering system (4WS) is an important part of active chassis technology. Since the 1990s, many automobile manufacturers have introduced concept vehicles with 4WS, and applied some mature four-wheel steering technology to its mass production vehicles. The main advantage of the four-wheel steering vehicle is that it can basically keep the center of gravity sideslip angle at zero when steering, and can greatly improve the dynamic response characteristics of the vehicle to the steering wheel input. In addition, the turning radius of the vehicle can be reduced at low speed.

Although researchers have done a lot of research on four-wheel steering from structure to control principle, no breakthrough has been made. The main reason is the production cost. Secondly, there are technical problems, mainly relying on experience, and the corresponding theoretical basis is still lacking. Therefore, it is of positive significance for promoting the development and application of four-wheel steering technology to carry out theoretical research on four-wheel steering technology [22].

6.3.1 Motion Stability Theory

1. System Stability

Stability is an extremely important characteristic of the system. The analysis of the stability of the system motion is an important part of the system and control theory. There are two definitions of system stability: (a) external stability characterized by input–output relationship, (b) internal stability characterized by state motion under zero input conditions.

For linear time-invariant systems

$$\dot{x} = Ax + Bu\,,\ x(0) = \text{known} \tag{6.98}$$

$$y = Cx + Du \tag{6.99}$$

For no input $u(t) = 0$, and the initial state is arbitrary, the state response is

$$x = \Phi(t, t_0)x_0 \to 0\,,\ t \to \infty \tag{6.100}$$

Such the system is internally stable and the system is internally asymptotically stable. As is well known, the sufficient and necessary condition for the asymptotic stability of differential equation $\dot{x} = Ax$ is that all eigenvalues of matrix A have negative real parts. If the linear time-invariant system is internally stable, it must be externally stable.

2. Stability Margin

In engineering applications, the stability margin of the system has been paid great attention. The stability margin is essentially a "dynamic" problem, which is related to various interference factors that affect the stability analysis of the actual system. In this section, the logarithmic decay rate is used to characterize the stability margin of the system. For the multi-degree of freedom system, let the j-th eigenvalue be $\lambda_j = -\tau_j + \mathrm{i}\omega_j$, the corresponding logarithmic decay rate is

$$\delta_j = \frac{2\pi\tau_j}{\omega_j} \tag{6.101}$$

Thus, the stability margin of the system is the minimum logarithmic decay rate $\delta_{\min} = \min(\delta_j)$.

6.3.2 Establishment of Driver-Four-Wheel-Steering-Vehicle Closed-Loop System Model

The lateral motion and yaw motion of the vehicle moving at a constant speed u are studied, and only small angle motion is performed. Therefore, a two degrees-of-freedom linear vehicle model is adopted. The linear tire model can be expressed as

$$\begin{cases} Y_1 = k_1\beta_1 \\ Y_2 = k_2\beta_2 \end{cases} \tag{6.102}$$

The state equation of the open-loop model of the two degrees-of-freedom linear vehicle system is

$$\begin{bmatrix} \dot{r} \\ \dot{\beta} \end{bmatrix} = \begin{bmatrix} a_{11} & a_{12} \\ a_{21} & a_{22} \end{bmatrix} \begin{bmatrix} r \\ \beta \end{bmatrix} + \begin{bmatrix} b_{11} & b_{12} \\ b_{21} & b_{22} \end{bmatrix} \begin{bmatrix} \delta_f \\ \delta_r \end{bmatrix} \tag{6.103}$$

where
$a_{11} = \frac{k_1a^2+k_2b^2}{I_zu}$, $a_{12} = \frac{k_1a-k_2b}{I_z}$, $a_{21} = \frac{k_1a-k_2b}{mu^2} - 1$, $a_{22} = \frac{k_1+k_2}{mu}$,
$b_{11} = -\frac{k_1a}{I_z}$, $b_{12} = \frac{k_2b}{I_z}$, $b_{21} = -\frac{k_1}{mu}$, $b_{22} = -\frac{k_2}{mu}$.
where Y_1, Y_2 are the lateral force on the front wheel and rear wheel, respectively. k_1, k_2 are the sum of the comprehensive lateral stiffness of the front and rear axle tires, respectively. β_1, β_2 are the front and rear axle sideslip angle, respectively. r is the yaw rate. $\dot{\beta}$ is the sideslip angle at the vehicle center of mass. δ_f, δ_r are the front and rear wheel steering angle, $\delta_r = R\delta_f$. R is the control parameter that the rear wheel steering angle changes with the change of the front wheel steering angle in the four-wheel steering system. $i\delta_f = \delta_{sw}$, i is the total transmission ratio of the steering system. δ_{sw} is the steering wheel angle. m is the total mass of the whole vehicle. I_z is the moment of inertia of the whole vehicle around the vertical axis. a, b are the distance from vehicle center of mass to front and rear axles, respectively.

From Fig. 6.1, the state Eq of driver four-wheel steering vehicle closed-loop system is derived as follows

$$\begin{bmatrix} \dot{r} \\ \dot{\beta} \\ \dot{y} \\ \dot{v}_a \\ \dot{x}_1 \\ \dot{x}_2 \end{bmatrix} = \begin{bmatrix} a_{11} & a_{12} & 0 & 0 & a_{15} & a_{16} \\ a_{21} & a_{22} & 0 & 0 & a_{25} & a_{26} \\ 0 & 0 & 0 & 1 & 0 & 0 \\ a_{41} & a_{42} & 0 & 0 & a_{45} & a_{46} \\ 0 & 0 & 0 & 0 & 0 & 1 \\ a_{61} & a_{62} & a_{63} & a_{64} & a_{65} & a_{66} \end{bmatrix} \begin{bmatrix} r \\ \beta \\ y \\ v_a \\ x_1 \\ x_2 \end{bmatrix} + \begin{bmatrix} 0 \\ 0 \\ 0 \\ 0 \\ 0 \\ b_6 \end{bmatrix} f_r \tag{6.104}$$

where a_{11}, a_{12}, a_{21} and a_{22} are the same with Eq. (6.103), and the other elements are as follows
$a_{15} = -\frac{k_1a-k_2bR}{iI_z}$, $a_{16} = -\frac{T_d}{2}a_{15}$, $a_{25} = -\frac{k_1+k_2R}{imu}$, $a_{26} = -\frac{T_d}{2}a_{25}$.

$a_{41} = u(1 + a_{21})$, $a_{42} = ua_{22}$, $a_{45} = ua_{25}$, $a_{46} = ua_{26}$.
$a_{61} = -\frac{4C_0T_c}{TT_hT_d}a_{41}$, $a_{62} = -\frac{4C_0T_c}{TT_hT_d}a_{42}$, $a_{63} = -\frac{4C_0}{T^2T_hT_d}$, $a_{64} = -\frac{4C_0(T+T_c)}{T^2T_hT_d}$.
$a_{65} = -\frac{2}{T_hT_d} - \frac{4C_0T_c}{TT_hT_d}a_{45}$, $a_{66} = -\frac{2T_h+T_d}{T_hT_d} - \frac{4C_0T_c}{TT_hT_d}a_{46}$, $b_6 = \frac{4C_0}{T^2T_hT_d}$

$$f_r = f(t+T) + T_c\dot{f}(t+T)$$

where v_a is the vehicle lateral speed. When $R = 0$, the system is transformed into the driver-vehicle closed-loop system model of traditional front wheel steering.

The system shown in Eq. (6.104) is a linear constant system. Therefore, the motion stability of the system depends on whether all eigenvalues of the system matrix have negative real parts [22].

6.3.3 Motion Stability Analysis of Four-Wheel Steering Vehicle

Taking the parameters of a certain vehicle as an example, the digital simulation is carried out. The parameters of the vehicle are: $m = 1818.2\text{kg}$, $I_z = 3885\text{kg} \cdot m^2$, $a = 1.463\text{m}$, $b = 1.585\text{m}$, $k_1 = -62618\text{N}/rad$, $k_2 = -110185\text{N}/rad$, $i = 20$. Parameters of skilled drivers are: $T_d = 0.2\text{s}$, $T_h = 0.1\text{s}$, $T = 1.05\text{s}$, $a = 0.8$.

If the eigenvalues are arranged in the order of the algebraic value of the real part λ_{rj} from small to large, the stability of the system can be judged only by examining the higher-order eigenvalues (e.g., the real part of the sixth-order eigenvalue). The real part of the sixth-order eigenvalue is considered for the motion stability of the vehicle, as shown in Figs. 6.19 and 6.20.

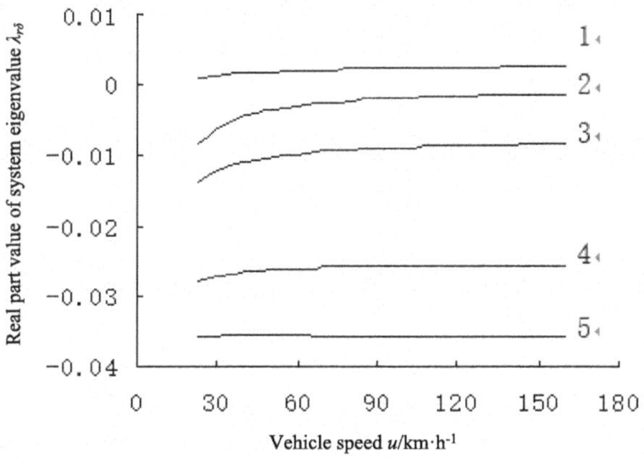

1-R=0.697; 2-R=0.696; 3-R=0.55; 4-R=0.2; 5-R=0

Fig. 6.19 Vehicle motion stability

Fig. 6.20 System stability margin

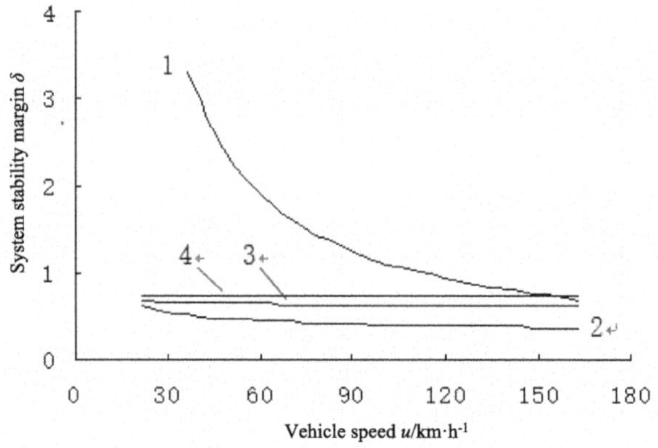

Figures 6.19 and 6.20 are the quantitative analysis results of the motion stability of the driver-vehicle closed-loop system. When $R = 0$, it is the result of the 4WS vehicle. When $R = 0$, the stability factor is about 0.002355, i.e., the steady-state steering characteristic is insufficient steering, and the characteristic vehicle speed is about 74.18 km/h. It can be seen from Fig. 6.19 that with the increase of R (i.e., the increase of rear-wheel steering), the closed-loop system is unstable when R is about 0.6965. Figure 6.20 shows that when R is less than 0.55, the stability margin of 4WS vehicle and 4WS vehicle is not much different in order of magnitude, and the stability margin tends to converge with the increase of vehicle speed. However, when R is large (i.e., $R = 0.696$, the stability margin changes significantly with the change of vehicle speed. Even if the closed-loop system does not have the risk of instability, it is difficult for the driver to obtain the handling stability feeling similar to the traditional 4WS vehicle. Therefore, when the skilled driver model is established, the motion stability of the four-wheel steering vehicle can be evaluated and optimized from the two aspects of the motion stability and stability margin of the closed-loop system.

Figures 6.21 and 6.22 are the quantitative analysis results of the motion stability of the driver-vehicle closed-loop system obtained when the total lateral stiffness of the rear wheels of the original vehicle decrease to −40185 N/rad. The reduction of the lateral stiffness can be understood as the reduction of the air pressure. When $R = 0$, it is the result of 4WS vehicle. When $R = 0$, the stability factor is about −0.00217. Therefore, the steady-state steering characteristic of the vehicle is excessive steering characteristic, and the critical speed is about 77.26 km/h. It can be seen from Figs. 6.21 and 6.22 that when approaching the critical speed, the driver-vehicle closed-loop system is unstable whether it is a 4WS vehicle or a four-wheel steering vehicle. When the vehicle speed is about 70 km/h, the system stability is significantly deteriorated. It should be noted that before reaching the speed of about 70 km/h, the values in Figs. 6.21 and 6.22 are not much different in order of magnitude. Therefore,

Fig. 6.21 Vehicle motion stability (lateral stiffness of the rear wheels decrease to −40185 N/rad)

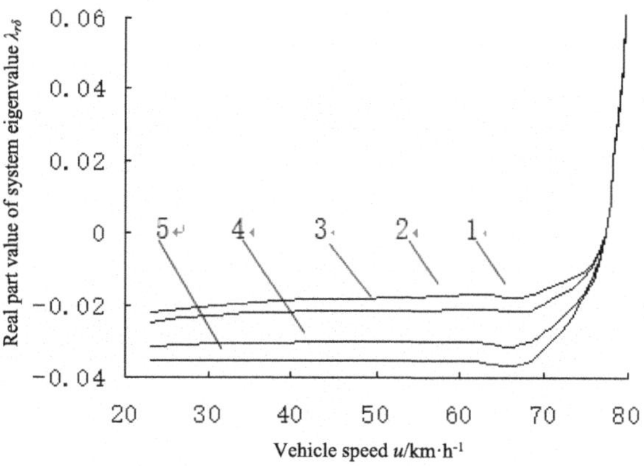

1-R=0.697; 2-R=0.696; 3-R=0.55;
4-R=0 2· 5-R=0

Fig. 6.22 System stability margin(lateral stiffness of the rear wheels decrease to −40185 N/rad)

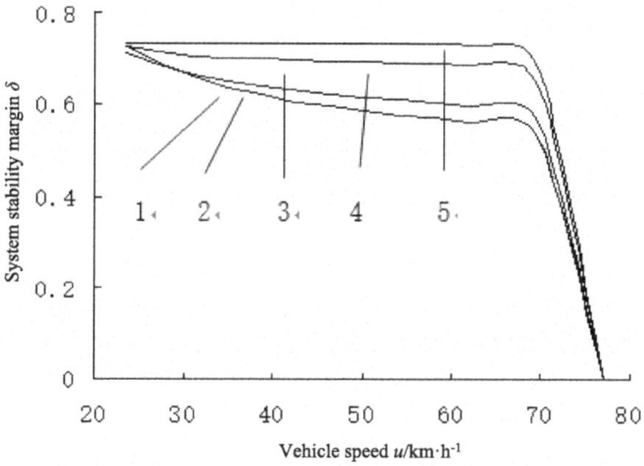

1-R=0.696; 2-R=0.55; 3-R=0.2; 4-R=0

when the vehicle has excessive steering, the four-wheel steering system is not very helpful to the vehicle's motion stability.

6.3.4 Conclusion

The four-wheel steering vehicle should also have moderate under-steering characteristics to ensure good stability. The control of rear wheel angle is crucial to the vehicle's motion stability. The change range of the rear wheel angle can be preliminarily determined by the analysis method of the driver-four-wheel-steering vehicle closed-loop system given in this paper, and the subjective evaluation of the driver on the handling stability should also be considered.

After the skilled driver model is established, the handling stability of the four-wheel steering vehicle can be ensured in the following two aspects in the vehicle design process: first, from the perspective of vehicle handling, according to the operation state of the vehicle under some typical driving conditions, the driver's handling of the vehicle and the driving safety of the vehicle are taken as indicators to comprehensively evaluate the vehicle handling. Secondly, from the perspective of vehicle motion stability, on the basis of the skilled driver model, the vehicle motion stability and stability margin are taken as indicators to comprehensively evaluate the vehicle motion stability.

6.4 Application of Matrix Perturbation Method in the Stability Analysis of Four Wheel Steering Vehicle

The natural characteristic problem of general vibration system is a generalized eigenvalue problem. Its eigenvalue and the first derivative of eigenvector have important applications in the analysis and design of engineering systems. Rogers first proposed the use of bi-normal condition in asymmetric feature systems.

As we all know, there are two methods to calculate the derivative of eigenvector: (1) algebraic method, (2) and modal method. This section gives the exact expression for calculating the eigenvector derivative in the case of modal truncation, in which the influence of the truncated higher-order modes is precisely given by the known lower order modes and the system matrix, and the required accuracy can be controlled according to the specific situation [23].

6.4.1 Matrix Perturbation Method for Eigenvalue Problems of Asymmetric Matrices

1. Full Mode Expansion Method

The natural vibration of the vibration system can be expressed as

$$Kx_i = \lambda_i M x_i, i = 1,\ 2,\ \cdots,\ n \tag{6.105}$$

$$K^{\mathrm{T}} y_i = \lambda_i M^{\mathrm{T}} y_i, i = 1,\ 2,\ \cdots,\ n \tag{6.106}$$

The bi-normal condition is

$$y_k^{\mathrm{T}} M x_i = \delta_{iik}, i, k = 1,\ 2,\ \cdots,\ n \tag{6.107}$$

$$x_i^{\mathrm{T}} M x_i = 1, i = 1,\ 2,\ \cdots,\ n \tag{6.108}$$

where, K and M are system matrices, respectively. λ_i is the eigenvalue. x_i and y_i are the right and left eigenvectors corresponding to λ_i.

When all N eigenvectors are known, the full mode expansion method can be used to calculate the derivative of the mode shape to the design variable b

$$\partial x_i/\partial b = \sum_1^n \alpha_{ik} x_k \tag{6.109}$$

$$\partial y_i/\partial b = \sum_1^n \gamma_{ik} y_k \tag{6.110}$$

When $i \neq k$, there is

$$\alpha_{ik} = \frac{y_k^{\mathrm{T}}(\partial K/\partial b - (\partial \lambda_i/\partial b) - M - \lambda_i(\partial M/\partial b))x_i}{\lambda_i - \lambda_k}$$

$$\gamma_{ik} = \frac{y_i^{\mathrm{T}}(\partial K/\partial b - (\partial \lambda_i/\partial b) - M - \lambda_i(\partial M/\partial b))x_k}{\lambda_i - \lambda_k}$$

When $i = k$, there is

$$\alpha_{ik} = -\frac{1}{2}\left(x_i^{\mathrm{T}}(\partial M/\partial b) + \sum_{j=1, j\neq i}^{n} \alpha_{ij} x_j^{\mathrm{T}}(M + M^{\mathrm{T}})x_i\right)$$

$$\gamma_{ik} = -y_i^{\mathrm{T}}(\partial M/\partial b)x_i - \alpha_{ii}$$

However, for large-scale vibration systems with many degrees of freedom, only L pair of low-order Eigen solutions can be accurately obtained, and $L \ll n$. Only the partial mode expansion method can be applied in this situation.

2. Partial Mode Expansion Method

When K is not singular, Eq. (6.109) can be rewritten into the following equation

$$\partial x_i/\partial b = \alpha_{ii} x_i + z_i \tag{6.111}$$

where

$$z_i = \sum_{j=1, j\neq i}^{n} \alpha_{ij} x_j \tag{6.112}$$

Equation (6.112) can be converted into

$$z_i = \sum_{j=1, j \neq i}^{n} \frac{y_j^{\mathrm{T}} F_i}{\lambda_i - \lambda_j} x_j \tag{6.113}$$

or

$$z_i = \sum_{j=1, j \neq i}^{L} \frac{y_j^{\mathrm{T}} F_i}{\lambda_i - \lambda_j} x_j + \sum_{j=L+1}^{n} \frac{y_j^{\mathrm{T}} F_i}{\lambda_i - \lambda_j} x_j \tag{6.114}$$

where

$$F_i = ((\partial K / \partial b) - (\partial \lambda_i / \partial b) M - \lambda_i (\partial M / \partial b)) x_i \tag{6.115}$$

It is assumed that $i \ll L$ from Eqs. (6.111) to (6.114), and the above eigenvalues are arranged from small to large. When the interval between the modes is large

$$\lambda_j - \lambda_i \approx \lambda_j, \; j > L \tag{6.116}$$

Equation (6.114) can be approximately rewritten as

$$z_i \approx \bar{z}_i = \sum_{j=1, j \neq i}^{L} \frac{y_j^{\mathrm{T}} F_i}{\lambda_i - \lambda_j} x_j + \sum_{j=L+1}^{n} \frac{y_j^{\mathrm{T}} F_i}{\lambda_j} x_j \tag{6.117}$$

Equation (6.117) can also be written as

$$\bar{z}_i = \sum_{j=1, j \neq i}^{L} \frac{y_j^{\mathrm{T}} F_i}{\lambda_i - \lambda_j} x_j + \sum_{j=1}^{n} \frac{y_j^{\mathrm{T}} F_i}{\lambda_j} x_j - \sum_{j=1}^{L} \frac{y_j^{\mathrm{T}} F_i}{\lambda_j} x_j \tag{6.118}$$

The bi-normal condition is written in the form of a matrix as

$$Y^{\mathrm{T}} K X = \Lambda \tag{6.119}$$

where

$\Lambda = diag(\lambda_1, \cdots, \lambda_n), X = [x_1, \cdots, x_n], Y = \left[y_1, \cdots, y_n\right]$.

or

$$X^{-1} K^{-1} \left(Y^{\mathrm{T}}\right)^{-1} = \Lambda^{-1}$$

Thus

$$K^{-1} = X \Lambda^{-1} Y^{\mathrm{T}} \tag{6.120}$$

or

$$K^{-1} = \sum_{j=1}^{n} \frac{x_j y_j^{\mathrm{T}}}{\lambda_j} \tag{6.121}$$

According to the spectral decomposition of Eq. (6.121), the second term of Eq. (6.118) becomes $-K^{-1}F_i$.

$$\overline{z}_i = \sum_{j=1, j \neq i}^{L} \frac{y_j^{\mathrm{T}} F_i}{\lambda_i - \lambda_j} x_j - K^{-1} F_i - \sum_{j=1}^{L} \frac{y_j^{\mathrm{T}} F_i}{\lambda_j} x_j \tag{6.122}$$

where term $-K^{-1}F_i$ is the contribution of truncated higher-order modes to the derivative of lower-order known modes. Thus, the improved partial mode expansion method of the eigenvector derivative can be expressed as

$$\partial x_i / \partial b = \overline{\alpha}_{ii} x_i + \overline{z}_i \tag{6.123}$$

where

$$\overline{\alpha}_{ii} = -\frac{1}{2}\left(x_i^{\mathrm{T}} (\partial M / \partial b) x_i + x_i^{\mathrm{T}} M \overline{z}_i + z_i^{\mathrm{T}} M x_i\right) \tag{6.124}$$

When the above eigenvalue distribution conditions are satisfied, Eq. (6.123) can obtain better results than modal truncation.

The derivative of the left eigenvector y_i to the design parameter b in the case of modal truncation can be obtained by the above method.

6.4.2 *Influence of Vehicle Parameters on Motion Stability [24]*

This section will conduct simulation research on the influence of vehicle structural parameters on vehicle motion stability. Firstly, the influence of single vehicle structural parameters on the motion stability (i.e., sensitivity) is studied. Then, the influence of the simultaneous changes of several vehicle structural parameters on the motion stability (i.e., the perturbation) is investigated. The method described in this section is applied to quantitatively reveal the internal law of the influence of single and multiple vehicle parameters on the motion stability of the driver-four-wheel-steering-vehicle closed-loop system, and provide search direction for the optimal design of four-wheel steering vehicles and 4WS vehicles, as well as the optimal design of vehicle dynamics and control system parameters.

Parameters of a certain vehicle: $m = 1818.2\,\mathrm{kg}$, $I_z = 3885\mathrm{kg} \cdot m^2$,$L_1 = 1.463\,\mathrm{m}$, $L_2 = 1.585\,\mathrm{m}$, $k_1 = -62618\,\mathrm{N}/rad$, $k_2 = -110,185\,\mathrm{N}/rad$, $i = 20$, $R = 0.35$. Skilled driver parameters: $T_d = 0.2\,\mathrm{s}$, $T_h = 0.1\,\mathrm{s}$, $T = 1.05\,\mathrm{s}$, $a = 0.8$.

If the eigenvalues are arranged in the order of the algebraic value of the real part from small to large, the stability of the system can be judged as long as the 6th-order eigenvalues are considered.

Figure 6.23a shows the analysis results of the sensitivity of the mass of the whole vehicle and the comprehensive lateral stiffness of the front and rear wheels to the motion stability. Figure 6.23b shows the analysis results of the sensitivity of the control parameters of the four-wheel steering system and the distance from the vehicle center of mass to the front and rear axles to the motion stability. It can be seen from Fig. 6.23 that vehicle structural parameters have an impact on the order of magnitude of motion stability sensitivity, which decreases as the order of magnitude of the parameter increases in general. In the optimization design, attention should be paid to the size of the sensitivity and the order of magnitude of the design parameters. Figure 6.23a shows that with the increase of vehicle speed, the sensitivity of vehicle mass m to motion stability tends to decrease after several large fluctuations. In the preliminary optimization design, it is sufficient to consider the impact of vehicle mass at low and medium vehicle speeds. In Fig. 6.23, the influence of the other five parameters on the sensitivity of motion stability does not monotonically change with the vehicle speed. However, the comprehensive lateral stiffness of the front and rear wheels $k_1$and k_2, the distance from the vehicle's center of mass to the front and rear axles L_1, L_2 have little impact on the sensitivity of motion stability, and the values are approximately symmetrical. This rule should be used in the optimization design.

Figure 6.24 shows the results of the perturbation analysis on the motion stability of FWS and 4WS vehicles when the comprehensive lateral stiffness of the front wheel increases by −10,000 N/rad, the comprehensive lateral stiffness of the rear wheel decreases by −20,000 N/rad, the distance between the vehicle's center of mass and the front axle increases by 0.061 m, the distance between the vehicle's center of mass and the rear axle decreases by 0.061 m, and the vehicle's mass increases by 200 kg. It can be seen from the figure that the influence of the change of the structural parameters of the whole vehicle on the movement stability of FWS and 4WS vehicles fluctuates with the change of vehicle speed. In terms of the overall effect, the movement stability of 4WS vehicles at low and medium speeds after the parameter change is better than that of FWS vehicles. Only at lower and higher speeds does the movement stability of 4WS vehicles tend to deteriorate. Therefore, the perturbation results of motion stability given in this method can be used for the optimization design of vehicle parameters in the preliminary design stage.

6.4.3 Conclusion

By applying the method in this section, the sensitivity of the vehicle structural parameters and the four-wheel steering system control parameters to the motion stability of the driver-four-wheel-steering vehicle closed-loop system can be obtained, thus providing the best search direction for the optimal design of the motion stability of the closed-loop system. The first-order perturbation can also be obtained when

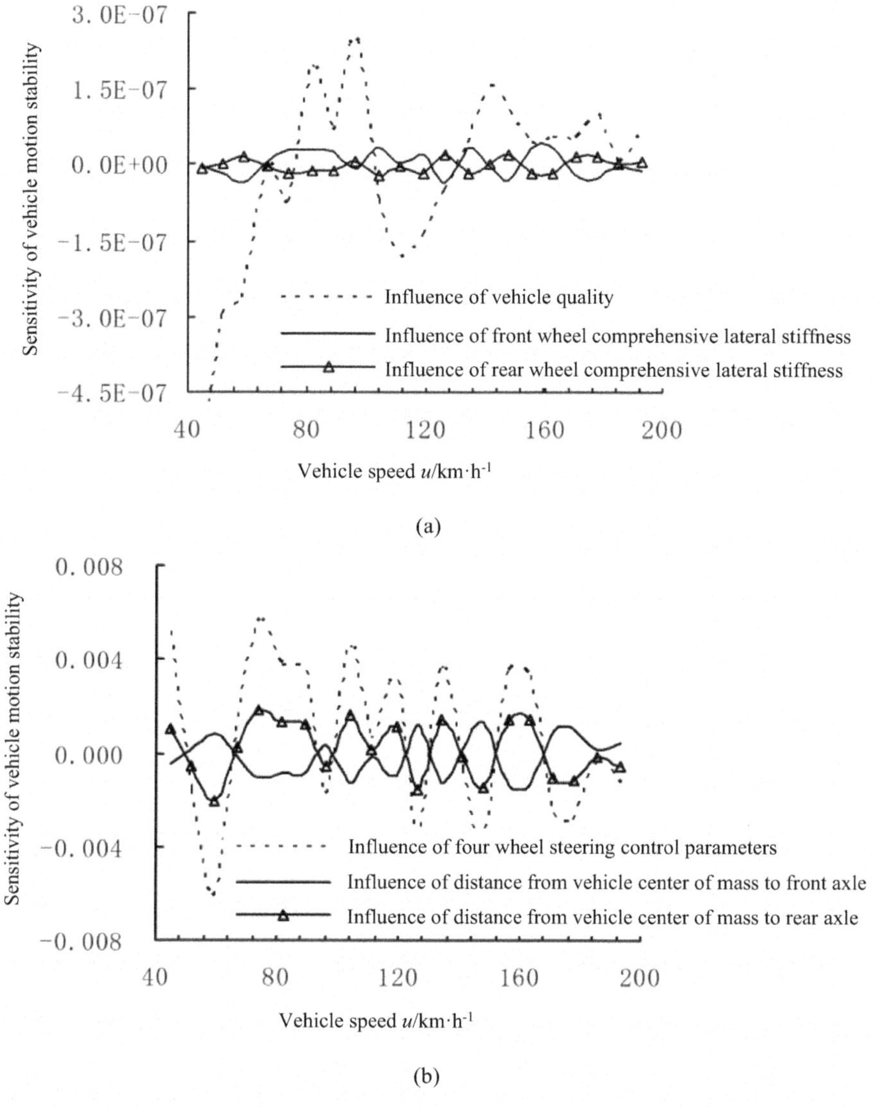

Fig. 6.23 Sensitivity of automotive structural parameters to kinematic stability

multiple vehicle structural parameters and four-wheel steering system control parameters change simultaneously; It provides a theoretical method for the qualitative and quantitative analysis of the motion stability of the driver-four-wheel-steering vehicle closed-loop system. The method in this section can be used for the optimization design of the FWS vehicle and 4WS vehicle.

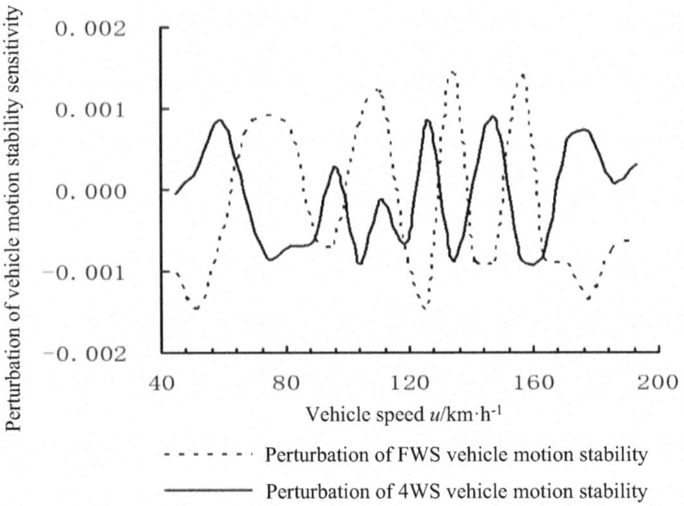

Fig. 6.24 Perturbation of motion stability when vehicle structural parameters change

References

1. Zhao Y, Guo K. A Research on analysis method of vehicle response. Automot Eng. 1999;21(3):134–139, 183.
2. Zhao Y, Guo K. Effect of driver statistic character on response of driver-vehicle closed-loop system and evaluation of vehicle active safety. Automot Eng. 1999;21(2):87–92.
3. Zhao Y, Guo K, Zhang G. Second order quadrature technique of driver vehicle closed loop system handling dynamics optimal design. Mech Sci Technol. 1999;18(5):737–8.
4. Zhao Y, Guo K, Zhang G. Fuzzy optimal design of vehicle active safety. Mech Sci Technol. 1999;18(6):919–21.
5. Zhao Y, Guo K. Development, significance of research and fundamental problems of motor vehicle maneuvering evaluation. Automob Technol. 1998;5:1–5.
6. Zhao Y, Guo K. Response analysis and handling safety evaluation for driver/vehicle/road closed-loop maneuvering system with random road input. Automot Eng. 1997;19(5):273–8.
7. Yin H. Modeling and simulation research on vehicle handling inverse dynamics. Nanjing University of Aeronautics and Astronautics; 2007.
8. Bai Y. Ease of Control Evaluation Via Stochastic Driver Model. Jilin University; 2012.
9. Lu Z. Study on Joint Optimization and Simulation of Handling and Stability of Four Wheel Steering Vehicle. Nanjing University of Aeronautics and Astronautics; 2007.
10. Wu Z. Research on active disturbance rejection control technology for automobile four-wheel steering application. Digit Technol Appl. 2014;(2):5–6, 8.
11. Li C, Luo W. Four-wheel steering vehicle optimal control research. Comput Technol Autom. 2013;32(4):20–2.
12. Song A. Research for Vehicle Stability Control of Driving simulation System. Chongqing University; 2012.
13. Zhao W. Joint Optimization Design of Vehicle Handling Dynamics and Motion Stability. Nanjing University of Aeronautics and Astronautics; 2005.
14. Chen H. Study on the Driver Model and Objective Evaluation of Tire Performance for Vehicle Handling. Jilin University; 2015.
15. Meng F. Study on Driver Modeling for Speed Control. Jilin University; 2007.

16. Zhang L. Study on Forward Collision Warning Algorithm for Curved Highway. Jilin University; 2004.
17. Xu J. Selection of Model Parameters for Different Types of Drivers Based on ADAMS. Nanjing University of Aeronautics and Astronautics; 2009.
18. Zhao Y. Research on Rollover Stability and Anti-Roll Control For Driver-Vehicle Closed-Loop System. Nanjing University of Aeronautics and Astronautics; 2016.
19. Cao J. Study on Vehicle Handling-Stability Evaluation Approach Based on Human-Vehicle-Road Closed-Loop System. Hanghai Jiao Tong University; 2015.
20. Li J. Study on Tire High Frequency Dynamic Properties Based on Brush Model to Road Unevenness. Jilin University; 2006.
21. Zhang X. Research on the Recognition Method of Drinking Driving State Based on Individual Driving Behavior. Beijing University of Technology; 2014.
22. Zhao Y, Wang L, He X, etc. Motion Stability Analysis of Vehicle with Four Wheel Steering System. China Mech Eng. 2003;14(14):1246–1248.
23. Zhao Y, Liu Z, Chen S. Accurate Truncated Modal Unfolding Method for Eigenvector Derivatives. Acta Astronaut. 1996;(2):52–58.
24. Zhao Y, Lin F, Guo K. Matrix perturbation method for motion stability analysis of vehicle with four wheel steering system. Mech Sci Technol Aerosp Eng. 2004;23(07):796–8.

Chapter 7
States and Parameters Estimation in the Vehicle System Dynamics

7.1 Vehicle Side-Slip Angle Estimation

In vehicle dynamic control, side slip angle is an important control variable of feedback control system. Vehicle dynamic control system mainly consists of estimator and controller. The function of the estimator is to estimate the side slip angle in real time. On this basis, the controller limits the side slip angle by increasing braking or driving force on the wheels to keep it below the characteristic value and generate a certain yaw moment to prevent the vehicle from losing control in the case of steering and emergency avoidance.

From the perspective of measurement, a non-contact speed sensor can be used to measure the longitudinal and lateral velocity vectors of the vehicle at the same time, and then the vehicle side slip angle can be estimated by $\beta = v/u$. This method is based on the optical principle to detect the speed with the help of the irregularity of the road surface, for the smooth surface of the road will often produce relatively large measurement error, in addition, the non-contact speed sensor is also very expensive, inconvenient to use in ordinary vehicles. Therefore, side slip angle is obtained by estimation method in practical application.

In this section, radial basis neural network (RBFNN) and adaptive Kalman filter (AKF) are applied to vehicle side slip angle estimation. A side slip angle estimator based on neural network and driver-vehicle closed loop system is proposed. At the same time, an estimator based on Kalman filter minimum variance is designed for the same road input, and the two estimators are compared.

Y. Zhao and F. Lin, *Vehicle System Dynamics*,
https://doi.org/10.1007/978-981-97-2019-4_7

7.1.1 Driver-Vehicle Closed Loop System Model

1. Driver-vehicle closed-loop system model and its state equation

According to the "optimal previewing curvature control theory" and "prediction-following theory" proposed by Prof Guo Konghui, the block diagram of driver-vehicle closed-loop system as shown in Fig. 7.1 [1] can be obtained. In the figure, $f(s)$ is the road input; y is the lateral displacement response of the vehicle; $V(s)$ is the transfer function of lateral acceleration ay and steering angle δ_{sw} of four-wheel steering vehicle. T_d is the driver's neural response lag time; T_h is the driver's manipulation response lag time; T is the driver's preview time; T_c is the correction time; C_0 is the correction parameter; $\ddot{y}^*$ is ideal lateral acceleration; δ_{sw}^* is the ideal steering wheel angle; δ_{sw} is the actual steering angle. The significance of the block diagram is as follows: The expected road input $f(s)$ becomes effective input f_e after the driver "digests" the road information ahead (preview) in advance. Effective input f_e and two negative feedbacks y and $T\dot{y}(s)$ form ideal lateral acceleration $\ddot{y}^*$, $\ddot{y}^*$ obtains ideal steering wheel angle δ_{sw}^* through driver's correction function $C(s) = C_0(1 + T_c s)$, δ_{sw}^* obtains actual steering wheel angle δ_{sw} through the driver's unavoidable neural response lag and steering response lag link $e^{-T_d s}/(1 + T_h s)$. δ_{sw} is derived from vehicle transfer function $V(s)$ to obtain vehicle lateral acceleration a_y, and a_y is derived from the actual vehicle lateral displacement response y through two integral links.

The equation of state of the driver-vehicle closed-loop system model can be derived from Fig. 7.1:

$$\begin{bmatrix} \dot{r} \\ \dot{\beta} \\ \dot{y} \\ \dot{v}_a \\ \dot{x}_1 \\ \dot{x}_2 \end{bmatrix} = \begin{bmatrix} a_{11} & a_{12} & 0 & 0 & a_{15} & a_{16} \\ a_{21} & a_{22} & 0 & 0 & a_{25} & a_{26} \\ 0 & 0 & 0 & 1 & 0 & 0 \\ a_{41} & a_{42} & 0 & 0 & a_{45} & a_{46} \\ 0 & 0 & 0 & 0 & 0 & 1 \\ a_{61} & a_{62} & a_{63} & a_{64} & a_{65} & a_{66} \end{bmatrix} \begin{bmatrix} r \\ \beta \\ y \\ v \\ x_1 \\ x_2 \end{bmatrix} + \begin{bmatrix} 0 \\ 0 \\ 0 \\ 0 \\ 0 \\ b_6 \end{bmatrix} f_r \tag{7.1}$$

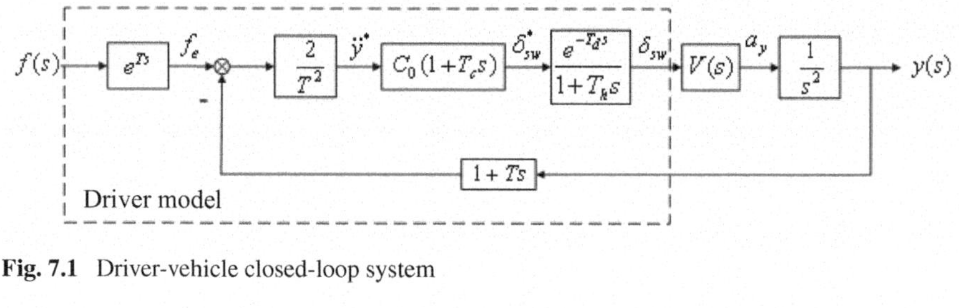

Fig. 7.1 Driver-vehicle closed-loop system

Output equation:

$$\begin{bmatrix} y \\ a_y \\ \delta_{sw} \\ \dot{\delta}_{sw} \\ S_1 \\ S_2 \end{bmatrix} = \begin{bmatrix} 0 & 0 & c_{13} & 0 & 0 & 0 \\ c_{21} & c_{22} & 0 & 0 & c_{25} & c_{26} \\ 0 & 0 & 0 & 0 & c_{35} & c_{36} \\ c_{41} & c_{42} & c_{43} & c_{44} & c_{45} & c_{46} \\ c_{51} & c_{52} & 0 & 0 & c_{55} & c_{56} \\ c_{61} & c_{62} & 0 & 0 & 0 & 0 \end{bmatrix} \begin{bmatrix} r \\ \beta \\ y \\ v_a \\ x_1 \\ x_2 \end{bmatrix} + \begin{bmatrix} 0 \\ 0 \\ 0 \\ d_4 \\ 0 \\ 0 \end{bmatrix} f_r \tag{7.2}$$

where $f_r = f(t+T) + T_c\dot{f}(t+T)$, r is the yaw rate, β is the slip side angle, v_a is the component of the absolute vehicle velocity in the Y direction of the absolute coordinate system, S_1 and S_2 are the lateral force coefficients of front and rear wheels respectively, x_1 and x_2 are the two state variables in the driver model. The calculation formula of a_{ij}, b_i, c_{ij} and d_i is as follows:

$$a_{11} = \frac{k_1a^2 + k_2b^2}{I_zu}, a_{12} = \frac{k_1a - k_2b}{I_z}, a_{21} = \frac{k_1a - k_2b}{mu^2} - 1, a_{22} = \frac{k_1 + k_2}{mu},$$

$$a_{15} = -\frac{k_1a}{iI_z}, a_{16} = -\frac{T_d}{2}a_{15}, a_{25} = -\frac{k_1}{imu}, a_{26} = -\frac{T_d}{2}a_{25},$$

$$a_{41} = u(1 + a_{21}), a_{42} = ua_{22}, a_{45} = ua_{25}, a_{46} = ua_{26}, a_{61} = -\frac{4C_0T_c}{TT_hT_d}a_{41},$$

$$a_{62} = -\frac{4C_0T_c}{TT_hT_d}a_{42}, a_{63} = -\frac{4C_0}{T^2T_hT_d}$$

$$a_{64} = -\frac{4C_0(T + T_c)}{T^2T_hT_d}, a_{65} = -\frac{2}{T_hT_d} - \frac{4C_0T_c}{TT_hT_d}a_{45}, a_{66} = -\frac{2T_h + T_d}{T_hT_d} - \frac{4C_0T_c}{TT_hT_d}a_{46}$$

$$b_6 = \frac{4C_0}{T^2T_hT_d}$$

$$c_{13} = 1, c_{21} = a_{41}, c_{22} = a_{42}, c_{25} = a_{45}, c_{26} = a_{26}, c_{35} = 1, c_{36} = -\frac{T_d}{2}$$

$$c_{41} = -\frac{T_d}{2}a_{61}, c_{42} = -\frac{T_d}{2}a_{62}, c_{43} = -\frac{T_d}{2}a_{63}, c_{44} = -\frac{T_d}{2}a_{64}, c_{45} = -\frac{T_d}{2}a_{65}.$$

$$c_{46} = 1 - \frac{T_d}{2}a_{66}, c_{51} = \frac{k_1a}{m_1u}, c_{52} = \frac{k_1}{m_1}, c_{55} = -\frac{k_1}{im_1}, c_{56} = -\frac{T_d}{2}c_{55}, c_{61} = -\frac{k_2b}{m_2u}$$

$$c_{62} = \frac{k_2}{m_2}, d_4 = -\frac{T_d}{2}b_6, m_1 = \frac{bm}{L}, m_2 = \frac{am}{L}$$

where a is the distance from the vehicle center of mass to the front axle, b is the distance from the vehicle center of mass to the rear axle, L is wheel base, m is vehicle mass, I_z is the moment of inertia of the whole vehicle around the plumb axis, k_1 is the comprehensive cornering stiffness of the front wheel, k_2 is the comprehensive cornering stiffness of the rear wheel, i is the total steering gear ratio, C_0 is the correction parameter, T_c is the correction time, T is previewing time, T_d is the neural response lag time, T_h is manipulation response lag time, u is longitudinal speed.

2. Calculation of parameters C_0 and T_c

The following briefly gives the derivation of driver correction parameters C_0 and correction time T_c in the 2DOF driver-vehicle–road closed-loop system:

[1] Calculation of correction parameters C_0.

According to the two-DOF linear open-loop vehicle state equation, there is:

$$\dot{x} = Ax + B\delta_{sw} \tag{7.3}$$

$$a_y = \dot{v} + ur = u\dot{\beta} + ur = U_1\dot{x} + U_2x \tag{7.4}$$

where $x = [r, \beta]^{\mathrm{T}}$, $U_1 = [0, u]$, $U_2 = [u, 0]$, $A = \begin{bmatrix} a_{11} & a_{12} \\ a_{21} & a_{22} \end{bmatrix}$, $B = \begin{bmatrix} a_{15} \\ a_{25} \end{bmatrix}$.

Laplace transform is applied to Eq. (7.3), we have:

$$sx(s) = Ax(s) + B\delta_{sw}(s) \tag{7.5}$$

so:

$$\delta_{sw}(s) = (sx(s) - Ax(s))B^{-1} \tag{7.6}$$

Laplace transform is applied to Eq. (7.4), we have:

$$a_y(s) = (sU_1 + U_2)x(s) \tag{7.7}$$

Therefore, the vehicle transfer function is:

$$V(s) = \frac{a_y}{\delta_{sw}}(s) = \frac{(sU_1 + U_2)x(s)}{(sx(s) - Ax(s))B^{-1}} = (sU_1 + U_2)(sI - A)^{-1}B \tag{7.8}$$

According to the prediction-following theory in literature [1], the driver correction function $C(s)$ is determined by Eq. (7.9):

$$C(s) \cdot e^{-T_d s} \cdot \frac{1}{1 + T_h S} \cdot V(s) = \frac{1}{1 + \sigma \frac{T}{3} s} \tag{7.9}$$

so:

$$C(s) = \frac{e^{T_d s}(1 + T_h s)}{V(s)\left(1 + \frac{\sigma T}{3} s\right)} \tag{7.10}$$

$$C_0 = C(0) = -\frac{1}{U_2 A^{-1} B} \tag{7.11}$$

[2] Calculation of correction time T_c

$$T_c = \frac{\dot{C}(0)}{C(0)} = T_d + T_h - \frac{\sigma T}{3} - \frac{\dot{V}(0)}{V(0)} \tag{7.12}$$

$$V(0) = -U_2 A^{-1} B,\ \dot{V}(0) = -U_1 A^{-1} B - U_2 A^{-1} A^{-1} B \tag{7.13}$$

$$T_c = T_d + T_h - \frac{\sigma T}{3} - \frac{U_1 A^{-1} B}{U_2 A^{-1} B} - \frac{U_2 A^{-1} A^{-1} B}{U_2 A^{-1} B} \tag{7.14}$$

3. Mathematical model of road input

The road input in this section adopts the path of double lane change test and slalom test which are commonly used in handling stability test.

[1] Double lane change test

The double lane change test reflects the handling performance of the vehicle when it is driving in a straight line at high speed to avoid obstacles or overtake vehicles. In the actual driving process, the ideal target trajectory of the driver should be the low-order smooth curve closest to the broken line in Fig. 7.2, which is shown in Fig. 7.3 after fitting, where the abscissa represents the forward direction of the vehicle, and the ordinate represents the left direction of the driver. Here, the x-axis and y-axis represent distances.

After cubic spline fitting, the function expression of the third order curve with continuous first derivative is [2]:

Fig. 7.2 The track of the double lane change test

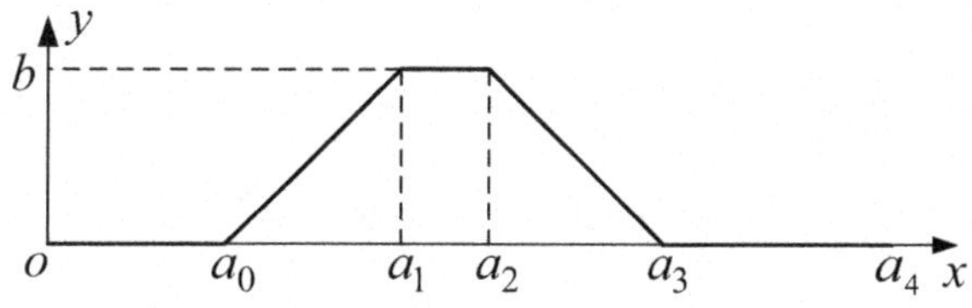

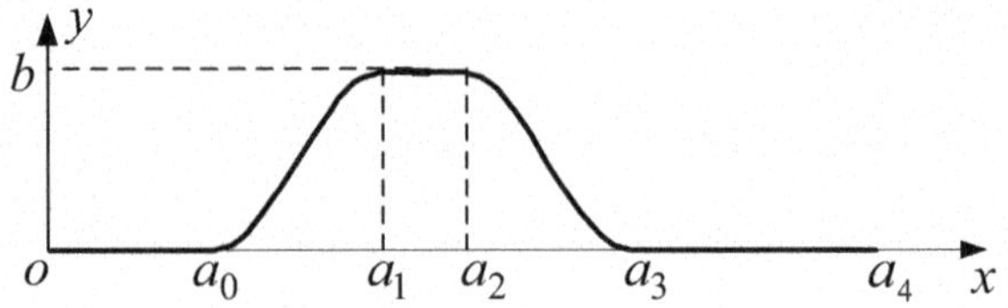

Fig. 7.3 The track of the double lane change test after fitting

$$f(x) = \begin{cases} 0 & (0 \le x < a_0) \\ e_0 + e_1 x + e_2 x^2 + e_3 x^3 & (a_0 \le x < a_1) \\ b & (a_1 \le x < a_2) \\ e_0' + e_1' x + e_2' x^2 + e_3' x^3 & (a_2 \le x < a_3) \\ 0 & (x \ge a_3) \end{cases} \tag{7.15}$$

where

$e_0 = a_0^2(3a_1 - a_0)b/d^3$, $e_1 = -6a_0 a_1 b/d^3$, $e_2 = 3(a_0 + a_1)b/d^3$, $e_3 = -2b/d^3$, $d = a_1 - a_0$.

$e_0' = a_3^2(3a_2 - a_3)b/d'^3$, $e_1' = -6a_3 a_2 b/d'^3$, $e_2' = 3(a_2 + a_3)b/d'^3$, $e_3' = -2b/d'^3$, $d' = a_2 - a_3$.

Because the state equation of the driver vehicle closed-loop system needs to be solved in the time domain, $f(x)$ needs to be transformed into a function of time t. When the vehicle starts from the origin and travels along the x-axis at a constant speed u, $x = ut$ is substituted into Eq. (7.15) to obtain:

$$f(t) = \begin{cases} 0 & (0 \le t < t_0) \\ g_0 + g_1 t + g_2 t^2 + g_3 t^3 & (t_0 \le t < t_1) \\ b & (t_1 \le t < t_2) \\ h_0 + h_1 t + h_2 t^2 + h_3 t^3 & (t_2 \le t < t_3) \\ 0 & (t_3 \le t \le t_4) \end{cases} \tag{7.16}$$

where

$t_0 = 2\text{s}$, $t_1 = 4\text{s}$, $t_2 = 5\text{s}$, $t_3 = 7\text{s}$, $t_4 = 12\text{s}$, $g_0 = e_0$, $g_i = e_i u^i (i = 1, 2, 3)$

$$h_0 = e_0'$$

$h_i = e_i' u^i$, $i = 1, 2, 3$

[2] Slalom test

Slalom lane change test is an important test to evaluate the vehicle handling stability. It can assess the vehicle handling performance under the condition of approaching sideslip or rollover, comprehensively evaluate the driving stability and riding comfort of the vehicle, and can also be used as a subjective perceptual test in the vehicle handling stability comparison test.

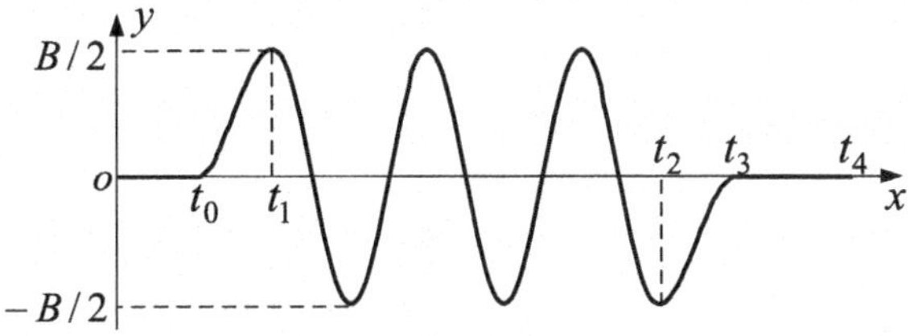

Fig. 7.4 The track of the slalom test after fitting

It is similar to the double lane change test, in the actual driving process of the slalom test, the ideal target trajectory should be a low-order continuous smooth curve as shown in Fig. 7.4.

After cubic spline fitting, the third order curve with continuous first derivative is:

$$f(t) = \begin{cases} 0 & (0 \le t < t_0) \\ B(1.0 + \sin \omega t)/2.0 & (t_0 \le t < t_1) \\ B \cos \omega t & (t_1 \le t < t_2) \\ B(-1.0 + \sin \omega t)/2.0 & (t_2 \le t < t_3) \\ 0 & (t_3 \le t \le t_4) \end{cases} \tag{7.17}$$

where $t_0 = 2\text{s}$, $t_1 = 4\text{s}$, $t_2 = 14\text{s}$, $t_3 = 16\text{s}$, $t_4 = 20\text{s}$.

7.1.2 Estimation of the Side Slip Angle Based on the Radial Basis Function Neural Network

1. Structure and principle of the radial basis function neural network

Radial basis function (RBF) neural network is an improved network for BP network which has two inherent defects, namely local minimum and slow convergence rate.

Figure 7.5 shows the basic structure of RBF neural network, which is composed of input layer, hidden layer of radial basis neuron and output layer of linear neuron [3]. RBF networks can approximate arbitrary nonlinear functions with arbitrary accuracy, and can be designed quickly to replace lengthy training.

As can be seen from Fig. 7.5, the input layer implements a nonlinear mapping as $x \to R_i(x)$, while the output layer implements a linear mapping from $R_i(x)$ to y_k.

$$y_k = \sum_{i=1}^{N} w_{ik} R_i(x), \quad k = 1, 2, \cdots, M \tag{7.18}$$

where w_{ik} is the connection weight between the i node in the hidden layer and the k node in the output layer, $R_i(x)$ is a nonlinear radial symmetric basis function.

The commonly used radial basis function is Gaussian function:

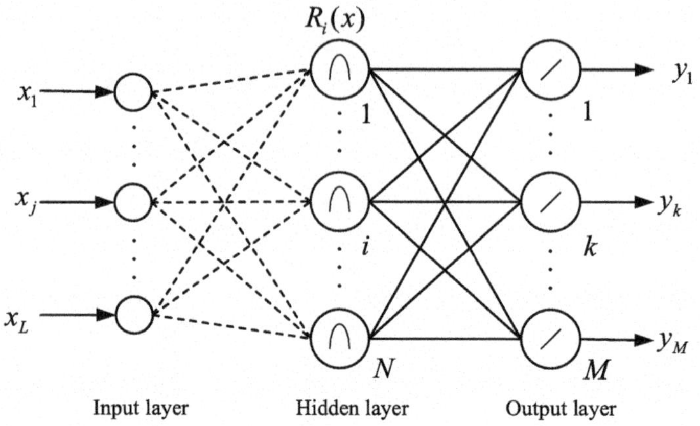

Fig. 7.5 The structure of RBF neural network

$$R_i(x) = \exp\left(-\frac{\|x - c_i\|^2}{2\sigma_i^2}\right), \quad i = 1, 2, \cdots, N \tag{7.19}$$

where c_i is the center vector of the basis function of the i node in the hidden layer, σ_i is the width of the i-th basis function around the center point, $\|\cdot\|$ is the Euclidean norm. So the network can map from input $x = [x_1, x_2, \cdots, x_L]^T$ to output $y = [y_1, y_2, \cdots, y_M]^T$.

The learning process of RBF neural network includes two stages: the unsupervised learning of hidden layer unit, and the parameters to be learned are the center c_i and width σ_i of basis function; Supervised learning of output layer units, the parameter to be learned is the weight w_{ik} of the output layer. In this chapter, K-means clustering algorithm is adopted to determine the center c_i and width σ_i, and then the least square method is used to determine the weight w_{ik}. The specific learning algorithm is as follows [4].

(1) Initialize the cluster center, and randomly select N input vectors x_i from the training sample as the initial cluster center c_i.
(2) The input vectors in the training samples are clustered according to the nearest neighbor rule, and the condition for x_q to be classified into class θ_i is:

$$\|x_q - c_i\| = \min_{j=1,\cdots,N} \|x_q - c_j\| \tag{7.20}$$

(3) Calculate the average value of input vectors in θ_i as the new cluster center c_i:

$$c_i = \frac{1}{M_i} \sum_{\mathbf{x}_j \in \theta_i} x_j \tag{7.21}$$

where M_i is the number of vectors in θ_i.

Repeat steps (1) to (3) until the distribution of cluster centers no longer changes. After the cluster center is determined, the parameter value of each hidden layer node σ_i^2 is taken as the average distance between each cluster center and other vectors of this class:

$$\sigma_i^2 = \frac{1}{M_i} \sum_{x_j \in \theta_i} \left\| x_j - c_i \right\|, \quad i = 1, 2, \cdots, N \tag{7.22}$$

(4) Output error of the network is

$$E = \sum_{k=1}^{M} (d_k - y_k)^2 \tag{7.23}$$

where d_k is the expected output of the output unit k, y_k is the actual output.

Substitute Eq. (7.18) into Eq. (7.23), and obtain:

$$E = \sum_{k=1}^{M} \left(d_k - \sum_{i=1}^{N} w_{ik} R_i(x) \right)^2 \tag{7.24}$$

The purpose of training output layer neurons is to find a set of weights w_{ik} to minimize the error E. By $\partial E / \partial w_{ik} = 0$ obtaining a system of linear equations with variables as w_{ik}, it can be solved.

2. Estimation model of the side slip angle based on the neural network

The neural network model for the side slip angle in this section is established based on the following assumptions: the side slip angle can be approximately expressed as a function of the yaw rate and lateral acceleration [5]:

$$\beta(k+1) = f(r(k), r(k-1), \cdots r(k-n), a_y(k), a_y(k-1), \cdots a_y(k-n)) \tag{7.25}$$

The side slip angle at the current $k + 1$ moment is a function of the time series consisting of yaw rate r and lateral acceleration a_y at the previous $n + 1$ moments. Then the mapping relationship between them is established by neural network. This n determines the size of the neural network. It is found that $n = 4$ is more suitable for considering both operation time and estimation accuracy.

The parameters describing the proficiency of drivers are neural response lag time T_d, manipulation response lag time T_h and preview time T. Due to the individual differences of different drivers, T_d, T_h and T have their own variation ranges $[\underline{T}_d, \overline{T}_d]$, $[\underline{T}_h, \overline{T}_h]$ and $[\underline{T}, \overline{T}]$. By combining the different values of the three parameters within their varying ranges, different drivers are simulated to drive the same vehicle, so as to obtain multiple groups of samples for training the network.

In theory, if the number of training samples is large enough and representative, the neural network can reveal any complex rules hidden in it. In this chapter, an efficient uniform experimental design method is adopted to select training samples, which not only greatly reduces the number of training samples, but also makes each group of training samples representative. The uniform design table was selected for simulation test design [6], that is, 5 factors, 15 levels, and 15 test times (Tables 7.1 and 7.2).

There are 3 factors T_d, T_h and T in this chapter. According to Table 7.2 [6], columns 1, 2 and 3 in Table 7.1 are selected to form a uniform test scheme.

The steps of establishing the neural network model for estimating the side slip angle are as follows.

(1) According to the test scheme given in Table 7.1, 15 groups of representative specific T_d, T_h and T are selected in the parameter given interval, and the state equation and output equation of the driver vehicle closed-loop system are solved according to the road input, so as to obtain 15 groups of yaw rate r, lateral acceleration a_y and the side slip angle β.

Table 7.1 Uniform design table $U_{15}(15^5)$

Levels	Factors				
	1	2	3	4	5
1	1	4	7	11	13
2	2	8	14	7	11
3	3	12	6	3	9
4	4	1	13	14	7
5	5	5	5	10	5
6	6	9	12	6	3
7	7	13	4	2	1
8	8	2	11	13	14
9	9	6	3	9	12
10	10	10	10	5	10
11	11	14	2	1	8
12	12	3	9	12	6
13	13	7	1	8	4
14	14	11	8	4	2
15	15	15	15	15	15

Table 7.2 Usage table of a $U_{15}(15^5)$

Factors	Column number			
2	1	4		
3	1	2	3	
4	1	2	3	5

(2) For the 15 groups of data in step (1), the white noise sequence of Gaussian distribution is added to r and a_y respectively to obtain r' and a'_y. The statistical characteristics of the noise are determined according to the actual vehicle test data.
(3) According to the description of Eq. (7.25), a mapping pair between β at $k+1$ time, r' and a'_y at the previous $n+1$ time is established, and the mapping pair is composed of 15 representative training samples.
(4) When training the RBF neural network, r' and a'_y are taken as the input and the corresponding sample β is taken as the output. When the training of the network is completed, the RBF network for estimating the vehicle side slip angle is established.

7.1.3 Estimation of the Side Slip Angle Based on Adaptive Kalman Filter

The state estimation based on Kalman filter is a soft sensor technology. Its essence is to transform the problem of parameter measurement into state observation and state estimation, that is, to obtain the estimated value of unobservable or difficult to observe variables from observable variables. In this section, the yaw rate r and lateral acceleration a_y of the measured information are known and the Kalman filter algorithm is used to obtain the estimated value of the side slip angle β, which is difficult to obtain.

Before the state estimation of Kalman filter, the state equation and observation equation should be discretized, the sampling period is set to 0.01 s, and the zero-order hold is used. There is system noise in w the discrete state equation and measurement noise v in the discrete measurement equation, and we assume that w and v are independent zero-mean random Gaussian white noise sequences.

A_d, B_d and C_d respectively represent the discretized system matrix, input matrix and measurement matrix, then the discrete state equation of the closed-loop system in Eq. (7.1) is:

$$\begin{bmatrix} r(k) \\ \beta(k) \\ y(k) \\ v_a(k) \\ x_1(k) \\ x_2(k) \end{bmatrix} = A_d \begin{bmatrix} r(k-1) \\ \beta(k-1) \\ y(k-1) \\ v_a(k-1) \\ x_1(k-1) \\ x_2(k-1) \end{bmatrix} + B_d f_r(k-1) + w(k-1) \tag{7.26}$$

The discrete observation equation of the closed-loop system with yaw rate and lateral acceleration as the observed quantities is:

$$\begin{bmatrix} r(k) \\ a_y(k) \end{bmatrix} = C_d \begin{bmatrix} r(k) \\ \beta(k) \\ y(k) \\ v_a(k) \\ x_1(k) \\ x_2(k) \end{bmatrix} + v(k) \tag{7.27}$$

Kalman filter can be divided into two parts: time update process (prediction) and measurement update process (correction). The time update process obtains the prior estimation of the next time according to the system state at the current time; In the measurement process, measurement and prior estimation are combined to obtain an improved system posterior estimation. The classical Kalman filter algorithm is as follows [7].

(1) Time update process

The state prediction equation is:

$$x_{k|k-1} = A_d x_{k-1|k-1} + B_d\, u_{k-1} \tag{7.28}$$

The error covariance prediction equation is:

$$P_{k|k-1} = A_d P_{k-1|k-1} A_d^T + Q \tag{7.29}$$

(2) Measurement update process

The Kalman gain equation is:

$$K_k = P_{k|k-1} C_d^T (C_d P_{k|k-1} C_d^T + R)^{-1} \tag{7.30}$$

The filtering equation is:

$$x_{k|k} = x_{k|k-1} + K_k (Z_k - C_d x_{k|k-1}) \tag{7.31}$$

The error covariance update equation is:

$$P_{k|k} = (I - K_k C_d) P_{k|k-1} \tag{7.32}$$

where Q and R are the covariance matrix of process noise and measurement noise, Z_k is the measured information yaw rate r and lateral acceleration a_y.

On the basic of the classical Kalman filter algorithm, the improved Sage-Husa adaptive Kalman filter algorithm is adopted in this section. For the specific algorithm process, please refer to the Ref. [5].

7.1.4 Comparison of Two Estimation Methods of Side Slip Angle

1. Real vehicle test

[1] Purpose of the test

The longitudinal velocity, lateral velocity, yaw rate, lateral acceleration and other test data are obtained through the real vehicle test of handling stability (double lane change test, slalom test). Among them, yaw rate and lateral acceleration are measured values, which can be used as the input of neural network trained by simulation test and as the observation value containing noise in Kalman filter state estimation. The test value of the side slip angle is obtained by the ratio of the lateral velocity and the longitudinal velocity.

[2] Test method

The IVRU-FX fiber optic gyroscope is installed on the vehicle to collect the yaw rate and lateral acceleration in real time. The LC-5100 non-contact speedometer and LC-1100 spatial filter speed sensor are used to measure the longitudinal and lateral velocity. This test uses a double non-contact speed sensor layout, that is, two speed sensors to measure the longitudinal speed and lateral speed of the vehicle curve motion.

The vehicle speed was approximate maintained at $u = 80$ km/h.

(1) the double lane change test steps and slalom test

The double lane change test is performed in accordance with ISO/TR3888-1975. The test steps are as follows:

(a) Mark piles are arranged on the ground of the test site, and the driving test route is drawn according to Fig. 7.2;
(b) Connect the test instrument, turn on the power supply of the instrument, and make it preheat to the normal operating temperature;
(c) Pass the test section at a constant speed of 80 km/h (± 3 km/h), and the driver is required to turn sharply into a lane parallel to the initial lane at the fastest speed and return to the initial lane as soon as possible. During driving, the dividing pile placed on the lane line is not allowed to be hit. The curves of the measured variables (the yaw rate, the lateral acceleration, the longitudinal velocity, the lateral velocity) and the time to pass the effective marker are recorded.

(2) the slalom test steps

The slalom lane change test shall be conducted in accordance with the test method for GB/T 6323–2014. The test steps are as follows:

(a) Arrange stakes on the test site and draw the driving test route according to Fig. 7.4;

(b) Connect the test instrument, connect the power supply of the instrument, and preheat it to the normal working temperature;
(c) Before the formal test, practice five round trips according to the route shown in Fig. 7.4;
(d) The test vehicle runs in a straight line at a stable speed close to half of the reference speed (65 km/h). Before entering the test section, record the zero line of each measured variable (the yaw rate, the lateral acceleration, the longitudinal speed and the lateral speed) through the notebook computer, and then pass the test section. Record the curves of each measured variable and the time of passing the effective stake;
(e) Increase the vehicle speed and repeat the process of step (d) with the reference vehicle speed close to 80 km/h for the test (it is invalid to knock down the stake).

2. Comparison of estimated values and experimental values

[1] The double lane change

The measured yaw rate and lateral acceleration signals including noise obtained by vehicle test are shown in Fig. 7.6.

Input the measured signal in Fig. 7.6 into the trained RBF neural network sideslip angle estimation model to obtain the vehicle side slip angle estimation curve β_{net}; At the same time, the measured signal in Fig. 7.6 is used as the measurement information of Kalman filter to obtain the estimation curve of vehicle sideslip angle β_{Kal}. The running time of the two algorithms is compared. The neural network method is 5.13 s and the Kalman filter method is 5.47 s. The side slip angle obtained by the real vehicle test method is recorded as β_{real}. Figure 7.7 shows the comparison results of the three, i.e., β_{net} obtained by neural network method, β_{Kal} obtained by Kalman filter method, and test value β_{real}.

It can be seen from Fig. 7.7 that the estimated values of the two methods are in good agreement with the test values in the trend. The estimated results are quite close to the actual vehicle test results, and the error is basically within the allowable range of engineering application.

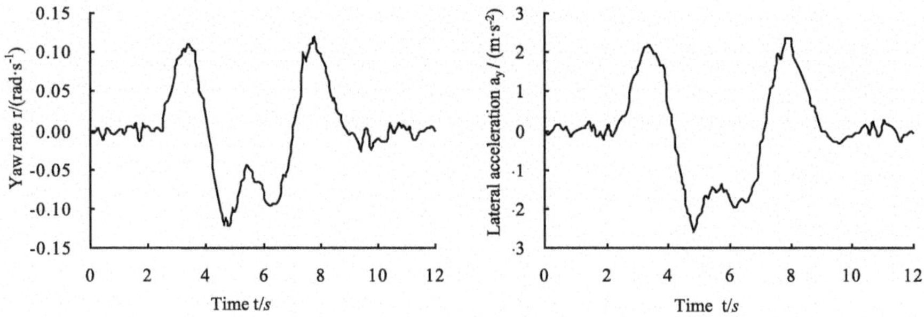

Fig. 7.6 Measured values of yaw rate and lateral acceleration (double lane change)

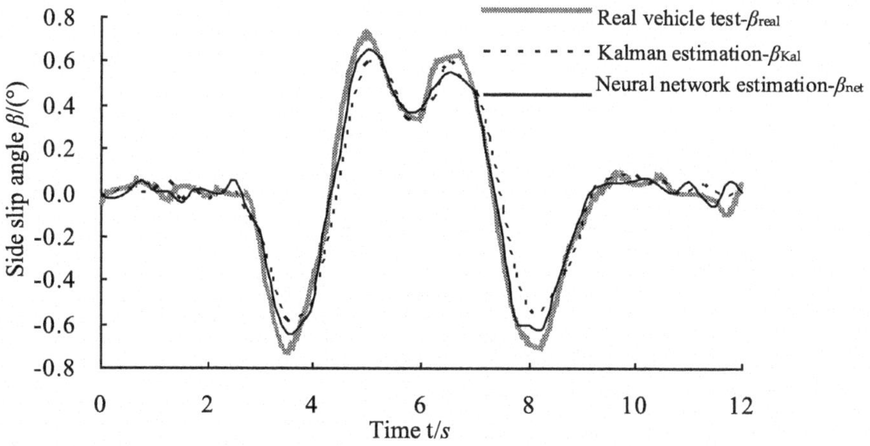

Fig. 7.7 Comparison between the estimated value and the test value of the two methods (the double lane change)

It can also be seen from Fig. 7.7 that the coincidence between β_{net} and β_{real} is obviously better. It can be seen the neural network model with 15 groups of training samples has good anti-noise interference ability. In order to improve the intuitively in comparing the accuracy of the two methods, the absolute estimation errors of the two methods are compared. It can be seen from Fig. 7.8 that the estimation result of RBF neural network is better than that of Kalman filter.

The main reasons for the error between the estimated value and the test value are as follows:

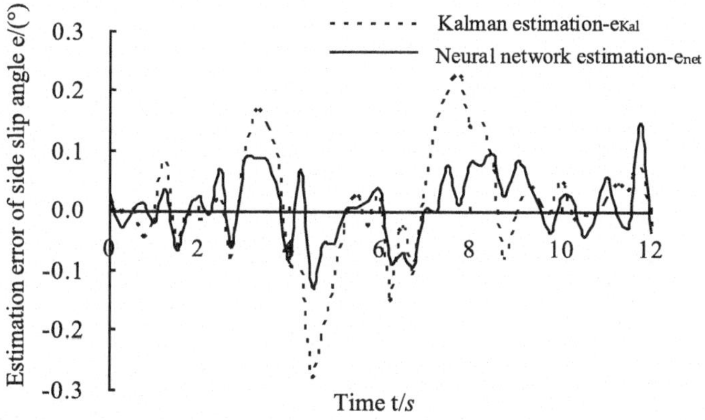

Fig. 7.8 Absolute error of two estimation methods (the double lane change)

(1) The two degree of freedom vehicle model used in the two estimation methods is relatively simple, and the nonlinear factors of the tire, suspension and other systems are not considered, and the influence of damping in the steering system, suspension and other systems is ignored;
(2) In the real vehicle test, the driver's own factors, such as the driver's driving skills and judgment of road conditions and vehicle driving conditions, will affect his input to the steering wheel;
(3) The test value of centroid sideslip angle also contains certain instrument measurement error.

[2] The slalom

The measured yaw rate and lateral acceleration signals obtained from the actual vehicle test on the serpentine are shown in Fig. 7.9.

The absolute errors of the estimated values and the test values of the two methods are also given, as shown in Figs. 7.10 and 7.11. It can be seen from the figure that the estimation error of side slip angle is also relatively large in the slalom lane change test with relatively complex control path. In addition, the estimation result of neural network is also better than that of Kalman filter.

In order to quantitatively compare the two estimation methods, Table 7.3 shows the error of the neural network method and Kalman filter method in estimating the vehicle sideslip angle relative to the test value. It can be seen from the table that the estimation error of the neural network method is smaller than that of the Kalman filter method under both conditions.

It can be seen the RBF neural network is simple and convenient to estimate the vehicle side slip angle, with clear ideas and high accuracy. Moreover, the uniform design method significantly reduces the number of simulation tests, and the response time of this method is also fast.

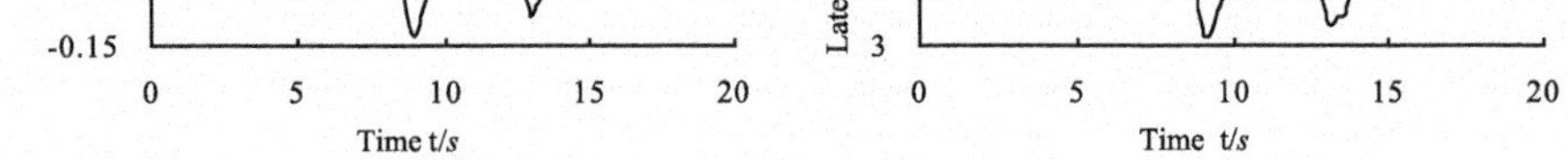

Fig. 7.9 Measured values of yaw rate and lateral acceleration (the slalom test)

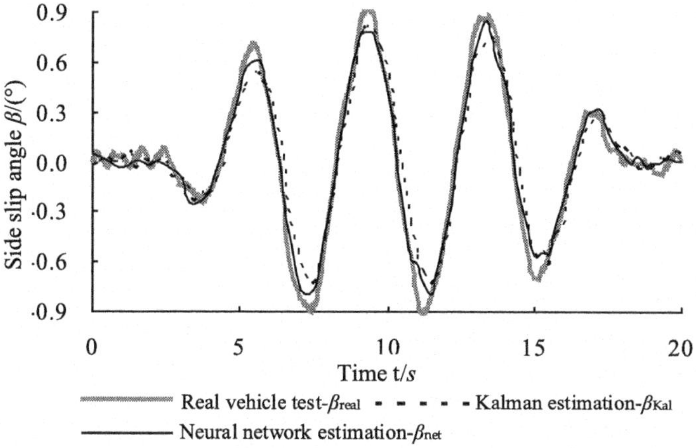

Fig. 7.10 Comparison between the estimated value and the test value of the two methods (the slalom lane change test)

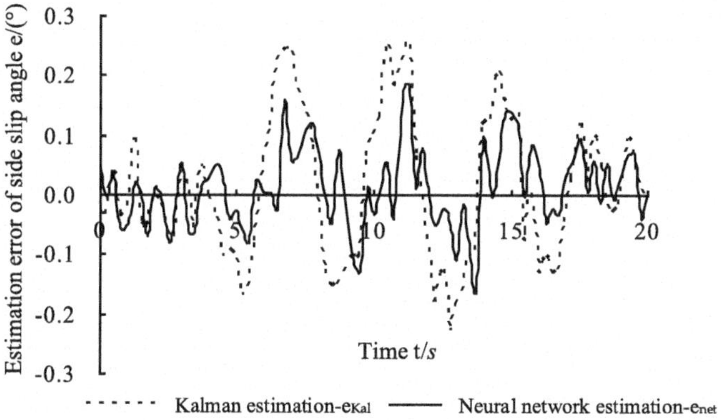

Fig. 7.11 Absolute error of two estimation methods (the slalom lane change test)

Table 7.3 Error of two methods for estimating sideslip angle

Index	The double lane change test		The slalom test	
	Neural network	Kalman Filter	Neural network	Kalman filter
Mean value of error/(°)	0.046333	0.069289	0.053323	0.097463
Standard deviation of error/(°)	0.057822	0.09576	0.068211	0.11844

7.2 Vehicle States Estimation Based on UKF Algorithm

7.2.1 Nonlinear Vehicle System with Noise

Normally, the state equation and observation equation of nonlinear discrete system have the following forms:

$$x_{k+1} = f(x_k, u_k, w_k)$$

$$y_k = h(x_k, v_k)$$

where x_k is the state vector of the system at time k, u_k is the system input, y_k is the observation vector, w_k and v_k are linearly independent process noise and measurement noise respectively, it is assumed that the two kinds of noise are Gaussian white noise with a mean value of zero, which satisfies $E(w_k w_j^{\mathrm{T}}) = Q_k \delta_{kj}$, $E(v_k v_j^{\mathrm{T}}) = R_k \delta_{kj}$, $f(\cdot)$ is a definite bounded nonlinear state transfer function, $h(\cdot)$ is a bounded nonlinear measurement function.

The state vector of the nonlinear vehicle system is set as:

$$x = [v, a_y, r, \beta]^{\mathrm{T}}$$

The system input u_k is the front wheel angle δ, the observation vector is:

$$y = \begin{bmatrix} r \\ a_y \end{bmatrix} = Cx = \begin{bmatrix} 0\,0\,1\,0 \\ 0\,1\,0\,0 \end{bmatrix} x$$

The process noise of the state estimation is set as the time invariant stationary Gaussian white noise sequence with the mean value of zero, and the variance matrix is set as follows:

$$Q = 0.01 \cdot |\max(\mathrm{state}(i)) - \min(\mathrm{state}(i))| \; (i = 1 \cdots 4, 0 \le t \le 10)$$

where $\max(\mathrm{state}(i))$ and $\min(\mathrm{state}(i))$ are the maximum and minimum values of the i-th state vector in the whole time.

The covariance matrix of the measurement noise is set as the time invariant stationary Gaussian white noise sequence with the mean value of zero, and the variance matrix is set as follows:

$$R = 0.05 \cdot |\max(y(i)) - \min(y(i))| \; (i = 1 \cdots 2, 0 \le t \le 10)$$

where $\max(y(i))$ and $\min(y(i))$ are the maximum and minimum values of the i-th state vector in the whole time.

7.2.2 UKF Algorithm for Vehicle States Estimation

UKF (unscented Kalman filter) is based on UT transform and adopts Kalman linear filter framework. The specific sampling form is deterministic sampling. The number of Sigma points sampled by UKF is small, and the specific number depends on the selected sampling strategy. The commonly used is $2n + 1$ sigma points symmetric sampling.

UKF algorithm has the following characteristics:

(1) UKF can accurately estimate the mean and covariance to the fourth order accuracy of Taylor series, while EKF (Extended Kalman filter) estimates the mean to the second order accuracy and the covariance to the fourth order accuracy;
(2) Through UT transformation, the tedious calculation of Jacobian matrix is avoided;
(3) The calculation amount is of the same order as that of EKF;
(4) The deterministic sampling strategy is adopted instead of the random sampling of particle filter to avoid the problem of particle decay.

In this section, the UKF algorithm is applied to the states estimation of running vehicles, and the appropriate sampling strategy and proportion correction algorithm are selected. The specific steps for deriving the UKF algorithm are as follows.

1. Symmetric sampling strategy of Sigma points

The key of UKF algorithm lies in the selection of Sigma point sampling strategy, that is, the number, position and weight determination method of Sigma points. It is necessary to ensure that the cost function approximating some performance indexes of the output is minimized while grasping the distribution characteristics of the input variables.

In this chapter, the symmetric sampling strategy [8] is used to select Sigma points. For n-dimensional random variables with mean value $\overline{x}$ and variance P_x, χ matrix is generated, which is composed of $2n + 1$ χ_i column vectors:

$$\chi_i = \begin{cases} \overline{x}, & i = 0 \\ \overline{x} + (\sqrt{(n+\kappa)P_x})_i, & i = 1, 2, \cdots, n \\ \overline{x} - (\sqrt{(n+\kappa)P_x})_{i-n}, & i = n+1, \cdots, 2n \end{cases} \tag{7.33}$$

where κ is a free parameter, which can be used to adjust the approximation degree of Sigma points to the distribution of random variables x. Document [8] points out that when the random variable x is Gaussian distribution, $\kappa = 3 - n$ can make the approximation error of the fourth-order term of the posterior variance be minimized. And the change of the value κ directly affects the distance between the Sigma point and the mean value $\overline{x}$. The larger the κ, the farther away the Sigma points are from $\overline{x}$, and the smaller the κ, the closer the Sigma points are to $\overline{x}$.However, if $\kappa < 0$, it is possible to make the posterior variance lose its semi positive nature.

It should be noted that $\sqrt{(n+\kappa)P_x}$ represents the i-th row or column of the matrix $(n+\kappa)P_x$. $\sqrt{P_x}$ can be obtained by Cholesky decomposition. When the variance matrix P_x is decomposed into $P_x = S^T S$ (upper triangular decomposition), $\sqrt{P_x}$ takes the row of S; When the variance matrix P_x is decomposed into $P_x = SS^T$ (lower triangular decomposition), $\sqrt{P_x}$ takes the column of S.

The weight of each Sigma point is:

$$W_i^m = W_i^c = \begin{cases} \kappa/(n+\kappa), & i = 0 \\ 1/2(n+\kappa), & i \neq 0 \end{cases} \tag{7.34}$$

It can be seen from the above that in symmetric sampling, except for the center point, the weights of other Sigma points are the same, and the distance to the center point is also the same. This shows that in symmetry sampling, all Sigma points except the center point have the same importance, and it can be seen from the distribution of Sigma points that Sigma points are spatially symmetric and axisymmetric. Symmetric sampling ensures that the approximation accuracy of arbitrary distribution reaches the second-order truncation of Taylor series expansion. This sigma point selection strategy makes the x odd central moment higher than the first order be 0. This makes it more consistent with the characteristics of Gaussian distribution. For Gaussian distribution, the third-order truncation of Taylor series expansion can be achieved.

2. Application of proportional correction algorithm

The main problem faced by the symmetric sampling strategy is that the distance from the Sigma point to the center becomes farther and farther with the increase of the dimension of x, which will produce the nonlocal effect of sampling. When the nonlinearity is strong (such as exponential function and trigonometric function), some serious problems will occur [9]. Although the selection $\kappa = 3-n$ can eliminate the influence of dimension increase on the distance, when $n > 3$, $\kappa < 0$, the weight $W_0 < 0$, which will cause the calculated covariance to be non-positive semidefinite. For this reason, we can consider the proportion correction of the original Sigma point set [9], and get the new Sigma point as:

$$\chi_i = \begin{cases} \overline{x}, & i = 0 \\ \overline{x} + (\sqrt{(n+\lambda)P_x})_i, & i = 1, 2, \cdots, n \\ \overline{x} - (\sqrt{(n+\lambda)P_x})_{i-n}, & i = n+1, \cdots, 2n \end{cases} \tag{7.35}$$

The weight of each sigma point is:

$$W_0^m = \frac{\lambda}{n+\lambda} \tag{7.36}$$

$$W_0^c = \frac{\lambda}{n+\lambda} + 1 - \xi^2 + \gamma \tag{7.37}$$

$$W_i^m = W_i^c = \frac{1}{2(n+\lambda)}, i = 1, 2, \cdots, 2n \tag{7.38}$$

where λ is the adjustment parameter, and $\lambda = \xi^2(n + \kappa) - n$, $\kappa \geq 0$ is selected to ensure the semi positive nature of the variance matrix, because the state quantity of the nonlinear vehicle system in this chapter is 4-dimensional ($x = [v, a_y, r, \beta]^\mathrm{T}$), so $\kappa = 0$ is selected. ξ is a scaling factor, which indicates the extension of Sigma point around $\overline{x}$. The distance between Sigma points and $\overline{x}$ can be adjusted by adjusting the value of ξ. This value is usually set to a small positive value ($10^{-4} \leq \xi \leq 1$) [10]. Considering that the transfer relationship of tire force and torque to tire sideslip angle in the modified Fiala model adopted is a case where tangent function and exponential function are coupled with each other, showing strong nonlinearity, in order to prevent non local sampling effect, Select the smaller $\xi = 0.01$. γ is used to contain higher-order moment information of x priori distribution. For Gaussian distribution, $\gamma = 2$ is optimal [10].

3. Steps of algorithms

UKF can deal with noise in two ways: extend and non-extended mode. The former expands the state noise and observation noise into states under the condition that the system model is unchanged. Only one Sigma point set needs to be generated in an iteration process, and the odd moment information can be transmitted, but the calculation amount is significantly increased; The non expansion method reduces the dimension of the system and the number of Sigma points. It has better real-time filtering performance and is more suitable for processing additive Gaussian noise. This chapter mainly adopts the non-extended UKF algorithm. For the nonlinear vehicle system in Sect. 7.2.1, assuming that the state noise and observation noise are Gaussian white noise and the variances are respectively Q and R, the whole filtering algorithm steps are as follows [10].

(1) Initialization of mean and variance

$$\overline{x}_0 = E(x_0) \tag{7.39}$$

$$P_0 = E((x_0 - \overline{x})(x_0 - \overline{x})^\mathrm{T}) \tag{7.40}$$

where $x_0 = [a_{y0}, r_0, v_0, \beta_0]^\mathrm{T}$, since the vehicle handling in this chapter is a straight line at the beginning, $\overline{x}_0 = [0\ 0\ 0\ 0]^\mathrm{T}$ can be set. In order to ensure the convergence of the algorithm, the initial error variance matrix P_0 is set as a fourth-order sparse square matrix with diagonal elements of 1.

(2) Calculate Sigma points ($k = 0, 1, 2, \cdots$).

Generate a matrix χ_k composed of $2n + 1$ Sigma points (column vectors) according to Eq. (7.35):

$$\chi_k = \left[x_{k|k}\ \ x_{k|k} + \sqrt{(n+\lambda)P_{k|k}}\ \ \ x_{k|k} - \sqrt{(n+\lambda)P_{k|k}} \right] \tag{7.41}$$

(3) Time update process

The nonlinear system state equation performs nonlinear transformation on each Sigma point:

$$\chi_{k+1|k} = f\left(\chi_{k|k}, u_k\right) \tag{7.42}$$

One step prediction value of the state is obtained by weighting:

$$x_{k+1|k} = \sum_{i=0}^{2n} W_i^m \chi_{i,k+1|k} \tag{7.43}$$

One step prediction value of variance matrix is obtained by weighting:

$$P_{k+1|k} = \sum_{i=0}^{2n} W_i^c (\chi_{i,k+1|k} - x_{k+1|k})(\chi_{i,k+1|k} - x_{k+1|k})^{\mathrm{T}} + Q \tag{7.44}$$

Nonlinear transformation of each Sigma point is performed by the observation equation:

$$\psi_{i,k+1|k} = h\left(\chi_{i,k+1|k}\right) \tag{7.45}$$

One step prediction observation value of the system is obtained by weighting:

$$y_{k+1|k} = \sum_{i=0}^{2n} W_i^m \psi_{i,k+1|k} \tag{7.46}$$

(4) Measurement update process

Find the variance matrix of the system output:

$$P_{y_{k+1}y_{k+1}} = \sum_{i=0}^{2n} W_i^c (\psi_{i,k+1|k} - y_{k+1|k})(\psi_{i,k+1|k} - y_{k+1|k})^{\mathrm{T}} + R \tag{7.47}$$

Calculate covariance matrix $P_{x_{k+1}y_{k+1}}$:

$$P_{x_{k+1}y_{k+1}} = \sum_{i=0}^{2n} W_i^c (\chi_{i,k+1|k} - x_{k+1|k})(\psi_{i,k+1|k} - y_{k+1|k})^{\mathrm{T}} \tag{7.48}$$

Calculate the filter gain matrix:

$$K_{k+1} = P_{x_{k+1}y_{k+1}} P_{y_{k+1}y_{k+1}}^{-1} \tag{7.49}$$

Obtain the filter value after status update:

$$x_{k+1|k+1} = x_{k+1|k} + K_{k+1}\left(y_{k+1} - y_{k+1|k}\right) \tag{7.50}$$

Solve the state posterior variance matrix:

$$P_{k+1|k+1} = P_{k+1|k} - K_{k+1}P_{y_{k+1}y_{k+1}}K_{k+1}^{\mathrm{T}} \tag{7.51}$$

The weights in the above equations are determined according to Eqs. (7.36) to (7.38).

7.2.3 Test Verification of Vehicle UKF Estimation Methods

This section uses the closed-loop control function in the virtual prototype software ADAMS/Car to drive the virtual prototype on the specified double lane change and slalom lane change test through the prototype controller. This examines the estimated accuracy and speed of the UKF algorithm used in this chapter for the critical state of the vehicle under two extreme conditions: rapid lane change and steering in the middle of the steering wheel. Using the data of the virtual test as a reference value has an important advantage over the real vehicle test, that is, many difficult to measure states can be obtained, which can more effectively examine the accuracy of the designed estimator to estimate these quantities.

1. Vehicle modeling in ADAMS/Car

Since the vehicle is an extremely complex mechanical system, if it is modeled according to the real structure of the vehicle, the workload and difficulty are very large. Therefore, according to the purpose of research, in the vehicle modeling according to the structure of the vehicle can be appropriately abstracted, simplified: in the suspension parts, except for the elastic parts, the rest of the parts are all regarded as rigid bodies; For the convenience of analyzing the problem, the body is regarded as a rigid body; Due to the complexity of the structure of the shock absorber, the Spring-damper force element provided by ADAMS is applied to simulate the role of the shock absorber and spring.

The automobiles studied in this chapter are mainly composed of double-wishbone independent front and rear suspension structures, rack steering structures, rigid chassis structures, etc. Firstly, the vehicle virtual prototype model is divided into front and rear suspension, body, steering system, brake system, front and rear wheels and other subsystems, and the corresponding subsystem models are established respectively; Then establish the input and output signal "Communicator" that exchanges information between the subsystems and between the subsystems and the experimental table provided by ADAMS/Car; The tire model uses the Fiala tire model that comes with the ADAMS/Car. Finally, the system is assembled into a vehicle virtual prototype test model as shown in Fig. 7.12 shown.

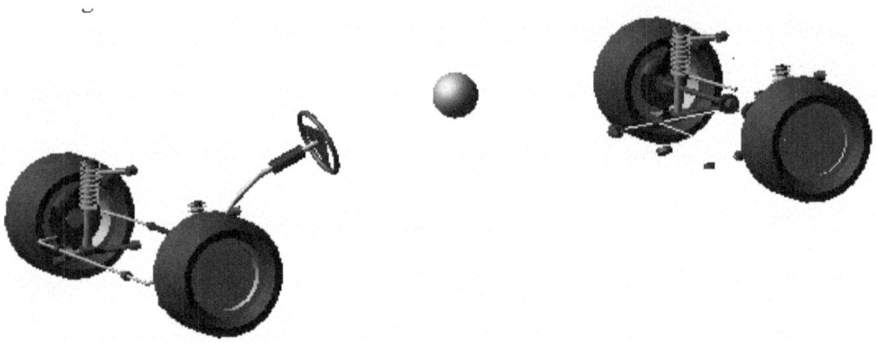

Fig. 7.12 The vehicle model in ADAMS

2. Closed-loop control settings

ADAMS/Car provides open-loop, quasi-static, and closed-loop analysis where open-loop and quasi-static analysis require only the required parameters to be entered on the user interface, and closed-loop analysis is more complex than open-loop analysis and quasi-static analysis, requiring the provision of control files required for analysis.

For closed-loop analysis, ADAMS/Car not only provides double lane change analysis and understeering characteristic analysis, designers can also design the required driver control file (*.dcf) to drive the vehicle according to specific test requirements, just like real-life tests. In the control file, you can set the speed and trajectory of the vehicle, and you can also add driver model information to make the vehicle drive intelligent.

Documents related to closed-loop control are included below.

(1) Driver Control File (*.dcf). The driver control file describes a series of actions that the vehicle model is prepared to perform, such as steering, braking, etc. In this file, you can set information such as the initial speed of the vehicle travel, the trajectory file, the control method, and the conditions at the end of the test (such as the distance traveled). Among them, the control mode has two cases: machine control and driver control. This virtual test adopts the machine control method, using the driver model to automatically control the operation of the vehicle, the input excitation is the steering wheel angle change, in order to match the time domain of the estimation algorithm, the test termination time is set to 10 s and the step size is 0.1 s in this file. In addition, the initial speed of the vehicle is adjusted accordingly according to the simulation program, and the other parameters remain at default values.

(2) Driver Control Data file (*.dcd). The driver control data file contains the data required by the Driver Control File. If the trajectory is used in *.dcf to control the movement of the vehicle, the main content of the *.dcd file is to set the coordinate data of the vehicle's driving trajectory. This virtual test gives the vehicle trajectory in the form of x, y coordinate data. It should be noted here

that the coordinate data should be as dense as possible, and the denser the vehicle, the less deviant it is to follow the specified path.

(3) Driver Input File (*.din). The driver input file specifies the values of various control parameters required by the ADAMS/Driver module, such as radial dynamic characteristics, lateral dynamic characteristics, vehicle structural characteristics, and control parameters of the learning file, which are used to guide the vehicle analysis.

3. Simulation results analysis

A virtual test is conducted in ADAMS/Car with a certain type of vehicle to verify the accuracy and efficiency of the UKF algorithm in the estimation of the state of the vehicle in this chapter. The parameters of the vehicle are as follows: vehicle mass $m = 1528\,\text{kg}$, $a = 1.48\,\text{m}$, $b = 1.08\,\text{m}$, $I_z = 2440\,\text{kg} \cdot \text{m}^2$, steering gear ratio $i = 26.7$.

[1] Double lane change test

The double lane change settings are as follows: $a_0 = 2u, a_1 = 4u, a_2 = 5u, a_3 = 7u$, $a_4 = 12u$.

To facilitate data processing, the simulation ends at 10 s. The comprehensive observation noise and measurement noise can obtain a visual measurement with noise interference r', a_y'.

(1) 80 km/h

Figure 7.13 is the lateral displacement and longitudinal displacement at the center of mass when the vehicle drives through the double lane change at a speed of 80 km/h.

Measurements of noise containing time-invariant statistical characteristics r', a_y' are shown in Fig. 7.14.

Figure 7.15 shows the comparison between the estimated values of the system state variables v, a_y r, β and the virtual test values when the vehicle passes through the double lane change test at a speed of 80 km/h.

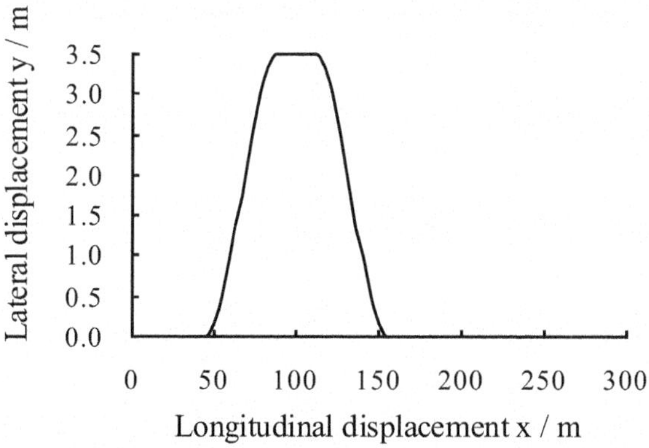

Fig. 7.13 Lateral and longitudinal displacement at 80 km/h (double lane change test)

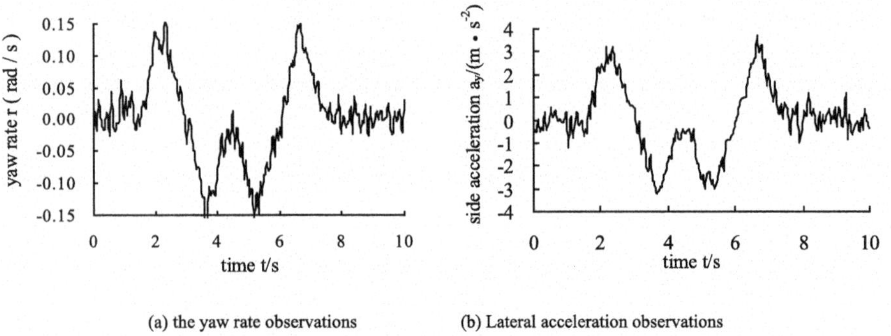

Fig. 7.14 Observations containing noise at 80 km/h (double lane change test)

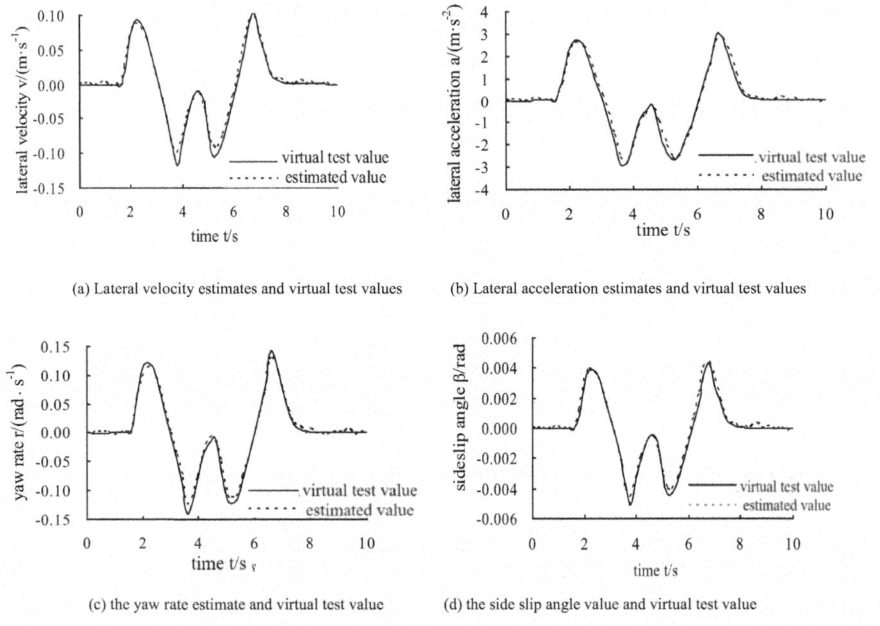

Fig. 7.15 Comparison of the estimated value of the UKF algorithm with the value of the virtual test at 80km/h (double lane change test)

As can be seen from Fig. 7.15, the estimates of the four state quantities at 80 km/h coincide well with the virtual test values. Lateral acceleration and the yaw rate follow slightly less at peaks and troughs.

(2) 120 km/h

With the improvement of road conditions and the development of automobile technology, it is very common for vehicles to travel at a speed of more than 100 km/h, and in general, the emergency of automobiles occurs at high speeds, so the design

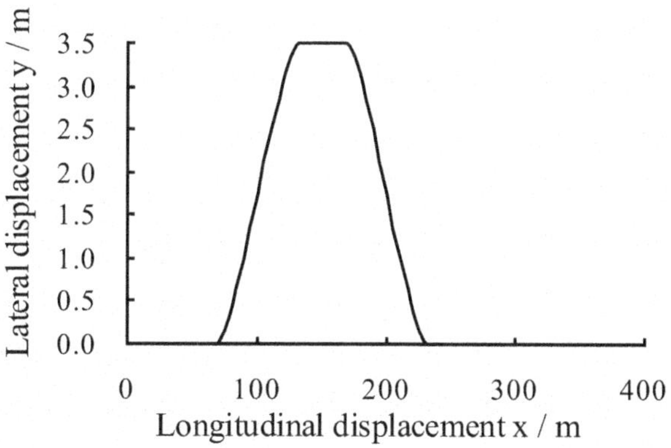

Fig. 7.16 Lateral displacement and longitudinal displacement at 120 km/h (double change lane test)

of the estimator in the dynamic control system of the vehicle should consider the working conditions at higher speeds. The following examines the accuracy of the state estimation algorithm in this chapter when a vehicle passes a double lane change test at a high speed of 120 km/h.

Figure 7.16 shows the lateral displacement and longitudinal displacement at the center of mass when the vehicle crosses the double lane change path at a speed of 120 km/h.

As can be seen from Fig. 7.16, the longitudinal displacement of the double change lane test completed in the same time increases as the speed of the vehicle increases.

Figure 7.17 shows the comparison of the system state variable, the estimate of the v, a_y, r, β and virtual test value when the vehicle passes through the double change lane test at a speed of 120 km/h.

From the coincidence of the estimated value of Fig. 7.17 and the virtual test value, it can be seen the overall speed increase still has a good estimation accuracy. At this moment, the lateral velocity and the side slip angle are significantly increased, and the estimation of the two state quantities does not match the virtual test value as well as the low speed, especially in the position of the peaks and troughs. This is due to the fact the change in the state quantity is more drastic at high speed, resulting in poor follow-up of the estimated value. In addition, the estimation accuracy of lateral acceleration and yaw rate is still high compared with low speed.

[2] Slalom

A more intense maneuver than the double change lane test is the slalom lane change test of halving the time course. The slalom lane change test is set as follows: $a_0 = u$, $a_1 = 2u$, $a_2 = 7u$, $a_3 = 8u$, $a_4 = 10u$.

(1) 80 km/h

Figure 7.18 shows the lateral and longitudinal displacement at the center of mass when the vehicle crosses the slalom lane change test at a speed of 80 km/h.

Measurements of time-constant statistical characteristic noise r', a_y' are shown in Fig. 7.19.

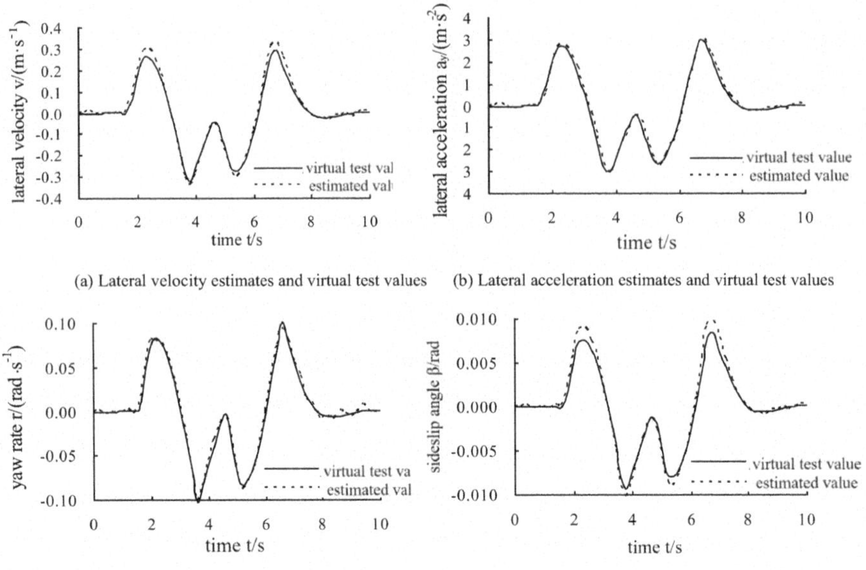

(a) Lateral velocity estimates and virtual test values (b) Lateral acceleration estimates and virtual test values

(c) Yaw rate estimate and virtual test value (d) the side slip angle value and virtual test value

Fig. 7.17 Comparison between the estimated value of UKF algorithm and the virtual test value at 120km/h (double change lane test)

Fig. 7.18 Lateral and longitudinal displacement at 80 km/h (slalom test)

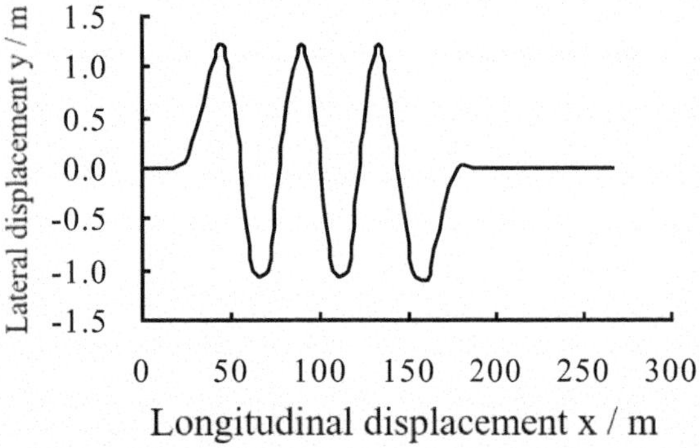

Figure 7.20 reflects the comparison between the estimated values of the system state variable v, a_y, r, β and the virtual test values when the vehicle passes through the slalom lane change test at a speed of 80 km/h.

As can be seen from Fig. 7.20, the maximum lateral acceleration has reached 8 m/s^2 when the vehicle is driven at a speed of 80 km/h under the slalom test which time is shortened by half, the tire has entered a nonlinear area. In general, the estimation of the four state variables is better in line with the virtual test value, and the estimated value of each state quantity is more error than that of the virtual test value on the peaks and troughs and other places.

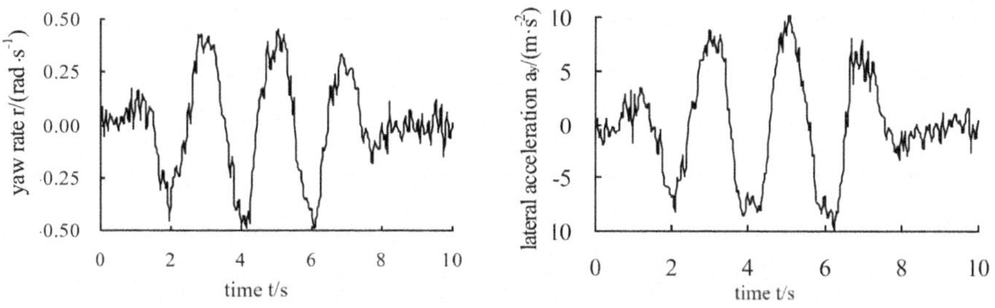

(a) the yaw rate observations (b) Lateral acceleration observations

Fig. 7.19 Measurements containing noise at 80 km/h (slalom test)

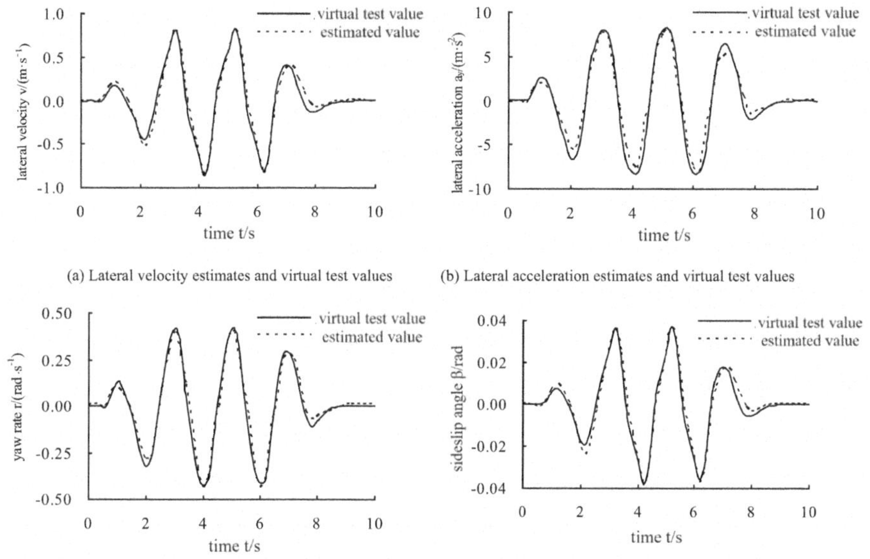

(a) Lateral velocity estimates and virtual test values (b) Lateral acceleration estimates and virtual test values

(c)the yaw rate estimate and virtual test value (d) the side slip angle value and virtual test value

Fig. 7.20 Comparison of UKF algorithm estimates and virtual test values at 80 km/h (slalom test)

(2) 120 km/h

Figure 7.21 shows the lateral displacement and longitudinal displacement at the centroid when the vehicle crosses the slalom lane change test at a speed of 120 km/h.

As can be seen from Fig. 7.21, the longitudinal displacement of the slalom lane change test completed in the same time increases as the speed of the vehicle increases.

Figure 7.22 shows the comparison of the system state variable, the estimate of the system state variable v, a_y, r, β and the simulation value obtained by the virtual experiment when the vehicle passes through the slalom lane change test at a speed of 120 km/h.

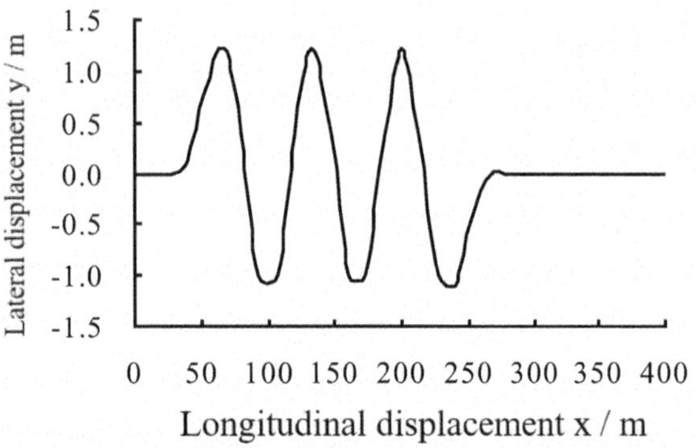

Fig. 7.21 Lateral and longitudinal displacement at 120 km/h (slalom lane change test)

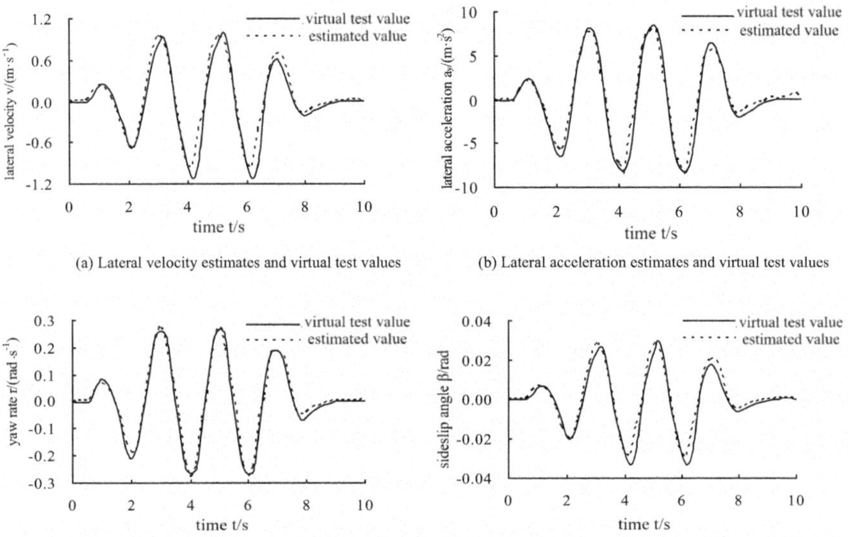

(a) Lateral velocity estimates and virtual test values

(b) Lateral acceleration estimates and virtual test values

(c) the yaw rate estimate and virtual test value

(d) the side slip angle value and virtual test value

Fig. 7.22 Comparison of UKF algorithm estimates and virtual test values at 120 km/h (slalom test)

From Fig. 7.22a, b, it can be seen the lateral speed of the vehicle has increased significantly with the increase of the speed, but the lateral acceleration has not changed much. The former reflects the increase in the side sliding trend of the vehicle after the speed is increased under the condition of constant lateral displacement, and it can be seen from the figure that the estimation accuracy of the UKF algorithm for the lateral speed is slightly lower than that of the low speed. The latter reflects that the lateral force of the vehicle tire does not change much under the two speeds, and it can be seen from the figure that the estimation accuracy of the lateral acceleration is not affected by the speed.

From Fig. 7.22c, it can be seen the yaw rate of the vehicle decreases at higher speeds, which indicates that the yaw torque of the whole vehicle is reduced, and the coincidence between the estimated angular velocity of the yaw rate and the virtual test value of the UKF algorithm at high speed is better than low speed. From Fig. 7.22d, it can be seen the amplitude of the side slip angle at higher speeds has also decreased, because after the speed is increased, the increase in lateral speed is not as large as the increase in longitudinal speed, and the estimation accuracy of the side slip angle is roughly the same as that at low speed.

In order to improve the intuitively and quantitatively in compare the state estimation accuracy of the UKF algorithm at different vehicle speeds, the mean absolute error (MAE) and root mean square error (RMSE) of the estimated value relative to the virtual test value are given below, as shown in Table 7.4. As shown in Table 7.4, the expressions for the two error indicators are as follows:

$$\mathrm{MAE}(x) = \frac{\sum_{k=1}^{N} \left| x_{k|k} - x_{\mathrm{actual}}(k) \right|}{N} \tag{7.52}$$

$$\mathrm{RMSE}(x) = \sqrt{\frac{\sum_{k=1}^{N} \left(x_{k|k} - x_{\mathrm{actual}}(k) \right)^2}{N}} \tag{7.53}$$

where $x_{k|k}$ and $x_{\mathrm{actual}}(k)$ are estimate and the virtual test value in the k time, respectively.

As can be seen from Table 7.4, the UKF algorithm has high state estimation accuracy under both typical limit conditions, and the average absolute error is within 10% of the state amplitude. In addition, due to the relatively high complexity of the path and the shorter manipulation time, the estimation error of the slalom test is higher than that of the double change lane test.

Table 7.4 MAE indicators and RMSE indicators of the UKF algorithm

Evaluation indicators	The amount of status	Double change lane test		Slalom lane change test	
		80 km/h	120 km/h	80 km/h	120 km/h
MAE	v/(m/s)	0.00405	0.0132	0.0555	0.0754
	a_y/(m/s2)	0.148	0.119	0.705	0.492
	r/ (rad/s)	0.00498	0.00373	0.0315	0.0178
	β/rad	0.000166	0.000491	0.00266	0.0023
RMSE	v/(m/s)	0.00595	0.0182	0.0777	0.121
	a_y/(m/s^2)	0.192	0.154	0.944	0.75
	r/(rad/s)	0.00695	0.00586	0.0436	0.0264
	β/rad	0.000245	0.000674	0.00365	0.00353

7.2.4 Comparison of UKF and EKF in Vehicle States Estimation

The EKF (Extended Kalman Filtering) algorithm is a representative of the traditional nonlinear estimation algorithm, the basic idea of which is to expand the nonlinear model around the state estimate of the first-order Taylor series, and then apply linear Kalman Filtering is estimated. This section compares the two algorithms, UKF and EKF, and examines their respective performance in nonlinear vehicle states estimation.

The nonlinear vehicle dynamics system containing pure additive noise is still studied, and the variance of the process noise and the measured noise of the system are Q and R, respectively. The steps of the EKF algorithm are as follows [7]:

(1) Initialization of the mean and variance.

$$\overline{x}_0 = E(x_0)$$

$$P_0 = E((x_0 - \overline{x})(x_0 - \overline{x})^{\mathrm{T}})$$

(2) Time update process.

State prediction equation:

$$x_{k+1|k} = f\left(x_{k|k}, u_{k+1}\right)$$

Error covariance prediction:

$$P_{k+1|k} = AP_{k|k}A_k^{\mathrm{T}} + Q$$

(3) Measure the update process.

Gain:

$$K_{k+1} = P_{k+1|k}C^{\mathrm{T}}(CP_{k+1|k}C^{\mathrm{T}} + R)^{-1}$$

Filtering equation:

$$x_{k+1|k+1} = x_{k+1|k} + K_{k+1}\left(y_{k+1} - Cx_{k+1|k}\right)$$

Error covariance update:

$$P_{k+1|k+1} = (I - K_{k+1}C)P_{k+1|k}$$

where A is the Jacobian matrix of f for x, C is the Jacobian matrix of h for x:

$$A = \begin{bmatrix} \frac{\partial f_1}{\partial x_1} & \cdots & \frac{\partial f_1}{\partial x_4} \\ \vdots & & \vdots \\ \frac{\partial f_4}{\partial x_1} & \cdots & \frac{\partial f_4}{\partial x_4} \end{bmatrix}$$

$$C = \begin{bmatrix} 0\,0\,1\,0 \\ 0\,1\,0\,0 \end{bmatrix}$$

Taking 120 km/h slalom test manipulation as an example, the estimation effects of the two algorithms are compared. The noise statistics characteristics of the system shall be set in accordance with the foregoing provisions. Figure 7.23 compares the estimated values of the four key state quantities v, a_y, r, β and the virtual trial values of the two algorithms.

From Fig. 7.23 It can be seen the estimates of the two algorithms are very close, and the accuracy of the estimation EKF algorithm at the peak of the lateral acceleration is not as good as the UKF algorithm, and the error indicators of the two estimation algorithms are given below, as shown in Table 7.5.

From Table 7.5, it can be seen under the same conditions, the estimation accuracy of the UKF algorithm is slightly higher than that of the EKF algorithm, which also supports the aforementioned "UKF can accurately estimate the mean to achieve."

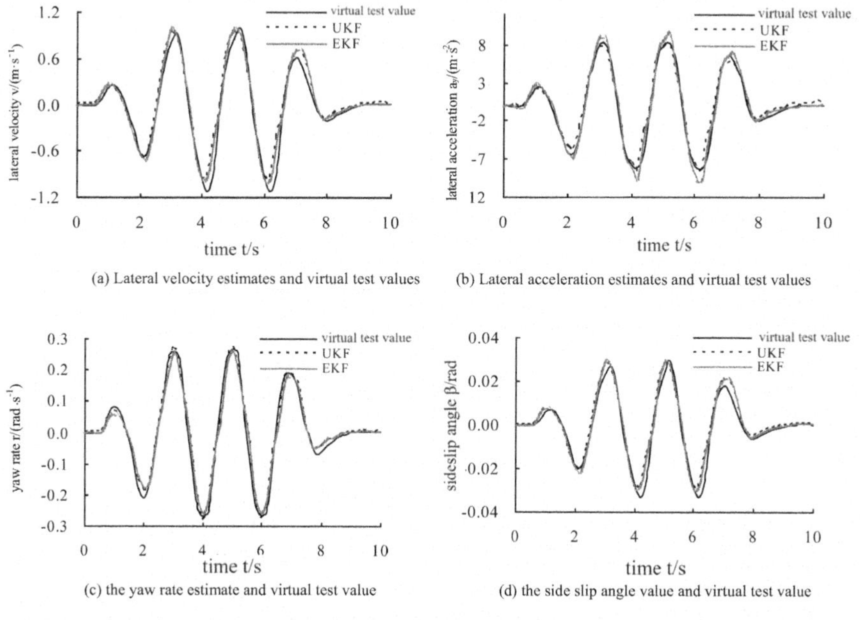

(a) Lateral velocity estimates and virtual test values

(b) Lateral acceleration estimates and virtual test values

(c) the yaw rate estimate and virtual test value

(d) the side slip angle value and virtual test value

Fig. 7.23 Comparison of UKF algorithm and EKF algorithm at 120km/h (slalom lane change test)

Table 7.5 Error comparison of UKF algorithm and EKF algorithm

Evaluation indicator	Status quantity	UKF	EKF
MAE	v/(m/s)	0.0754	0.0787
	a_y/(m/s^2)	0.492	0.546
	r/(rad/s)	0.0178	0.0208
	β/rad	0.0023	0.00243
RMSE	v/(m/s)	0.121	0.123
	a_y/(m/s^2)	0.75	0.772
	r/(rad/s)	0.0264	0.0295
	β/rad	0.00353	0.00358

The Taylor series has a fourth-order precision, while the EKF estimates the mean to achieve second-order precision" conclusion.

The running time of the two algorithm programs is compared at the same time, and the running time of the UKF algorithm is 3.735 s and the running time of the EKF algorithm is 6.468 s in the same environment. It can be seen although the two computation quantities are of the same order, in the MATLAB environment, because the EKF algorithm needs to calculate the cumbersome Jacobian matrix through symbolic operations, it occupies a lot of running time. It can be seen if the program in this chapter is directly used for hardware real-time control through the Dspace rapid prototyping control development system, the real-time nature of the UKF method is higher.

In summary, the UKF algorithm has advantages over the widely used EKF algorithm in terms of estimating accuracy and real-time, and can fully meet the software performance requirements of the automotive state estimator [11].

7.3 Vehicle States Estimation Based on DEAKF

7.3.1 *Nonlinear Seven Degrees of Freedom Vehicle Dynamics Model*

1. Vehicle model

The estimation algorithm presented in this section is based on the seven-degree-of-freedom nonlinear vehicle dynamics model shown in Fig. 7.24, where the seven degrees of freedom include longitudinal, lateral, yaw, and the rotational motion of the four wheels.

The differential equations of motion for automobiles are as follows.

Longitudinal:

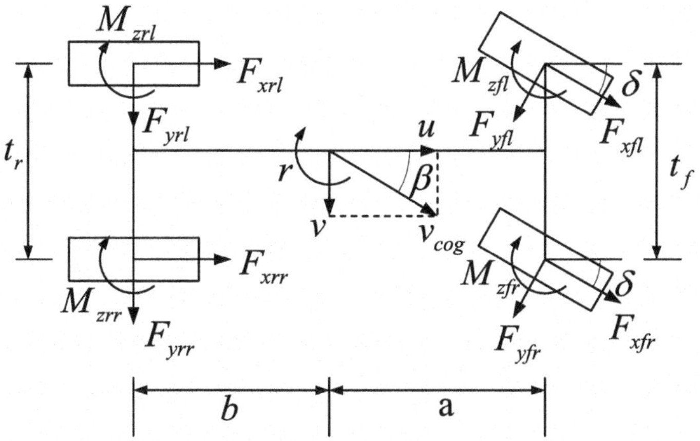

Fig. 7.24 Schematic diagram of a seven-degree-of-freedom vehicle dynamics model

$$\dot{u} = a_x + vr \tag{7.54}$$

$$\begin{aligned} a_x = (&F_{xfl}\cos\delta + F_{xfr}\cos\delta + F_{xrl} + F_{xrr} \\ &-F_{yfl}\sin\delta - F_{yfr}\sin\delta)/m \end{aligned} \tag{7.55}$$

Lateral:

$$\dot{v} = a_y - ur \tag{7.56}$$

$$\begin{aligned} a_y = (&F_{xfl}\sin\delta + F_{xfr}\sin\delta + F_{yfl}\cos\delta \\ &+F_{yfr}\cos\delta + F_{yrl} + F_{yrr})/m \end{aligned} \tag{7.57}$$

Yaw:

$$\begin{aligned} \Gamma = &\frac{t_f}{2}F'_{xfl} - \frac{t_f}{2}F'_{xfr} + \frac{t_r}{2}F_{xrl} - \frac{t_r}{2}F_{xrr} \\ &+ aF'_{yfl} + aF'_{yfr} - bF_{yrl} - bF_{yrr} \\ &+ M_{zfl} + M_{zfr} + M_{zrl} + M_{zrr} \end{aligned} \tag{7.58}$$

$$\dot{r} = \frac{\Gamma}{I_z} \tag{7.59}$$

where

$$F'_{xij} = F_{xij}\cos\delta - F_{yij}\sin\delta$$

$$F'_{yij} = F_{yij} \cos\delta + F_{xij} \sin\delta$$

Ignoring the inertia resistance even moment and air lift action, the vertical load of each tire when the vehicle turns right is

$$F_{zfl,zfr} = \left(\frac{mg}{2} \pm ma_y \frac{h}{t_f}\right)\frac{b}{L} - \frac{1}{2}ma_x\frac{h}{L} \tag{7.60}$$

$$F_{zrl,zrr} = \left(\frac{mg}{2} \pm ma_y \frac{h}{t_r}\right)\frac{a}{L} + \frac{1}{2}ma_x\frac{h}{L} \tag{7.61}$$

The slip angle of each tire can be obtained by using the lower male type:

$$\alpha_{fl,fr} = \delta - \arctan\frac{v + ar}{u \pm \frac{t_f}{2}r} \tag{7.62}$$

$$\alpha_{rl,rr} = -\arctan\frac{v - br}{u \pm \frac{t_r}{2}r} \tag{7.63}$$

The side slip angle is

$$\beta = \arctan\frac{v}{u} \tag{7.64}$$

The slip rate of each tire

$$s_{ij} = \frac{r_e\omega_{ij} - u_{\omega ij}}{u_{\omega ij}} \tag{7.65}$$

where, the center speed of the wheel $u_{\omega ij}$ can be obtained by the following:

$$u_{\omega fl,\omega fr} = v_{cog} \pm r\left(\frac{t_f}{2} \pm a\beta\right) \tag{7.66}$$

$$u_{\omega rl,\omega rr} = v_{cog} \pm r\left(\frac{t_r}{2} \mp b\beta\right) \tag{7.67}$$

In the above equations, u is the longitudinal velocity, v is the lateral velocity, v_{cog} is the centroid velocity, a_x is longitudinal acceleration, a_y is lateral acceleration, r is the yaw rate, β is the side slip angle of the whole vehicle, Γ is the yaw moment for the whole vehicle around the z-axle, δ is the steering angle of front wheel, F_{xij} is the longitudinal forces on each wheel, F_{yij} is the lateral force on each wheel, F_{zij} is the vertical loads on each wheel, M_{zij} is the self aligning moment on each wheel, m is the mass of the whole vehicle, I_z is the moment of inertia for the whole vehicle around the plumb shaft, a and b are the distance from the center of mass of the vehicle to the front and rear axles, $L = a + b$ is the wheelbase, t_f and t_r are front and rear track, respectively, h is centroid height, α_{ij} is slip angle of each tire, s_{ij} is the slip rate of each tire, r_e is the effective rolling radius for the wheel, ω_{ij} is angular velocity for wheel rotation, $u_{\omega ij}$ is the center speed of the wheel.

2. Tire model

The Pacejka nonlinear tire combined condition model [12], the model input variables are tire vertical load, tire side angle, tire slip rate, can be used in the uniform form of Eqs. (7.68)–(7.70) Calculating the tire lateral force, longitudinal force and self aligning moment:

$$y(x) = D\sin(C\arctan(Bx - E(Bx - \arctan Bx))) \tag{7.68}$$

$$Y(X) = y(x) + s_v \tag{7.69}$$

$$x = X + s_h \tag{7.70}$$

In Eq. (7.69), the output variable Y represents the tire longitudinal force F_x, lateral force F_y and positive returning moment M_z.The input variable X in Eq. (7.70) can represent the tire slip rate s (used for calculating longitudinal force) and the tire slip angle α (used for calculating lateral force and self aligning moment), respectively. Specific expressions of parameters B, C, D, E, s_v and s_h under different outputs can be referred to Ref. [12]. In this chapter, $B = 0.237$, $C = 1.65$, $D = 3610.5$, $E = 0.707$, $s_v = 40.379$, $s_h = 0.0473$ under the vertical load of 3.16 kN.

7.3.2 Dual Extended Adaptive Kalman Filter Algorithm

Based on the Dual Extended Kalman Filtering (DEKF) algorithm [10], this chapter proposes a dual extended adaptive Kalman filter [13]. (DEAKF) algorithm.

1. Improved filtering algorithm

The DEKF algorithm uses two decoupled extended Kalman filters to work in parallel, and the states estimation is embedded between the prediction and correction of the parameters, which can estimate the states and parameters of the system online. The algorithm is based on the conventional extended Kalman filtering algorithm, so it is impossible to estimate the state parameters of the situation where the noise statistical characteristics are unknown, and the statistical characteristics of the process noise and observation noise of the vehicle as a complex system are often unknown during driving. To this end, this chapter improves on the DEKF algorithm, and integrates the robust filtering algorithm and the DEKF algorithm to obtain the DEAKF algorithm, which can perform parallel estimation of states and parameters of systems with unknown noise statistical characteristics.

The equations of state and observation for nonlinear discrete systems are as following:

$$x_{k+1}^{s} = f\left(x_{k}^{s}, u_{k}, x_{k}^{p}\right) + w_{k} \tag{7.71}$$

$$y_{k} = h\left(x_{k}^{s}, x_{k}^{p}\right) + v_{k} \tag{7.72}$$

where x^{s} is the state vector, x^{p} is a parameter vector, u is input for the system, y is observation vector, w and v are process noise and measurement noise, respectively.

State and parameter parallel estimation of time-varying noise statistical nonlinear systems, according to the DEKF algorithm, parameter prediction process:

$$x_{k+1|k}^{p} = x_{k|k}^{p} \tag{7.73}$$

$$P_{k+1|k}^{p} = P_{k|k}^{p} + Q^{p} \tag{7.74}$$

States prediction process:

$$x_{k+1|k}^{s} = f\left(x_{k|k}^{s}, u_{k+1}, x_{k+1|k}^{p}\right) + q_{k}^{s} \tag{7.75}$$

$$P_{k+1|k}^{s} = A_{k}^{s} P_{k|k}^{s} A_{k}^{s\mathrm{T}} + Q_{k}^{s} \tag{7.76}$$

States correction process:

$$K_{k+1}^{s} = P_{k+1|k}^{s} C^{s\mathrm{T}} \left(C^{s} P_{k+1|k} C^{s\mathrm{T}} + R_{k}^{s}\right)^{-1} \tag{7.77}$$

$$\varepsilon_{k+1} = y_{k+1} - C^{s} x_{k+1|k}^{s} - r_{k}^{s} \tag{7.78}$$

$$x_{k+1|k+1}^{s} = x_{k+1|k}^{s} + K_{k+1}^{s} \varepsilon_{k+1} \tag{7.79}$$

$$P_{k+1|k+1}^{s} = \left(I - K_{k+1}^{s} C^{s}\right) P_{k+1|k}^{s} \tag{7.80}$$

Add a noise statistical estimator with a forgetting factor [14] to the state estimation process, and the state estimation Eqs. (7.75)–(7.80) Form a two-segment coupling adaptive Kalman filter algorithm that estimates the mean and variance of unknown noise online, the noise statistical estimator is

$$q_{k+1}^{s} = (1 - d_{k}) q_{k}^{s} + d_{k}\left(x_{k+1|k+1} - f\left(x_{k|k}^{s}, u_{k+1}, x_{k+1|k}^{p}\right)\right) \tag{7.81}$$

$$Q_{k+1}^{s} = (1 - d_{k}) Q_{k}^{s} + d_{k}\left(K_{k+1}^{s} \varepsilon_{k+1} \varepsilon_{k+1}^{\mathrm{T}} K_{k+1}^{s\mathrm{T}} + P_{k+1|k+1}^{s} - A_{k}^{s} P_{k|k}^{s} A_{k}^{s\mathrm{T}}\right) \tag{7.82}$$

$$r_{k+1}^s = (1 - d_k)r_k^s + d_k(y_{k+1} - C^s x_{k+1\,k}^s) \tag{7.83}$$

$$R_{k+1}^s = (1 - d_k)R_k^s + d_k(\varepsilon_{k+1}\varepsilon_{k+1}^{\mathrm{T}} - C^s P_{k+1|k}^s C^{s\mathrm{T}}) \tag{7.84}$$

where $d_k = (1 - e)/(1 - e^{k+1})$, e is the forgetting factor is used to limit the memory length of the filter and strengthen the effect of recent data on estimation. Typically, $0.95 \leq e \leq 0.995$ [14], in this section e is taken as 0.97.

Finally, add the parameter correction process:

$$K_{k+1}^p = P_{k+1|k}^p C^{p\mathrm{T}}(C^p P_{k+1|k}^p C^{p\mathrm{T}} + R_k^p)^{-1} \tag{7.85}$$

$$x_{k+1|k+1}^p = x_{k+1|k}^p + K_{k+1}^p \varepsilon_{k+1} \tag{7.86}$$

$$P_{k+1|k+1}^p = (I - K_{k+1}^p C^p)P_{k+1|k}^p \tag{7.87}$$

Thus, Eqs. (7.73)–(7.87) establish the DEAKF algorithm for nonlinear discrete systems as shown in Eqs. (7.71) and (7.72). In the above formulas, Q^p and Q^s are the process noise covariance matrix in parameter estimation and state estimation respectively, R^p and R^s are the measurement noise covariance matrix in parameter estimation and state estimation respectively, P^p and P^s are the estimation error covariance matrix in parameter estimation and state estimation, q^s is the mean value of process noise in state estimation, r^s is the mean value of the measurement noise in the state estimation.

According to the seven-degree-of-freedom nonlinear vehicle model, the state vector is set to

$$x^s = [u, v, a_x, a_y, r, \beta, \Gamma]^{\mathrm{T}} \tag{7.88}$$

The parameter vector is

$$x^p = [m, I_z, a]^{\mathrm{T}} \tag{7.89}$$

The system input is

$$u = [\delta, \omega_{fl}, \omega_{fr}, \omega_{rl}, \omega_{rr}]^{\mathrm{T}} \tag{7.90}$$

The observation vector is

$$y = [r, a_y, u]^{\mathrm{T}} \tag{7.91}$$

The dynamic equation $f(x_k^s, u_k, x_k^P)$ and measurement equation $h(x_k^s, x_k^p)$ of the system can be obtained by using the differential equation of motion of the vehicle.

Like the general EKF algorithm, in DEAKF algorithm, A^s is the Jacobian matrix of f to x^s, C^s is the Jacobian matrix of h to x^s:

$$A^s = \begin{bmatrix} \frac{\partial f_1}{\partial x_1^s} & \cdots & \frac{\partial f_1}{\partial x_7^s} \\ \vdots & & \vdots \\ \frac{\partial f_7}{\partial x_1^s} & \cdots & \frac{\partial f_7}{\partial x_7^s} \end{bmatrix} \tag{7.92}$$

$$C^s = \begin{bmatrix} 0\,0\,0\,0\,1\,0\,0 \\ 0\,0\,0\,1\,0\,0\,0 \\ 1\,0\,0\,0\,0\,0\,0 \end{bmatrix} \tag{7.93}$$

In addition, according to the literature [10], the matrix in the parameter estimation C^p can be obtained by Eq. (7.94).

$$C^p = C\frac{\partial f(x^s, x^p)}{\partial x^p} = \begin{bmatrix} \frac{\partial r}{\partial m} & \frac{\partial r}{\partial I_z} & \frac{\partial r}{\partial a} \\ \frac{\partial a_y}{\partial m} & \frac{\partial a_y}{\partial I_z} & \frac{\partial a_y}{\partial a} \\ \frac{\partial u}{\partial m} & \frac{\partial u}{\partial I_z} & \frac{\partial u}{\partial a} \end{bmatrix} \tag{7.94}$$

2. Filter parameter settings

This section assumes that the process noise in the measurement noise and state estimation has unknown time-varying statistical characteristics, and the process noise in the parameter estimation has known invariant noise statistical characteristics.

Since the parallel parameter estimator and state estimator in this chapter adopt the same output y, $R^p = R^s$, and the online calculation is conducted according to Eq. (7.84) when the initial value is given; The process noise covariance matrix Q^s in the state estimation is calculated online according to Eq. (7.82) when the initial value is given; The process covariance matrix Q^p in parameter estimation is calculated according to Eq. (7.95):

$$Q^p = \begin{bmatrix} S_m^2 & 0 & 0 \\ 0 & S_{I_Z}^2 & 0 \\ 0 & 0 & S_a^2 \end{bmatrix} \tag{7.95}$$

where S_m, S_{IZ} and S_a can be defined as 1% of the approximate value of the vehicle parameter m, I_z and a, after trying, such setting helps to improve the accuracy of the parameter estimation.

7.3.3 Three Methods of Vehicle States Estimation and Their Virtual Test Verification

As mentioned earlier, due to its good real-time performance, EKF has become the most used algorithm in the estimation of the states of vehicle, and it is also an important tool for the estimation of the state of nonlinear systems. The formula for estimating the state of the vehicle based on the EKF algorithm is stated in Sect. 7.2 and will not be repeated here. Although the EKF algorithm is widely used, its shortcomings are also relatively obvious. Because it is only an estimate of the system state, the uncertainty of the vehicle parameters does not have an adaptive adjustment function, so once the vehicle parameters fluctuate, the robustness of the estimator is poor, and even causes filter divergence.

The DEKF algorithm was proposed by Wan and Nelson, and because of the parallel parameter estimation, it solves the problem of filter divergence caused by fluctuations in the parameters of the system structure. The essence of the DEKF algorithm is that parameter estimation and state estimation are coupled and updated. The specific DEKF algorithm has been stated earlier and will not be repeated here.

In the actual application process, because the measured value of the sensor is often interfered with by external conditions, the time-varying nature of the noise cannot be ignored, and the DEKF algorithm may have divergence when estimating the state of the nonlinear system with noise time-varying. In view of the above situation, this chapter proposes the DEAKF algorithm on the basic of the DEKF algorithm, which is robust to the time-varying nature of noise and the uncertainty of parameters. The following is a comparison of the application of three nonlinear estimation methods of EKF, DEKF and DEAKF in the estimation of the state of vehicle.

1. Double lane change condition

In order to compare the advantages and disadvantages of the three algorithms, a simulation test of a certain type of vehicle was conducted on the virtual prototype software ADAMS. The parameters of the vehicle are: $m = 1528\,\text{kg}$, $a = 1.48\,\text{m}$, $b = 1.08\,\text{m}$, $h = 0.432\,\text{m}$, $I_z = 2440\,\text{kg} \cdot m^2$, $t_f = 1.52\,\text{m}$, $t_r = 1.594\,\text{m}$, $r_c = 0.33\,\text{m}$. In order to simulate the vehicle handling response under extreme operating conditions, this chapter uses a driving prototype to generate the required vehicle driving path in the vehicle simulation, so that the vehicle travels along the double change lane test shown in Fig. 7.3. The entire maneuver lasted 10 s and the sampling time was 0.01 s.

A smooth Gaussian white noise sequence with the time-varying statistical characteristics of the state estimation set to the mean zero, the variance array changes as follows:

$$Q^s = 0.01 \cdot |\max(\text{state}(i)) - \min(\text{state}(i))| \cdot \sin\left(\frac{\pi}{10}t\right)$$

$$i = 1, 2, \cdots, 7,\ 0 \le t \le 10$$

where max(state(i)) and min(state(i)) are the maximum and minimum values of the i-th state quantity in the whole time, respectively. The process noise covariance matrix Q^p of parameter estimation is determined according to Eq. (7.95).

A stationary Gaussian white noise sequence with the measured noise covariance matrix set to the time-varying statistical characteristics of the mean zero, the variance array changes according to the following law:

$$R^s = 0.05 \cdot |\max(y) - \min(y)| \cdot \sin\left(\frac{\pi}{10}t\right),\ 0 \le t \le 10$$

The observation noise with noise interference r', a_y' and u' can be obtained by synthesizing the observation noise and the measurement noise, as shown in Fig. 7.25.

In order to investigate the estimation ability of the DEAKF algorithm on nonlinear systems with inaccurate model parameters and statistical characteristics of unknown time-varying noise, this section compares three nonlinear Kalman filtering algorithms. In all three algorithms, inaccurate vehicle model parameters $m_0 = 1250\text{kg}$, $I_{z0} = 2047\text{kg} \cdot m^2$, $a_0 = 1.26\text{m}$ are used as the initial values. Since the DEKF algorithm and the DEAKF algorithm have parameter correction estimation functions, this set of automotive parameters is constantly being corrected online.

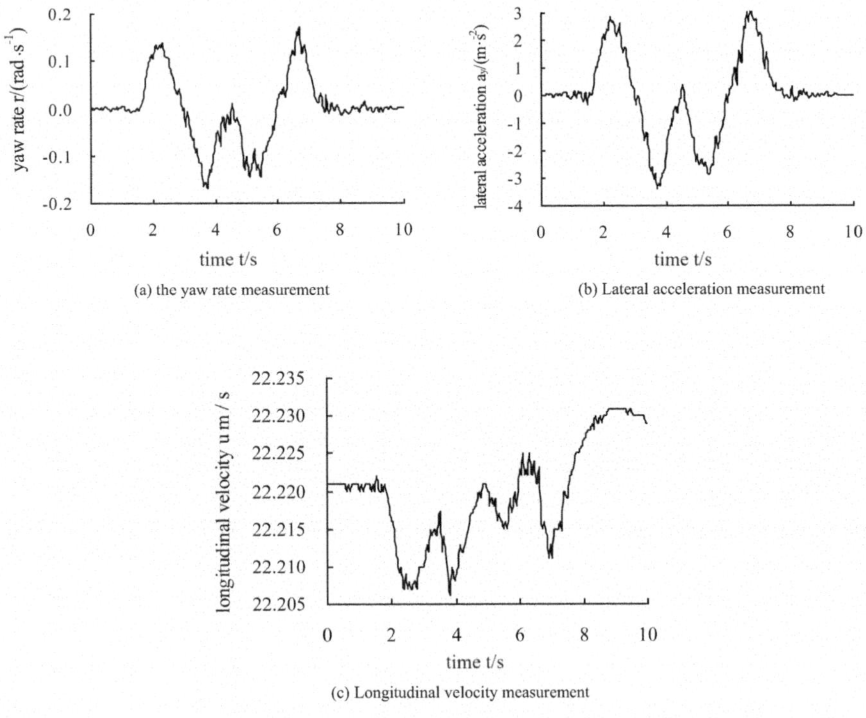

(a) the yaw rate measurement

(b) Lateral acceleration measurement

(c) Longitudinal velocity measurement

Fig. 7.25 Measurement of noise containing time-varying statistical characteristic noise (double change lane test)

For the process noise covariance matrix of state estimation, all the three algorithms take the following initial values:

$$Q_0^s = 0.001 \cdot |\max(\text{state}(i)) - \min(\text{state}(i))|$$

Q_0^p is determined according to Eq. (7.95) (EKF algorithm does not have this value).

For the measured noise covariance matrix, all three algorithms take the following initial values:

$$R_0^s = R_0^p = 0.01 \cdot |\max(y) - \min(y)|$$

Figure 7.26 compares the estimated values of six key state quantities u, v, a_x, a_y, r, β with the virtual experimental values (the actual values in the figure) by the EKF, DEKF, and DEAKF algorithms.

From the overall point of view of Fig. 7.26, the estimated value obtained by the DEAKF algorithm is higher than that of the EKF algorithm and the DEKF algorithm. From Fig. 7.26c, it can be seen the difference between the three calculation methods is more obvious in the estimation of longitudinal acceleration, and the deviation between the EKF calculation method and the virtual test value is significantly higher than that of the other two the arithmetic is bigger. The same trend can be seen from Fig. 7.26e: since the model adopts inaccurate initial parameters, while the EKF algorithm does not have a parameter correction estimator, its estimation accuracy is significantly lower than that of the DEKF algorithm and DEAKF algorithm, where there is a case of severe distortion in estimating the state quantity of the yaw rate.

In DEAKF algorithm, the state estimation and parameter estimation are conducted in parallel. In order to investigate the correction ability of the algorithm for the model with inaccurate parameters, the changes of m, I_z and a in the parameter vector in the whole time course are shown in Figs. 7.27, 7.28 and 7.29.

As can be seen from Figs. 7.27, 7.28 and 7.29, the three vehicle parameters to be estimated in the DEAKF algorithm gradually tend to the actual value with the increase of time, although there is still some deviation from the actual value, but it can basically be considered DEAKF The algorithm can compensate well for the impact caused by the error of model parameters, so that the state estimation is based on a more accurate model.

2. slalom test conditions

A more intense maneuver than the double change lane test is the slalom lane change test of halving the time course. The following examines the accuracy of the DEAKF algorithm in the estimation of the state of the vehicle under more extreme operating conditions.

In the vehicle simulation, the driving prototype is used to generate the required vehicle driving path, so that the vehicle travels along the slalom lane change test shown in Fig. 7.4. The entire maneuver lasted 10 s and the sampling time was 0.01 s.

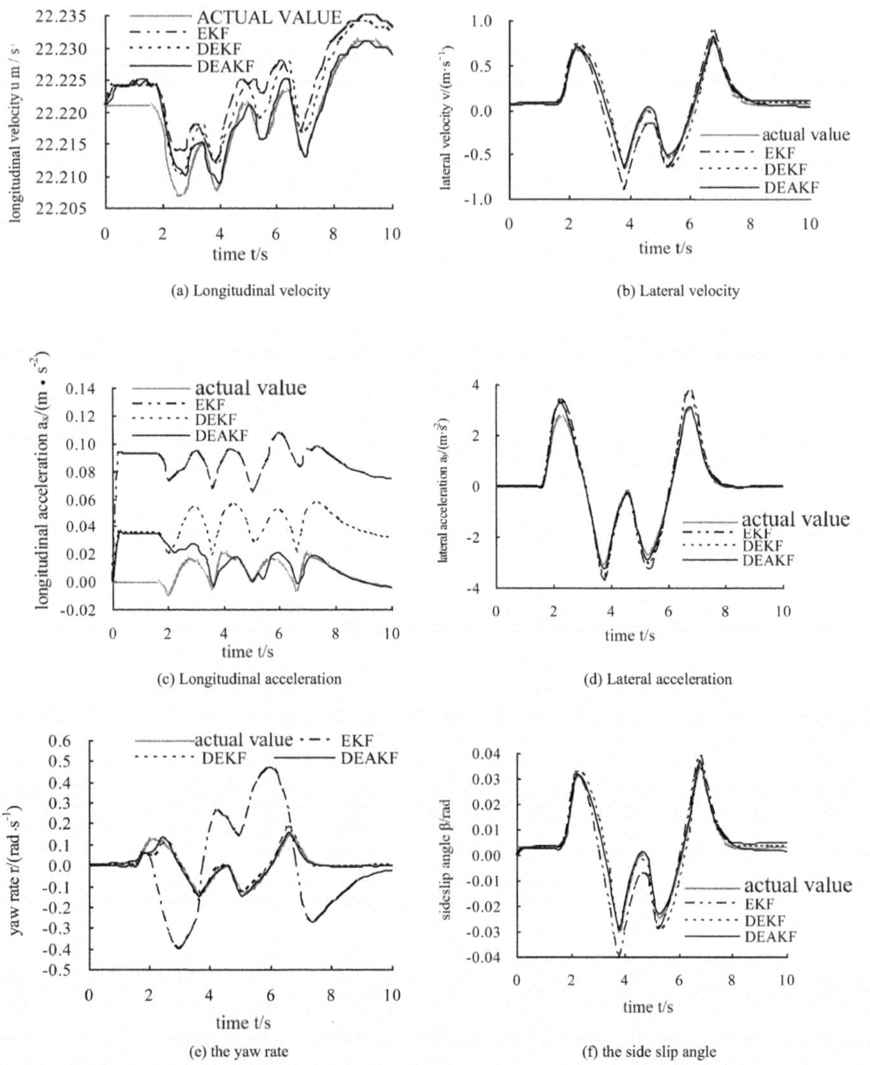

Fig. 7.26 Comparison of three algorithm estimates with virtual experimental values (double change lane test)

The process noise of the state estimation and the measurement noise are set up in the same case as the double lane change test, which will not be repeated here. The observation noise with noise interference can be obtained by synthesizing the observation noise and the measurement noise, as shown in Fig. 7.30.

The initial values of the process noise covariance matrix for state and parameter estimation and the covariance matrix for measuring noise are set to the same as the double lane change test.

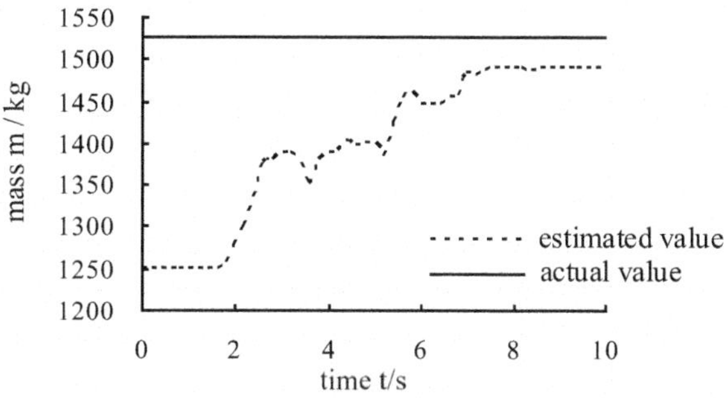

Fig. 7.27 Estimated and actual values of m (double lane change test)

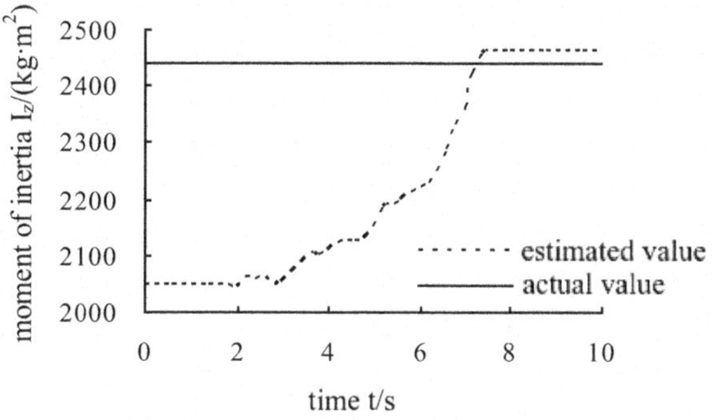

Fig. 7.28 Estimated and actual values of I_z (double lane change test)

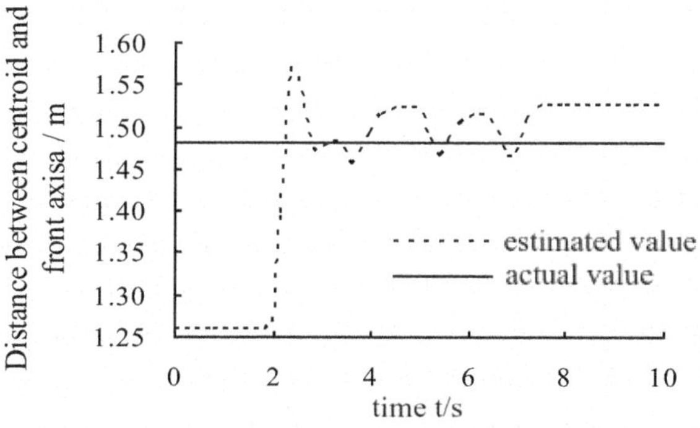

Fig. 7.29 Estimated versus actual values of a (double lane change test)

Figure 7.31 compares the estimated values of six key state quantities u, v, a_x, a_y, r, β with the virtual test values (the actual values in the figure) of the three algorithms of EKF, DEKF, and DEAKF.

As can be seen from Fig. 7.31, under the condition of more intense serpentine line manipulation, the lateral acceleration of the vehicle reaches about 0.8 g, at which time the vehicle has entered a strong nonlinear region, and DEAKF The algorithm still maintains a high estimation accuracy. In addition, the tracking ability of the DEAKF algorithm is also higher than that of the EKF algorithm and the DEKF algorithm.

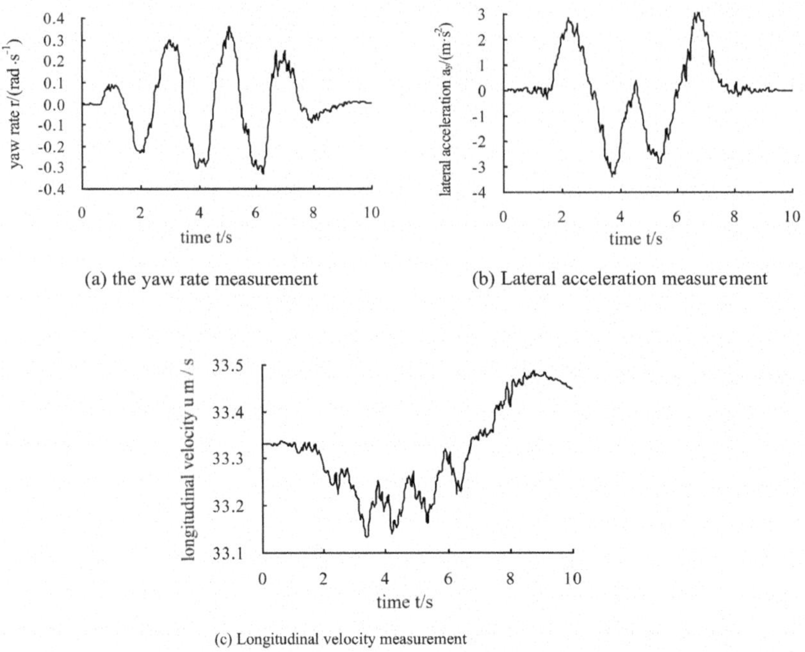

Fig. 7.30 Observational measurements (slalom lane change test) containing noise from time-varying statistical characteristics

Figures 7.32, 7.33 and 7.34 shows the change of the parameter estimation m, I_z, a in the entire slalom test.

From Figs. 7.32, 7.33 and 7.34, it can be seen under the condition of more intense operation of the slalom test, the estimated value of the structural parameters of the vehicle can also quickly converge to the real value. But as the manipulation increases, so does the oscillation of the parameter estimate in the initial stages, as is evident from the comparison of Figs. 7.34 and 7.29.

In order to quantitatively compare the estimation accuracy of the three algorithms, MAE and RMSE of the estimated value relative to the virtual test value are given, as shown in Table 7.6 shown.

From the two commonly used error analysis indicators in Table 7.6, it can be more intuitively seen that the estimation accuracy of the DEAKF algorithm proposed in this chapter is significantly higher than that of the first two algorithms. In addition, due to the relatively high complexity of the path and the shorter manipulation time, the estimation error of the slalom lane change test is higher than that of the double change lane test.

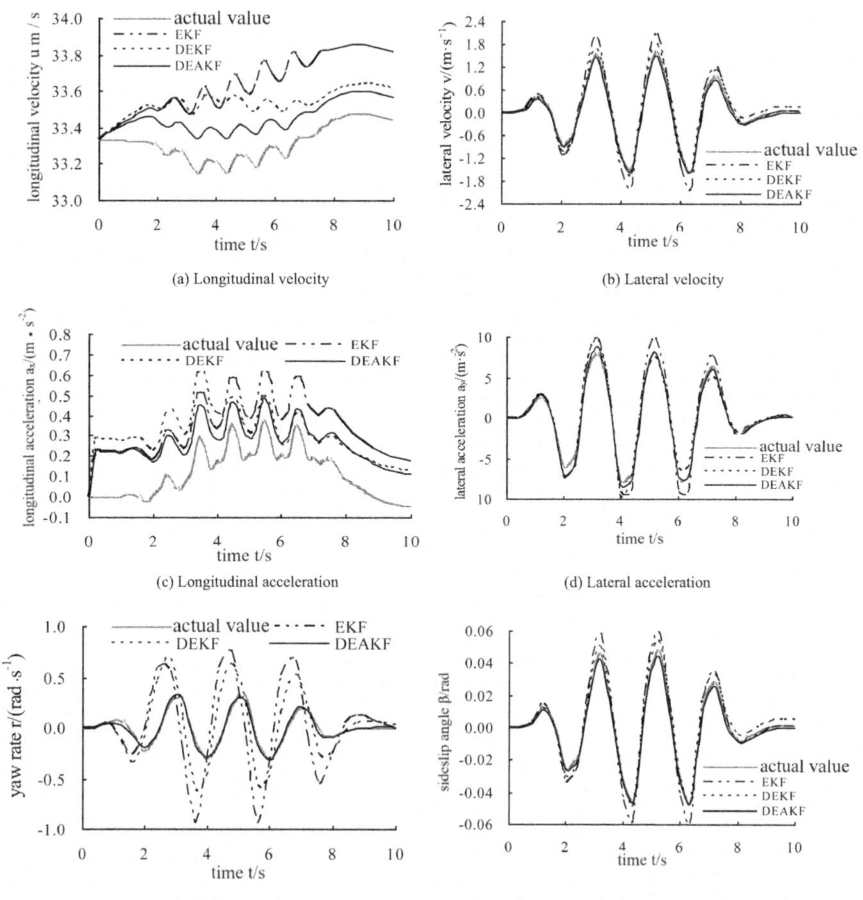

(a) Longitudinal velocity

(b) Lateral velocity

(c) Longitudinal acceleration

(d) Lateral acceleration

(e) the yaw rate

(f) the side slip angle

Fig. 7.31 Comparison of three algorithm estimates with virtual experimental values (slalom test)

Fig. 7.32 Estimated and actual values of parameter m (slalom lane change test)

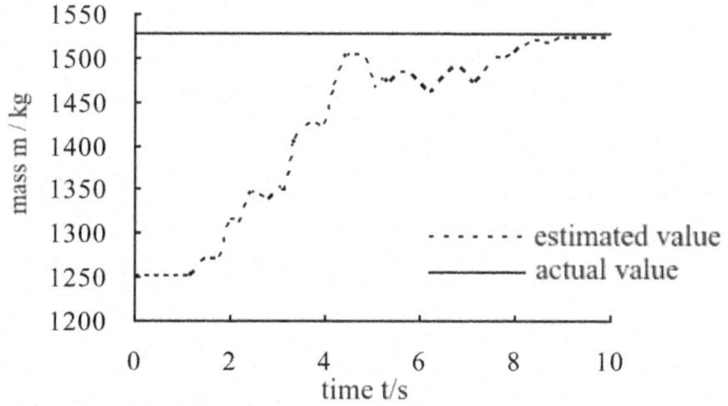

Fig. 7.33 Estimated and actual values of parameter I_z (slalom test)

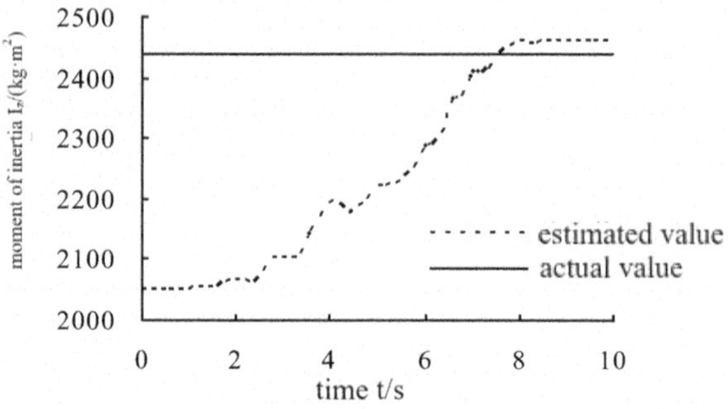

Fig. 7.34 Estimated and actual values of parameter *a* (slalom test)

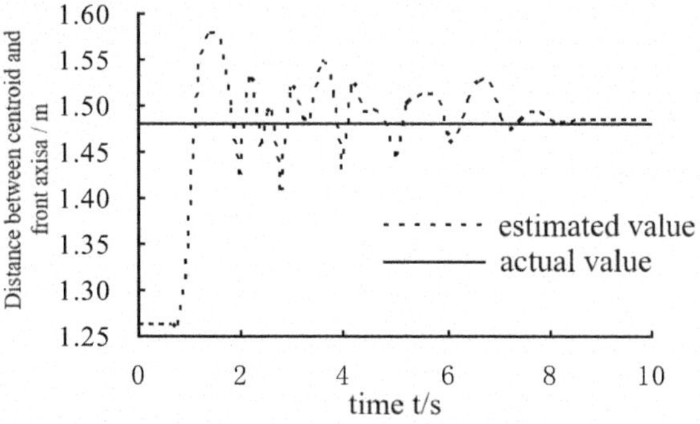

Table 7.6 MAE indicators and RMSE indicators for the three algorithms

Appraise index	States	Double lane change test			Slalom test		
		EKF	DEKF	DEAKF	EKF	DEKF	DEAKF
MAE	u/(m/s)	0.00447	0.0032	0.00147	0.324	0.220	0.140
	v/(m/s)	0.0702	0.0448	0.0274	0.183	0.101	0.0481
	a_x/ (m/s^2)	0.0798	0.0336	0.0111	0.228	0.2	0.161
	a_y/ (m/s^2)	0.227	0.188	0.156	0.834	0.647	0.348
	r/(rad/s)	0.188	0.0129	0.00835	0.309	0.207	0.0183
	β/rad	0.00312	0.00202	0.00124	0.0052	0.00289	0.00143
RMSE	u/(m/s)	0.0047	0.00323	0.002	0.348	0.236	0.144
	v/(m/s)	0.101	0.0677	0.0306	0.248	0.12	0.0511
	a_x/(m/s^2)	0.0803	0.034	0.0172	0.23	0.217	0.166
	a_y/(m/s^2)	0.337	0.244	0.15	1.104	0.94	0.505
	r/(rad/s)	0.242	0.0206	0.0139	0.406	0.263	0.0234
	β/rad	0.00455	0.00305	0.00138	0.00694	0.00339	0.00155

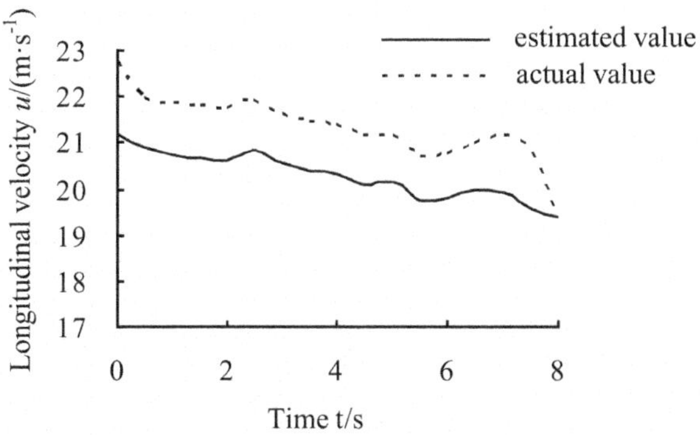

Fig. 7.35 The estimated value and the real vehicle test value of longitudinal velocity

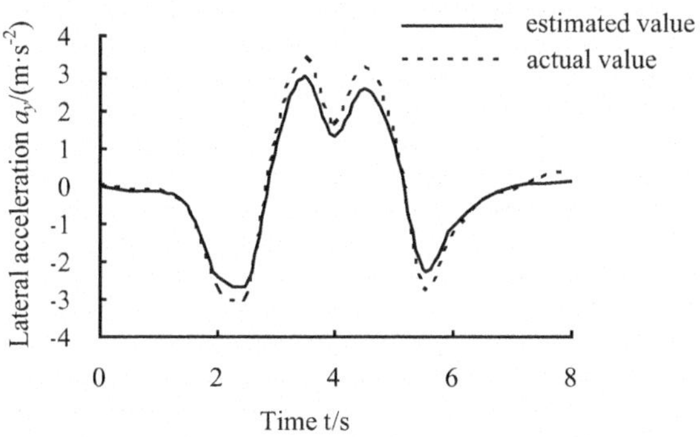

Fig. 7.36 The estimated value and the real vehicle test value of Lateral acceleration

7.3.4 Real Vehicle Test Verification

In order to further verify the effectiveness of the DEAKF algorithm proposed in this chapter in the vehicle state estimation, a double lane change test of a certain vehicle is carried out. The test shall be carried out in accordance with ISO/TR 3888-1975. The test speed is 80 km/h (±3 km/h). The test preparation, test standards and test procedures have been described in the previous chapters and will not be repeated here. A gyroscope is installed on the vehicle to collect the yaw rate and lateral acceleration of the vehicle in real time. The non-contact speed sensor is used to measure the longitudinal speed of the vehicle. In addition, the wheel speed sensor collects the angular velocity of four wheels. The steering wheel angle tester measures the steering wheel angle.

Figures 7.35, 7.36 and 7.37 are the comparison between the estimated value of DEAKF algorithm and the real vehicle test value of three state variables u, a_y, r.

From the above comparison, it can be seen although there is a certain error, the estimated value basically keeps the trend consistent with the test value. There is a big difference between the test value and the estimated value of the longitudinal velocity, because the estimation of the longitudinal velocity in the model is affected by the

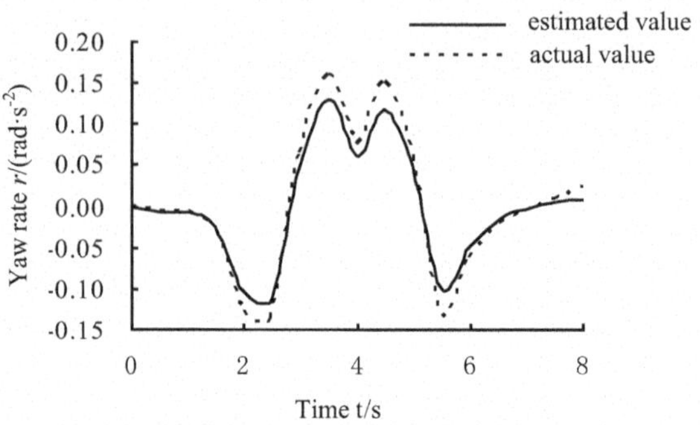

Fig. 7.37 The estimated value and the real vehicle test value of yaw rate

longitudinal acceleration, lateral velocity and yaw rate; In addition, the historical value of the previous time is used in the process of discretization, resulting in the cumulative effect of error.

The Pacejka tire model used in this chapter still has some deviation when simulating the mechanical characteristics of real vehicle tires. In addition, the measurement error and installation position of the sensor are also important reasons for the deviation between the estimated value and the test value.

7.4 Vehicle States Estimation Based on Fuzzy Logic

7.4.1 Fuzzy Kalman Filter and S-AKF Algorithm for State Estimation

1. Fuzzy Kalman Filter (FKF) algorithm

The traditional Kalman filtering algorithm assumes that the observed noise is Gaussian white noise with zero mean in the filter calculation, and the observed noise covariance matrix R is known and fixed. During the driving of the vehicle, if it encounters certain interference, it will have a certain impact on the sensor; At this time, the noise level of the sensor will change, if the fixed observation noise covariance matrix R is still used at this time, it will affect the accuracy of the Kalman filter. Here, a fuzzy controller is established to adjust the observed noise covariance matrix online to increase the robustness of the Kalman filter.

The estimated value of the observed noise covariance at time k is defined as $\hat{R}_k$, and is defined

$$\hat{R}_k = \alpha \hat{R}_{k-1} \tag{7.96}$$

where, α is the adaptive adjustment factor.

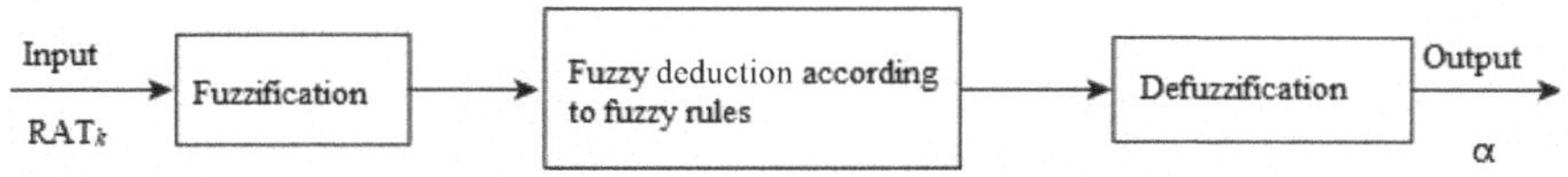

Fig. 7.38 Fuzzy_*R* fuzzy controller

The ratio of the actual variance to the theoretical variance of the innovation at time k is defined as

$$\mathrm{RAT}_k = \mathrm{Act}_k/\mathrm{The}_k \tag{7.97}$$

where $\mathrm{Act}_k = \tilde{z}_{k/k-1}\tilde{z}_{k/k-1}^{\mathrm{T}}$, is the actual variance of innovation, $\mathrm{The}_k = H_k P_{k/k-1} H_k^{\mathrm{T}} + R_k$ is the theoretical variance of innovation.

2. Design of Adaptive Fuzzy Controller for Observation Noise Covariance Matrix

In this section, the ratio of the actual variance to the theoretical variance of the innovation is selected as the input of the fuzzy controller, and the adaptive adjustment factor α is the output. Theoretically, under the condition of optimal state estimation, the theoretical variance of the innovation should be approximately equal to its actual variance. If the observed noise level changes significantly, this will cause the theoretical variance to deviate from the actual variance. At this time, the theoretical variance needs to be adjusted to bring it back close to the actual variance. According to the above idea, the ratio RAT_k of the actual variance to the theoretical variance of the innovation is used as the input of the fuzzy controller, and the adaptive adjustment factor α is used as the output, and the fuzzy controller is designed with the help of the Fuzzy toolbox of MATLAB.

The composition of the observation noise covariance matrix adaptive fuzzy controller (Fuzzy_*R* fuzzy controller) is shown in Fig. 7.38.

The Fuzzy_*R* fuzzy controller is programmed based on the MATLAB fuzzy toolbox. The specific steps are as follows.

(1) The fuzzy subset of the input is defined as: L_1 indicates less than 1; E_1 indicates approximately equal to 1; M_1 indicates greater than 1. The fuzzy subset of the output is defined as the same.
(2) The membership function of the input and output adopts the triangle membership function, as shown in Figs. 7.39 and 7.40 shown.

(3) The fuzzy logic rules are

$$\begin{aligned}&\text{if } \mathrm{RAT}_k \in L_1 \text{ then } \alpha \in L_1\\&\text{if } \mathrm{RAT}_k \in E_1 \text{ then } \alpha \in E_1\\&\text{if } \mathrm{RAT}_k \in M_1 \text{ then } \alpha \in M_1\end{aligned}$$

(4) Defuzzification adopts the centroid method

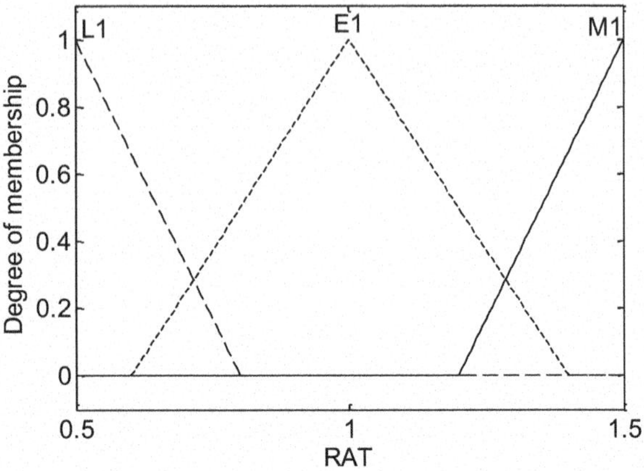

Fig. 7.39 Input membership function

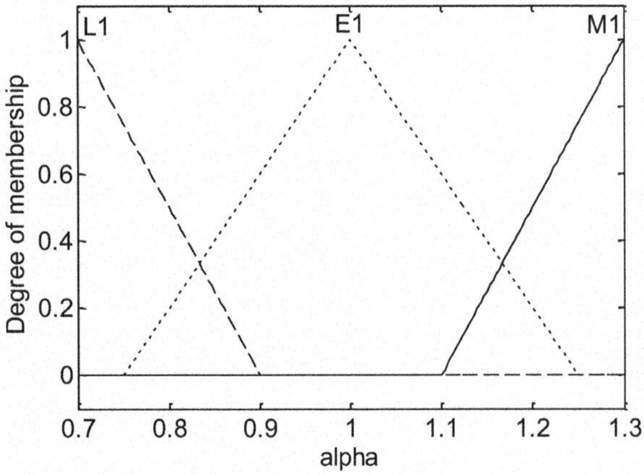

Fig. 7.40 Output membership function

3. Program implementation of FKF combined with S-AKF [15]

In order to increase the robustness of the algorithm, while using the fuzzy controller to estimate the observation noise covariance matrix online, the S-AKF algorithm is added, so that the estimation error covariance matrix is adjusted in real time, which increases the universality of the algorithm. Considering that the fuzzy controller is developed based on Simulink, and the Kalman filter program is more intuitive to implement in the general M language, so the mixed programming of M file and Simulink is used to realize the above algorithm.

Simulink simulation environment is diagram type, program readable, flowsheet structure clear, modeling rapidly, but the implementation cycle such as complicated operation, M code programming environment execution efficiency is high, easy to modify, and more conform to the habits of thinking, but the modeling efficiency is low, need to transfer between processing parameters and assignment, program structure is complex. This section combines the advantages of the two, the main structure of the program is written using M files, and the fuzzy controller is built using Simulink block diagrams and called by the main program. The specific program structure is shown in Fig. 7.41.

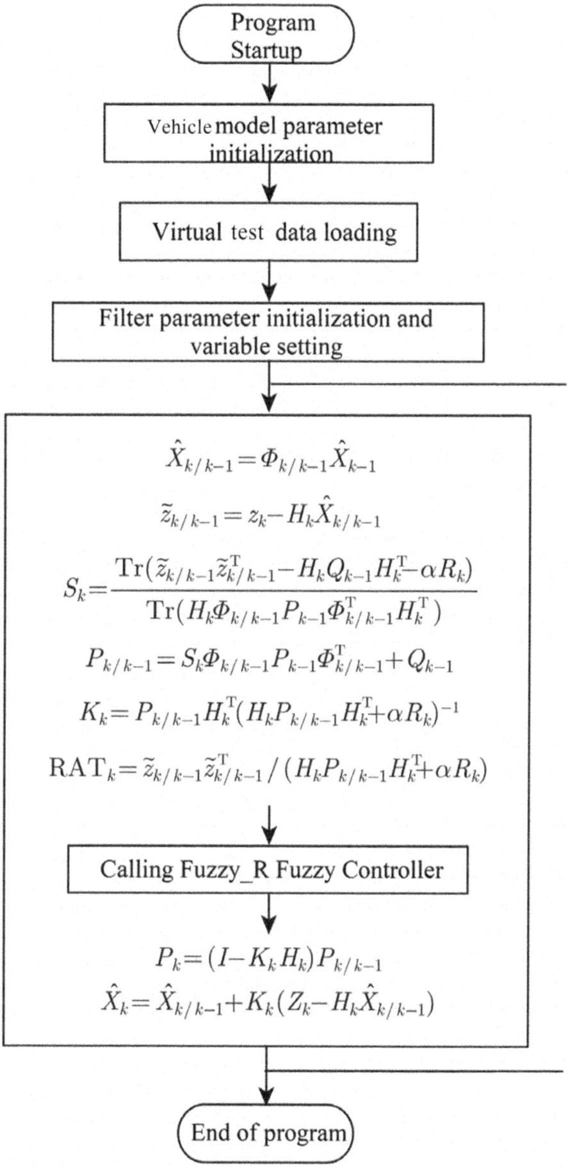

Fig. 7.41 Mixed programming flow of FKF and S-AKF

7.4.2 Virtual Test Verification

In order to verify the correctness of the above S-AKF and FKF algorithms, the vehicle model is established in the software ADAMS/CAR, and in order to simulate the vehicle handling response under extreme working conditions, the driving prototype is used to generate the double lane change path of the vehicle in the simulation, as shown in Fig. 7.42. For the simulation of double lane change, the parameters of the vehicle are as follows:$m = 1528\,\text{kg}$, $I_z = 2440\,\text{kg} \cdot \text{m}^2$, $a = 1.48\,\text{m}$, $b = 1.08\,\text{m}$, $k_1 = -226{,}180\,\text{N/rad}$, $k_2 = -294{,}390\,\text{N/rad}$. The test speed was 80 km/h, and the whole process lasted 10 s.

1. Verification and analysis of S-AKF algorithm

Derive the observations of the virtual experiment from ADAMS and add Gaussian white noise a_y with a mean of zero and a variance of 0.001 to the observations. Kalman filter parameter settings: $Q_k = I_{2\times2}$, $R_k = 0.001$. According to the above conditions and filter parameter settings, the S-AKF algorithm is used to estimate the yaw rate and the sideslip angle of the vehicle.

[1] Kalman filter (KF) state estimation without any adaptive algorithm

As can be seen from Figs. 7.43 and 7.44, the estimation result of the Kalman filter without any adaptive algorithm is closer to the virtual experimental value, and the KF estimated value can better track the virtual test value. However, compared with the sideslip angle, the KF estimate of the yaw rate shows a certain error near the peak value.

[2] State estimation of the S-AKF algorithm

Compared with the above results, Fig. 7.45 compared to Fig. 7.43, the estimation effect of the yaw rate is better, and the estimation error is relatively reduced.

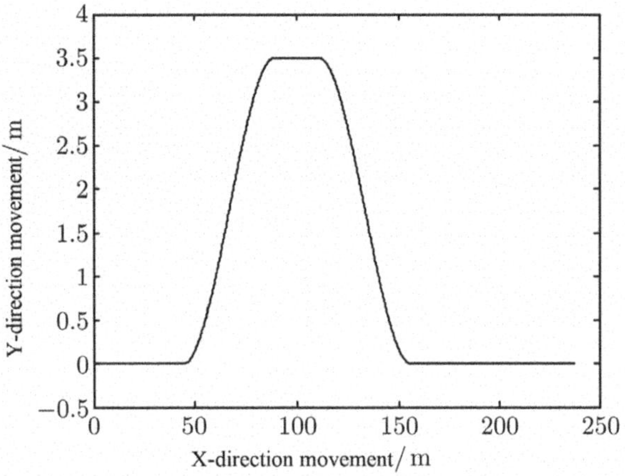

Fig. 7.42 Double lane change driving path

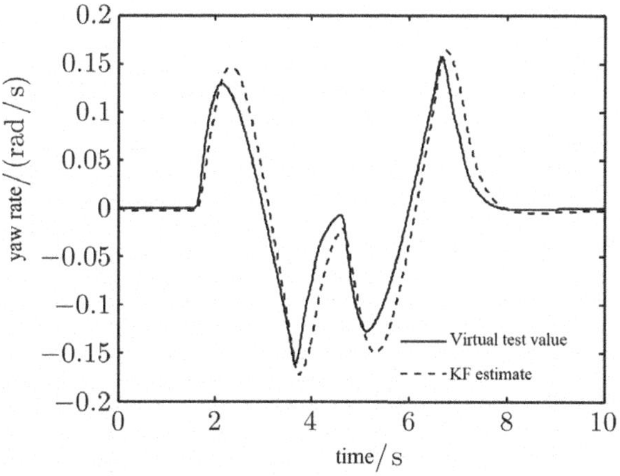

Fig. 7.43 Estimation of the yaw rate

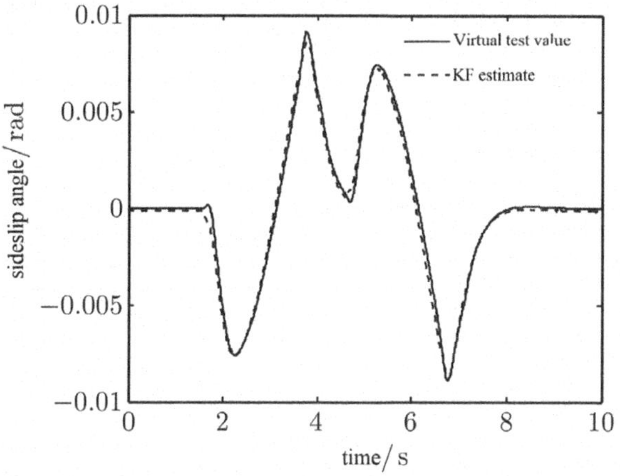

Fig. 7.44 Estimation of sideslip angle

From Figs. 7.44 and 7.46, the estimated value of the sideslip angle did not change significantly.

At the same time, it is found in the simulation that a similar effect of the S-AKF algorithm can be achieved by changing the system process noise covariance matrix Q_k. Originally $Q_k = I_{2\times 2}$, modify it to the following form:

$$Q_{k1} = \begin{bmatrix} 1 & 0 \\ 0 & 10^{-4} \end{bmatrix} \tag{7.98}$$

When the process noise covariance matrix is Q_{k1}, the KF estimation without the adaptive algorithm is shown in Figs. 7.47 and 7.48.

The process noise covariance in the second state equation where the sideslip angle is the element of the right lower corner in the Eq. (7.98). If the value is about 10^{-4}

Fig. 7.45 S-AKF estimate of yaw rate

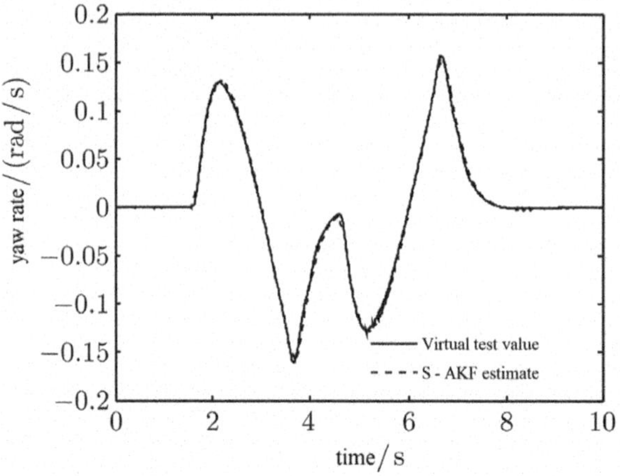

Fig. 7.46 S-AKF estimate of the sideslip angle

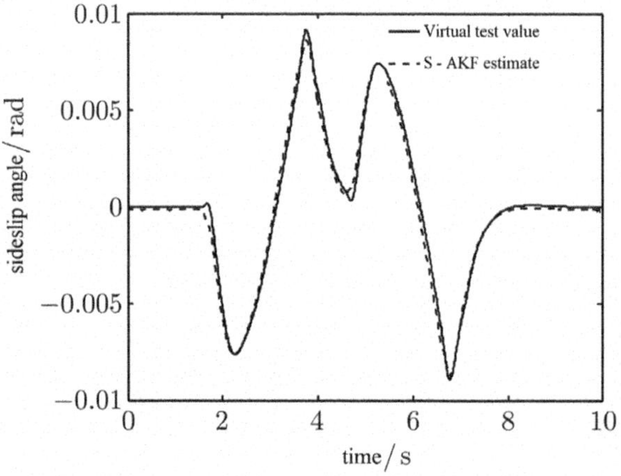

Fig. 7.47 Estimated value of yaw rate (Q_{k1})

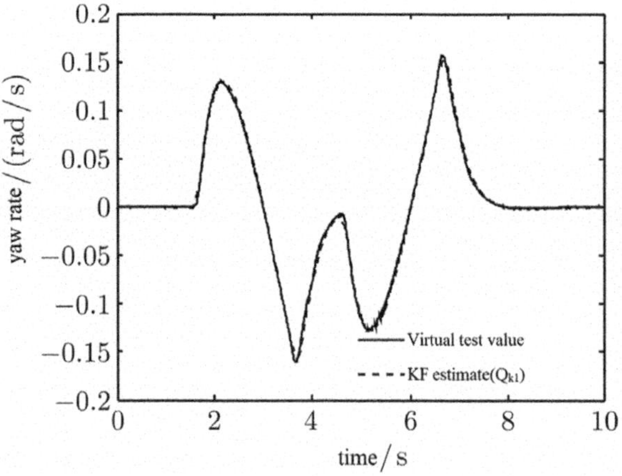

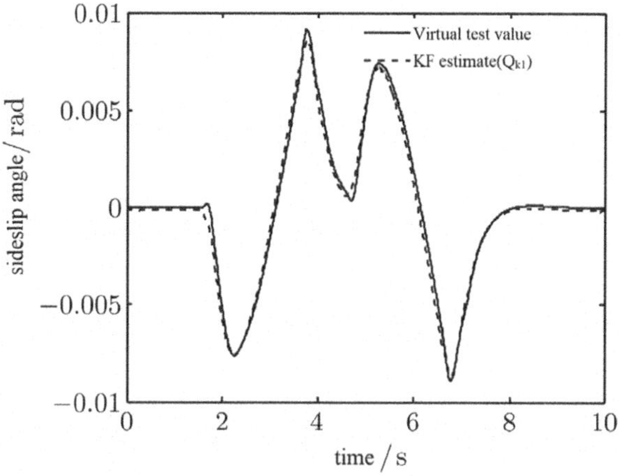

Fig. 7.48 Estimated value of the sideslip angle (Q_{k1})

or smaller than the left upper corner element "1", it can achieve the same effect as S-AKF when the process noise is not accurately known. Conversely, S-AKF can reduce the error of the estimate and improve the robustness of the process noise in the case of inaccurate statistical characteristics of the process noise.

[3] Comparison of indicators of each algorithm

It can be seen from the previous simulation that the S-AKF algorithm reduces the error of the state estimation by simply weighting the covariance matrix of the estimation error to achieve the purpose of self-adaptation. In order to quantitatively compare the performance of each algorithm, the mean absolute error (MAE) and root mean square error (RMSE) of filtering under various conditions are compared, as shown in Table 7.7.

In Table 7.7, the column "KF" indicates that the KF algorithm is adopted and the process noise covariance matrix is Q_k; "KF(Q_{k1})" column indicates that the KF algorithm is adopted but the process noise covariance matrix is Q_{k1}; The column "S-AKF" indicates that the S-AKF algorithm is used and the process noise covariance matrix is Q_k, and the other parameter settings are the same. As can be seen from the data in the table, MAE and RMSE indexes of the two states of S-AKF algorithm and "KF (Q_{k1})" are close, and MAE and RMSE indexes of yaw rate are far better than the estimated result indexes of "KF", while the indexes of sideslip angle are slightly larger than those of "KF", but there is little difference. It can be explained

Table 7.7 Comparison of three types of algorith indicators

Algorithm	MAE indicator		RMSE indicator	
	Yaw rate	Sideslip angle	Yaw rate	Sideslip angle
KF	0.0181	2.9636×10^{-4}	0.0249	3.8950×10^{-4}
KF (Q_{k1})	0.0022	3.1613×10^{-4}	0.0031	4.1203×10^{-4}
S-AKF	0.0023	3.1575×10^{-4}	0.0033	4.1150×10^{-4}

that the S-AKF algorithm can reduce the estimated error overall when the process noise covariance matrix is unknown.

2. Verification and analysis of FKF + S-AKF algorithm

In Sect. 7.4.1, a fuzzy controller is designed to provide an online estimate of the covariance matrix of the observed noise, and this section adds the virtual test values a_y to the time-varying Gaussian white noise, as shown in Table 7.8, the obtained observation of lateral acceleration with time-varying noise is shown in Fig. 7.49. Kalman Filter Parameter Settings: $Q_k = I_{3\times3}$, $R_0 = 0.001$.

Before the adaptive algorithm is added, the ordinary KF algorithm is used for state estimation, and the estimation results of yaw rate and sideslip angle obtained are shown in Figs. 7.50 and 7.51 respectively.

As can be seen in Figs. 7.50 and 7.51, due to the influence of noise, the errors between the estimated yaw rate and the sideslip angle are larger than that in Figs. 7.43 and 7.44.

The FKF algorithm and the FKF + S-AKF algorithm are applied to the state estimation respectively, and the comparison of different estimation results are obtained is shown in Figs. 7.52 and 7.53.

It can be seen from Figs. 7.52 and 7.53 that when the FKF algorithm is used alone, the estimated value of the sideslip angle can smoothly track the results of the virtual experimental value; However, there is still a certain error in the estimation of yaw rate; Because the estimation accuracy of the states is not only related to the observed noise, but also to the process noise, and the process noise covariance matrix is Q_k

Table 7.8 Time-varying noise variance of observed values

Sampling time/s	0–2	2–4	4–6	6–8	8–10
Noise variance	0.001	0.01	0.1	0.05	0.001

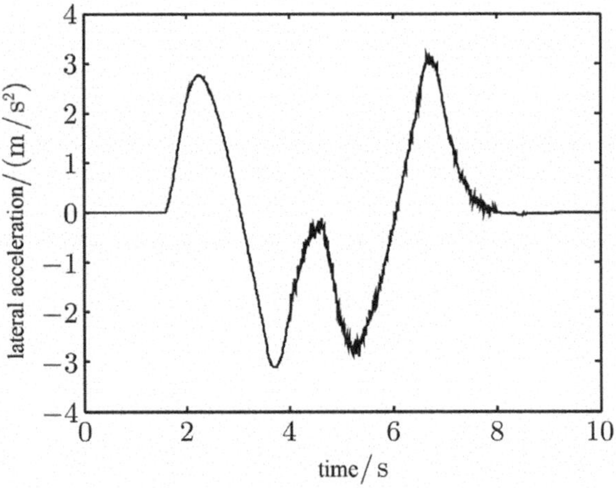

Fig. 7.49 Observation of lateral acceleration with time-varying noise

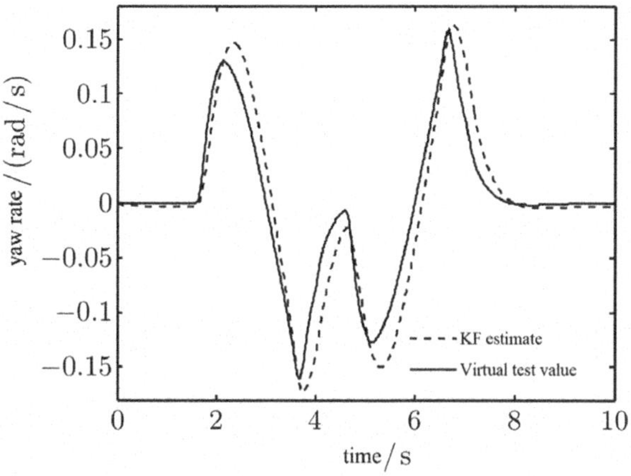

Fig. 7.50 Estimation of yaw rate without adaptive algorithm

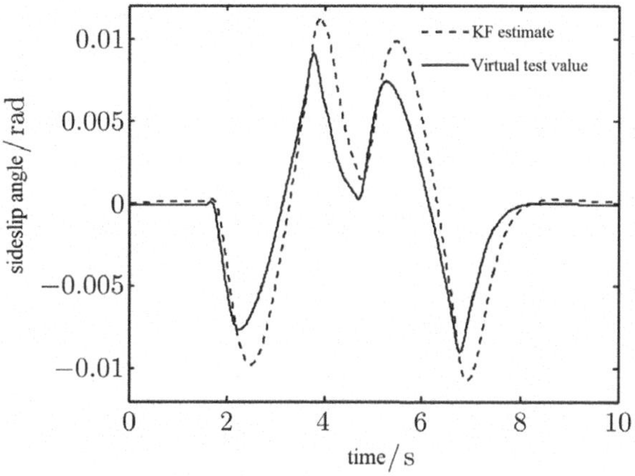

Fig. 7.51 Estimation of sideslip angle without adaptive algorithm

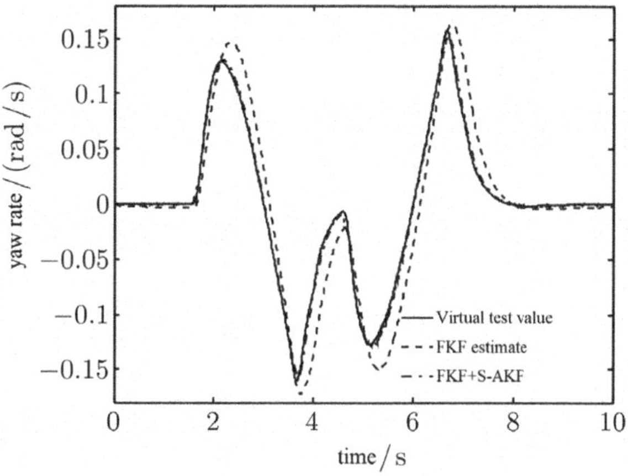

Fig. 7.52 Yaw rate estimation of different adaptive algorithms

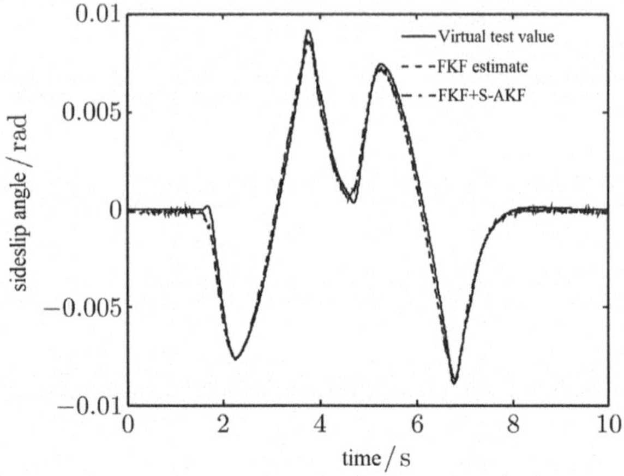

Fig. 7.53 Estimation of sideslip angle of different adaptive algorithms

in the simulation, it can be seen from the simulation results of the S-AKF algorithm that the inaccurate process noise covariance matrix leads to the estimation error of the yaw rate.

When S-AKF is added into the algorithm, the estimation results of FKF + S-AKF algorithm are greatly improved compared with the previous algorithm, and the estimated values of the yaw rate and the sideslip angle can better track the results of the virtual test value, indicating that the addition of S-AKF can make up for the shortcomings of the FKF algorithm, as to obtain better robustness and estimation accuracy.

Similarly, in order to quantitatively compare the performance of each algorithm, the mean absolute error (MAE) and root mean square error (RMSE) of the estimated values of each algorithm relative to the actual value are given, as shown in Table 7.9.

As can be seen from Table 7.9, MAE and RSME indexes of yaw rate gradually improved with the improvement of the algorithm; The MAE and RSME index of the sideslip angle show a downward trend, although the FKF + S-AKF corresponding index rebounds slightly relative to FKF, but the difference is very small. From the overall effect of the algorithm, the FKF + S-AKF algorithm is superior and can realize the adaptive filtering.

Table 7.9 Comparison of index of different adaptive algorithms

Algorithm	MAE index		RMSE index	
	Yaw rate	Sideslip angle	Yaw rate	Sideslip angle
KF	0.0187	0.0015	0.0257	0.0021
FKF	0.0181	2.9629×10^{-4}	0.0249	3.8738×10^{-4}
FKF + S-AKF	0.0024	3.1851×10^{-4}	0.0034	4.1053×10^{-4}

7.5 Parallel Estimation of Vehicle States and Parameters Based on EKF and RLS

At present, for the estimator design in the vehicle stability control system, it is generally assumed in the algorithm that vehicle parameters (such as mass, etc.) are known and fixed. However, in the process of vehicle driving, these parameters will change according to the different working conditions, such as the impact of empty and full load of passenger vehicles on the vehicle mass is more obvious. If the initial parameters of the vehicle are not accurate, it will have a great impact on the accuracy of the vehicle dynamics model, which directly affect the working effect of the model-based estimator, and even lead to errors in the estimation of the control logic. This will seriously affect the timing and intensity of intervention of the automotive active safety controller. Therefore, it is very important to consider the uncertainty of the vehicle parameters during the driving process of the vehicle, and only by parallel correction of the initial inaccurate vehicle parameters at the same time as the state estimation can more accurate vehicle driving information be obtained, as to impose more accurate external control on the vehicle, and the estimation-control frame system can maximize its performance.

Based on the above ideas, a parallel estimation algorithm combining EKF and RLS is proposed, which reasonably sets the driving identification conditions and estimation-identification process in the algorithm framework, and it is applied to the vehicle states estimation with initial inaccurate mass parameters, as to realize the parameter adaptation in the process of vehicle state estimation. Virtual tests and real-vehicle tests are carried out to verify the methods proposed in this section can accurately estimate key states of the vehicle such as the sideslip angle, and the identification of mass parameters can quickly converge to the true value.

7.5.1 *Nonlinear Vehicle Dynamic Model*

1. State estimation model

Consider the Bicycle model in which the vehicle has only lateral movement along the Y-axis and yaw movement around the Z-axis, as shown in Fig. 7.54.

The differential equation of vehicle motion is

$$\dot{\omega}_r = 2(aF_{yf}\cos\delta - bF_{yr} + M_{zf} + M_{zr})/I_z \tag{7.99}$$

$$\dot{v}_y = (2F_{yf}\cos\delta + 2F_{yr})/m - v_x\omega_r \tag{7.100}$$

The sideslip angle of vehicle centroid is

$$\beta = \arctan(v_y/v_x) \approx v_y/v_x \tag{7.101}$$

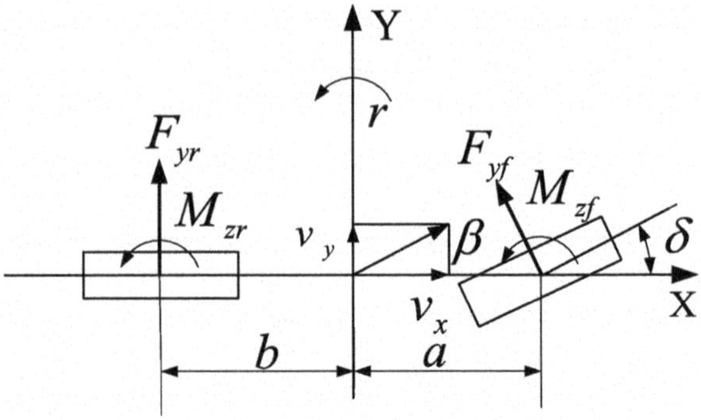

Fig. 7.54 Model of vehicle dynamics

According to Eqs. (7.100) and (7.101), it can be obtained

$$\dot{\beta} = (2F_{yf} \cos \delta + 2F_{yr}) / (mv_x) - \omega_r \tag{7.102}$$

Equations (7.99) and (7.102) are used as state equations.
The measurement equation is

$$a_y = (2F_{yf} \cos \delta + 2F_{yr}) / m \tag{7.103}$$

The sideslip angle of the front and rear wheels is:

$$\alpha_f = -(\delta - \arctan (v_y + a\omega_r) / v_x)$$

$$\alpha_r = \arctan (v_y - b\omega_r) / v_x$$

where, m is vehicle mass, I_z is moment of inertia for the vehicle's axle z, δ for the steering angle of front wheel, α is the slip angle of tire, a_y is lateral acceleration, β is the sideslip angle, ω_r is the yaw rate, a, b is the distance from the centroid to the front and rear axle respectively, v_x, v_y is the longitudinal and lateral velocity at the centroid, F_{yf}, F_{yr} are lateral forces of front and rear wheels, M_{zf}, M_{zr} are the front and rear wheel self aligning torque.

2. Modified Fiala tire model

A modified Fiala model is used for the description of the nonlinear cornering characteristics of tires because it is more accurate in simulating lateral force [16]. The fitting formula for the modified Fiala model is as follows:

$$\begin{cases} F_y = -M_u F_z(\theta - 0.0668\theta^2 - 0.1032\theta^3), & \theta < 1.54 \\ F_y = -M_u F_z \cdot 1.0063, & \theta \geq 1.54 \end{cases}$$

$$\begin{cases} M_z = M_u F_z L_r (0.2749\theta - 0.095\theta^2 - 0.0872\theta^3 + 0.0353\theta^4), & \theta \leq 2 \\ M_z = 0.2258 \cdot \mu F_z e / (\theta - 0.32), & \theta > 2 \end{cases}$$

where, θ is the dimensionless sideslip angle, $\theta = C_\alpha \tan\alpha / (\mu |F_z|)$, C_α is the cornering stiffness of the tire in the linear domain, M_u is the friction coefficient of road, F_z is the vertical load of the tire, L_r is the length of the tire imprint, e is the pneumatic trail.

3. Parameter identification model

Considering that without adding a sensor, that is, without introducing new variables, the Eq. (7.103) is used as the identification equation of m, and the Eq. (7.103) is written as follows according to the RLS identification algorithm:

$$ma_y = 2F_{yf}\cos\delta + 2F_{yr} \tag{7.104}$$

This chapter only considers the identification of parameter m, so it is assumed that the increase and decrease of vehicle mass m occurs at the center of mass position, and the axle load distribution remains unchanged.

7.5.2 EKF and RLS Parallel Algorithm [17]

1. EKF algorithm filtering process

The EKF algorithm is a minimum variance estimation, which expands the nonlinear function of the equation of state and the measurement equation at the optimal estimation point by performing a Taylor series and taking its linear term, thus obtaining an approximate linear model. The specific implementation process of the EKF algorithm in this chapter is as follows.

(1) State and measurement equations:

$$x_k = f(x_{k-1}, k-1) + \Gamma(x_k, k-1)w_{k-1} \tag{7.105}$$

$$z_k = h(x_k, k) + \upsilon_k \tag{7.106}$$

(2) Linearization of nonlinear model:

$$\Phi_k = \frac{\partial f}{\partial x} = \begin{bmatrix} \partial f_1/\partial x_1 & \partial f_1/\partial x_2 \\ \partial f_2/\partial x_1 & \partial f_2/\partial x_2 \end{bmatrix}$$

$$H_k = \partial h/\partial x = \begin{bmatrix} \partial h_1/\partial x_1 & \partial h_1/\partial x_2 \end{bmatrix}$$

where, x is the state vector, and $x = [\omega_r, \beta]^{\mathrm{T}}$, f_1, f_2 is the state equations after discretization, h_1 is a discretized measurement equation, Φ_k, H_k are the Jacobian matrix of the state equations and the measurement equations on the state variables.

(3) Statistical characteristics of noise.

w_k and υ_k are normally distributed with zero mean and variance Q_k and R_k, respectively, and the process noise is assumed to be uncorrelated with the measurement noise. There are

$$E\{w_k\} = 0, E\{\upsilon_k\} = 0, \mathrm{Cov}(w_k w_j) = Q_k\delta_{kj}, \mathrm{Cov}(\upsilon_k \upsilon_j) = R_k\delta_{kj}, \mathrm{Cov}(w_k \upsilon_j) = 0$$

where, δ_{kj} is the Kronecker function, which is

$$\delta_{kj} = \begin{cases} 1, & k = j \\ 0, & k \neq j \end{cases}$$

(4) The recursive process of the algorithm.

The specific recursion process consists of five sets of prediction and correction formulas, which are shown in the left dotted box in Fig. 7.55.

2. RLS algorithm identification process

RLS is a recursive form of least square method, suitable for online parameter identification of dynamic models, and the calculation method does not require any known

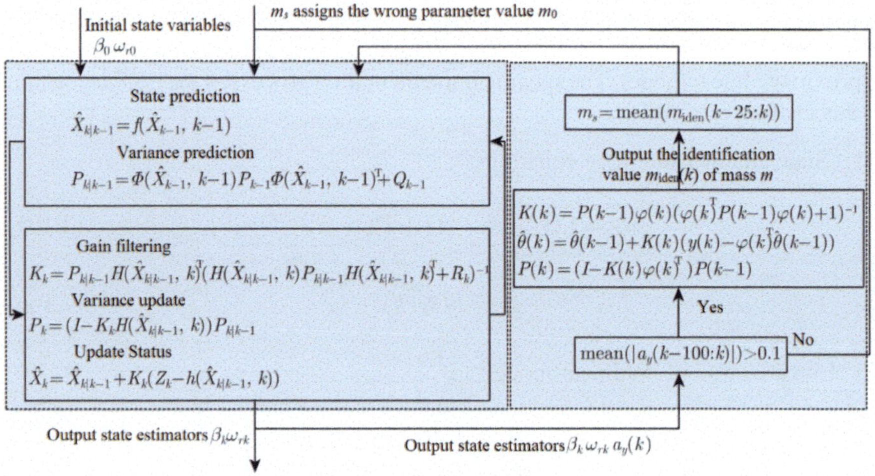

Fig. 7.55 Flow chart of parallel algorithm for EKF and RLS

pre-test statistical characteristics, and the calculation amount is small and the convergence speed is fast. The use of RLS in this section can achieve rapid convergence of mass estimation, and in the parallel estimation process, the computational burden is basically not increased, which can meet the requirements of the control system for real-time and accuracy. The specific implementation process of the RIS algorithm is as follows [17]:

(1) Recursive equation for input and output:

$$y(k) + \sum_{i=1}^{n} a_i y(k-i) = \sum_{i=1}^{n} b_i u(k-i) + e(k) \tag{7.107}$$

where, $u(k)$ is the input, $y(k)$ is the output, $e(k)$ is the error caused by measuring noise or modeling, $a_i, b_i (i = 1, 2 \cdots, n)$ are the parameters to be identified.

The parameters to be identified are denoted in vector form:

$$\theta = (a_1, a_2, \cdots a_n, b_1, b_2, \cdots b_n)^{\mathrm{T}}$$

If $\varphi(k) = \left[-y(k-1), \cdots, -y(k-n), -u(k-1), \cdots, -u(k-n)\right]^{\mathrm{T}}$, then $y(k) = \varphi(k)^{\mathrm{T}}\theta + e(k)$. When combined with Eq. (7.104), it can be obtained

$$\theta = m$$

$$\varphi(k) = a_y$$

$$y(k) = 2F_{yf} \cos\delta + 2F_{yr}$$

(2) The recursive process of the algorithm.

The specific recursion process is composed of three formulas: parameter recognition gain, parameter identification update, and identification error update, and the specific formula is shown in the dotted box on the right side of Fig. 7.55.

(3) The initial parameter settings for identity.

The RLS algorithm needs to be implemented by setting the identification error covariance matrix. The size of this initial value affects the convergence time and error of the identification process; If it is set larger, the convergence time will be shortened, but the identification process will have a large error fluctuation; If it is set smaller, the effect is reversed.

Combining the above factors, and since the RLS is estimated in parallel with the EKF, so in this chapter, P_0 is 1.

(4) Condition setting of drive identification.

Combined with the selected identification equation, when the vehicle has no obvious lateral and yaw motion, the input and output of the identification are very small, considering the signal-to-noise ratio of the sensor, in this case, the noise signal will have a great impact on the identification result, so the judgment conditions is set here to exclude the above situation.

In this chapter, the lateral acceleration a_y is the measured value, and the average value of absolute value of 100 points between $k - 100$ and the current moment k, namely $\text{mean}(|a_y(k - 100 : k)|)$, is taken as the judgment value. When it is greater than 0.1, the identification program is driven; otherwise, the identification program is skipped. In this way, the error of identification result can be reduced effectively.

3. Parallel operation of EKF and RLS

The parallel operation implementation steps for EKF and RLS are as follows [17].

(1) State vector and process noise, measurement noise, estimation error covariance matrix initialization; vehicle mass was initialized (the vehicle mass was assigned m_0), and the identification error variance matrix P_0 was assigned 1.
(2) If the driving identification condition is satisfied, that is, when $\text{mean}(|a_y(k - 100 : k)|)$ is greater than 0.1, the state variables with errors are estimated by EKF to identify m. Under the excitation of the state variables with errors, the identification value of m tends to close to the true value. If the drive identification condition is not satisfied, the m_s assignment is unchanged.
(3) When the identified mass m_{iden} data points reach 30, intercept the average of the latest 20 to 30 points, namely $\text{mean}(m_{\text{iden}}(k - 25 : k))$, pass it to the state estimation.
(4) In the above process, after driving the identification procedure, the identified mass m_{iden} begins to move closer to the truth value, driven by the output state estimators β_k and ω_{rk}.
(5) In the transition process from the emergence of this trend to convergence tends to be stable, the identified mass is transmitted to the state estimation, so that the mass transmitted to the state estimate is a gradually changing variable, effectively avoiding the sudden transmission of a recognition value after convergence stabilization, resulting in an instantaneous mutation of the state quantity, and this can be significantly shorten the time for state estimation and parameter identification to converge gradually, and the real-time performance is better.

The specific parallel algorithm flowchart is shown in Fig. 7.55.

7.5.3 Virtual Test Verification and Analysis

In order to test the correctness of the algorithm in this section, a virtual test is conducted on a vehicle in the software ADAMS/Car, and in order to simulate the operation response of the vehicle under extreme working conditions, the slalom path

Fig. 7.56 Slalom test

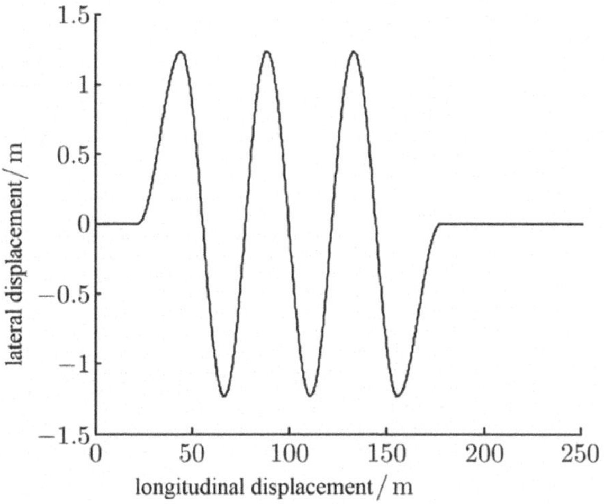

of the vehicle is generated by driving the prototype in the simulation, as shown in Fig. 7.56.

The virtual test speed is 80 km/h, the sampling time is 0.01 s, and the Gaussian white noise with a variance of 0.001 is added to the measured value a_y, $Q_k = 10I_{3\times3}$, $R = 0.001$.

The parameters of the vehicle are as follows: the mass of the vehicle $m_0 = 1528\,\text{kg}$, $I_z = 2440\,\text{kg}\cdot\text{m}^2$, $a = 1.48\,\text{m}$, $b = 1.08\,\text{m}$. Add 300 kg at the center of mass, then the real vehicle mass in this test is $m_{\text{act}} = 1828\,\text{kg}$.

The parameters in the Fiala tire model are as follows: $C_{\alpha f} = -46,792\,\text{N}/\text{rad}$, $C_{\alpha r} = -61,024\,\text{N}/\text{rad}$, $\mu = 1.0$, $L_r = 70\text{mm}$, $e = 31.7\,\text{mm}$.

The algorithm structure shown in Fig. 7.55 is implemented in MATLAB according to the algorithm described in this chapter.

As can be seen from Fig. 7.57, EKF (without RLS for mass adaptation, the same below) and EKF + RLS (parallel estimation of state parameters, the same below). The results of the two algorithms and the virtual test value are very small, indicating that the inaccurate vehicle mass has less influence on the estimation of the yaw rate.

From Fig. 7.58, it can be seen that before about 2.2 s, the estimated value of the sideslip angle has a certain error compared with the virtual test value; the vehicle mass shown in Fig. 7.59 is driven by the identification program at about 1.7 s, the error of the sideslip angle is due to the wrong vehicle mass participating in the iterative operation, after the identification program is driven, the identified mass begins to gradually approach the true value, and the identification value at this time is transmitted to the state estimation for iteration so that the error of the side slip angle is reduced, and the final states and parameters are gradually converged to a higher precision at the same time. And the relative error of the vehicle mass identification result is only 1.2%.

Fig. 7.57 Estimated yaw rate

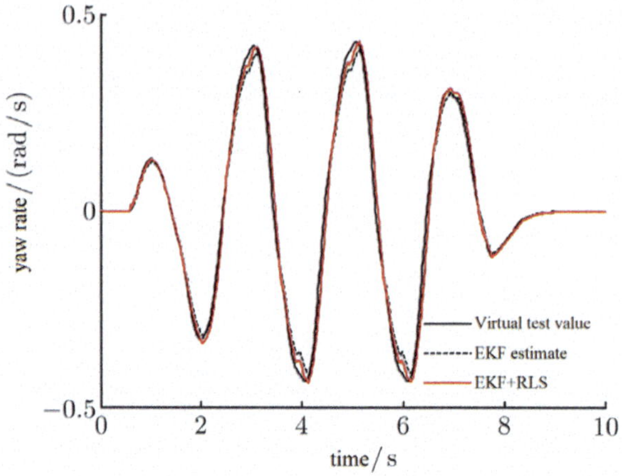

Fig. 7.58 Estimated sideslip angle of the centroid

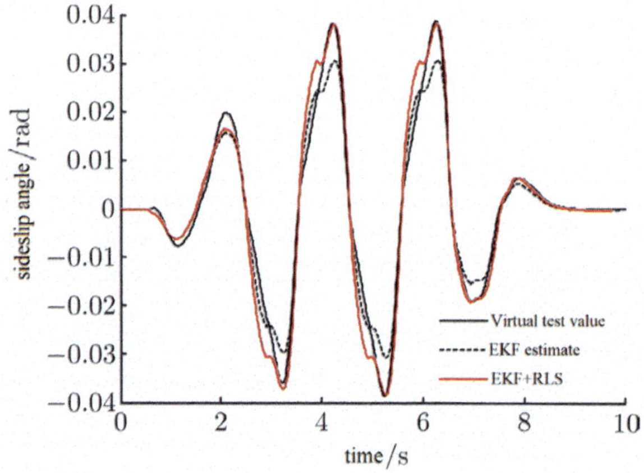

Fig. 7.59 Identification of vehicle mass

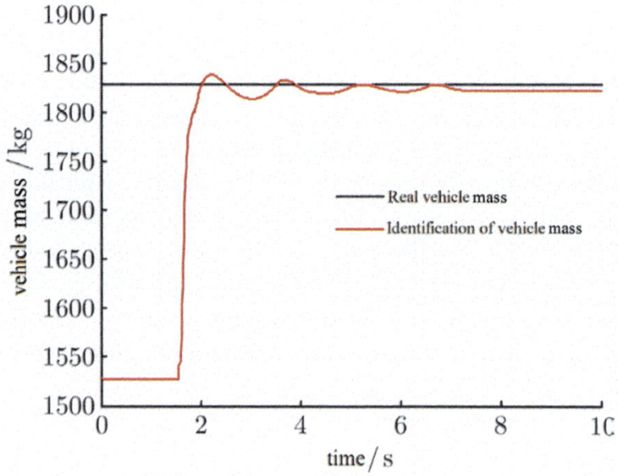

7.5.4 Real Vehicle Test Verification and Analysis

1. Slalom test

The test of a certain type of off-road vehicle was conducted in accordance with the standard of "Vehicle Handling Stability Test Method" (GB/T 6333-2014). A gyroscope is installed on the vehicle to collect the yaw rate and lateral acceleration of the vehicle in real time. A two-axis non-contact photoelectric velocity measurement instrument is used to measure the longitudinal velocity and the sideslip angle of the vehicle, and a steering wheel angle tester is used to measure the steering wheel Angle. The test speed was 65 km/h (±3 km/h). The sampling time is 0.05 s, the process noise covariance matrix $Q = 10I_{3\times3}$, and the measurement noise covariance $R = 0.001$.

The parameters of the vehicle are as follows: the mass of the vehicle $m_0 = 4460\,\text{kg}$, $a = 2.02\,\text{m}$, $b = 1.28\,\text{m}$, $I_z = 12,259\,\text{kg}\cdot\text{m}^2$. Add 200 kg of sandbag weight to the vehicle, then the true mass of the vehicle in this test is $m_{\text{act}} = 4660\,\text{kg}$.

The algorithm flow is implemented in MATLAB according to the arithmetic framework diagram described earlier.

Figures 7.60 and 7.61 are the steering wheel angle and lateral acceleration measured in the real vehicle test, respectively. It can be seen from Fig. 7.62 that the estimation results of the yaw rate of the two algorithms are in good agreement with the measured results, and the estimation accuracy of the EKF + RLS algorithm is slightly higher than that of the EKF method. As can be seen from Fig. 7.63, about 3 s before, the EKF estimate and the EKF + RLS estimate have a certain error from the real vehicle test; the identification value of the vehicle mass of Fig. 7.64 is in a relatively large fluctuation, and the accuracy of the mass identification value transmitted to the state estimate is low. After 3 s, the identification value of the vehicle mass gradually tends to converge smoothly, and the final relative error is 0.95%; This reduces the error of the EKF + RLS estimate, which can effectively track the measured value of the test, while the EKF estimate still shows a large error.

2. Real vehicle test and analysis of double lane change

The test of a cross-country vehicle under double lane change was conducted according to ISO/TC22/TR 3888 standard. The speed is 65 km/h (±3 km/h), and the vehicle parameters and filter parameters are set the same as the parameters of the previous slalom lane change test. Figures 7.65 and 7.66 are the steering wheel angle and lateral acceleration measured in the real vehicle test, respectively.

It is similar to the previous analysis results, the identification value of the vehicle mass before about 4 s is in a relatively large fluctuation, in Figs. 7.67 and 7.68, before 4 s, EKF + RLS estimates show a certain error to the measured values of the test, after 4 s, with the estimated value of the mass gradually approaches the true value, and the error gradually decrease (Fig. 7.69). Figure 7.68 The estimated value of the sideslip angle in Fig. 7.68 has a certain error at the peak, which is mainly caused by the longitudinal speed change in the test, and the deviation of the tire model in

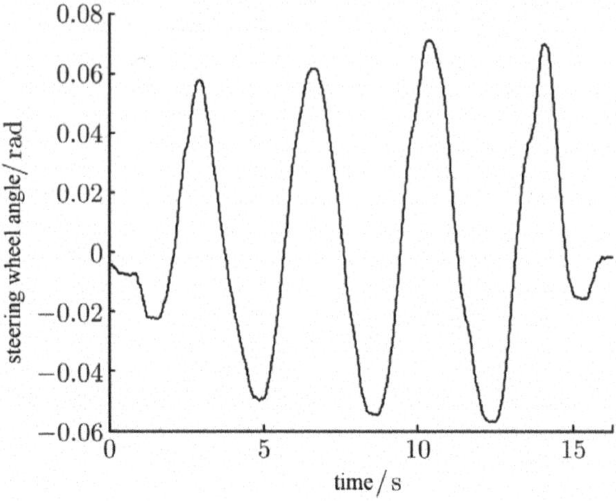

Fig. 7.60 Steering wheel angle (slalom condition)

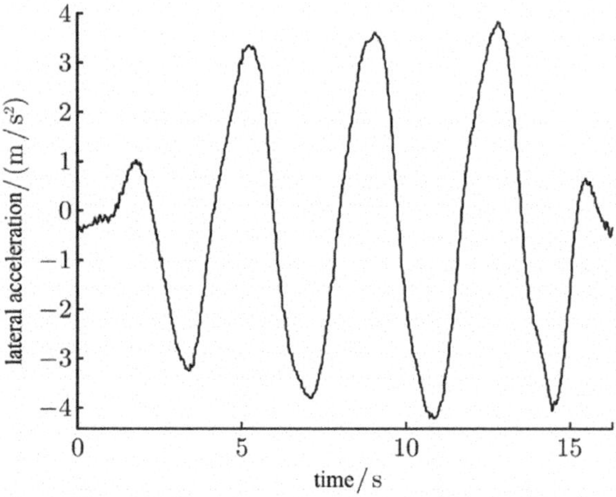

Fig. 7.61 Lateral acceleration measurement (slalom condition)

simulating the real tire mechanical characteristics is also one of the reasons for the estimation error.

From the previous analysis results, it can be seen the EKF process using RLS parallel can realize the adaptation of mass parameter and effectively improve the estimation accuracy.

Compared with the parallel algorithms of state estimation and parameter recognition proposed in the literature [13], the parallel algorithms in this section have better real-time performance in parameter identification; In the literature [13] after 8 s, the parameters converge to near the true value, and the algorithm proposed in this section sets the drive identification conditions, and the mass value transmitted to the state estimation after satisfying the driving conditions is a variable that gradually approaches the true value, reflecting better real-time performance.

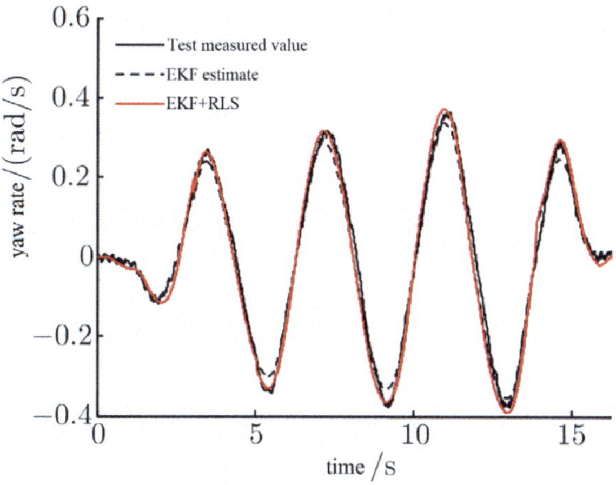

Fig. 7.62 Estimation of the yaw rate (slalom condition)

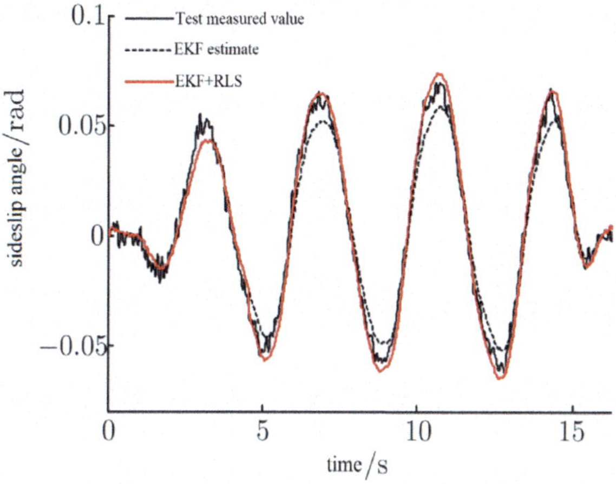

Fig. 7.63 Estimation of the sideslip angle (slalom condition)

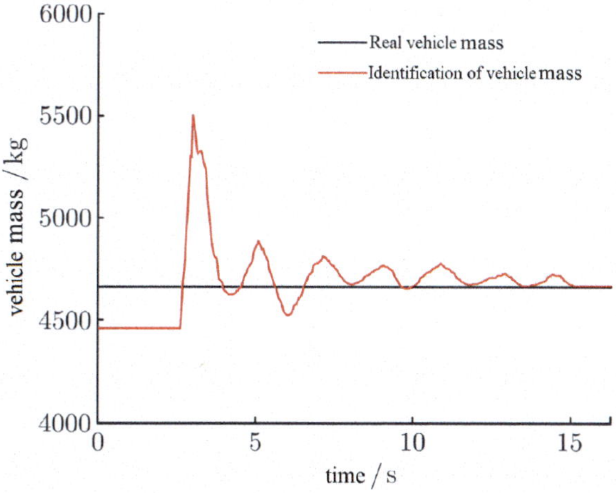

Fig. 7.64 Identification of vehicle mass (slalom condition)

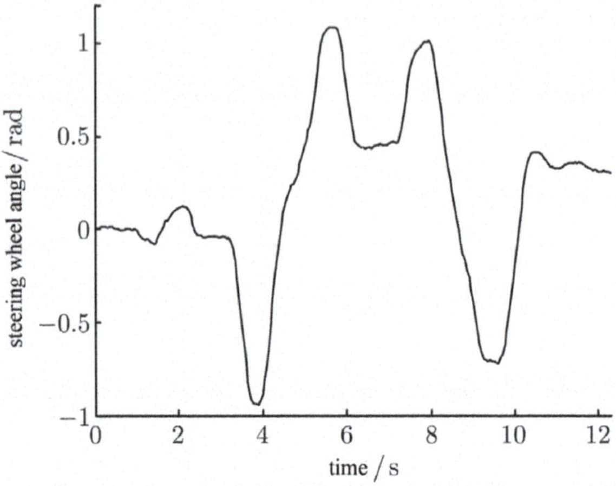

Fig. 7.65 Steering wheel angle (double lane change)

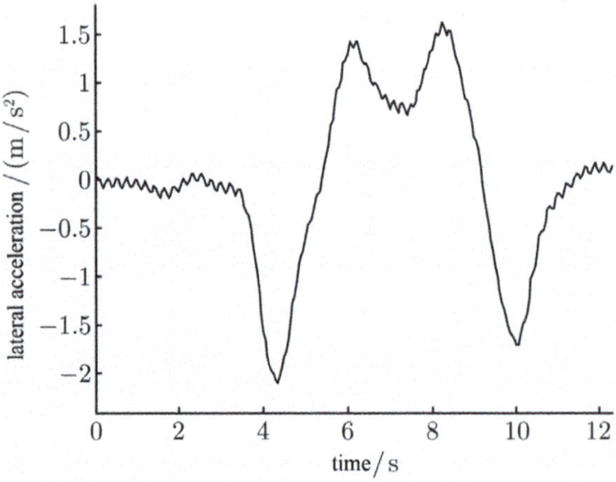

Fig. 7.66 Lateral acceleration measurement (double lane change)

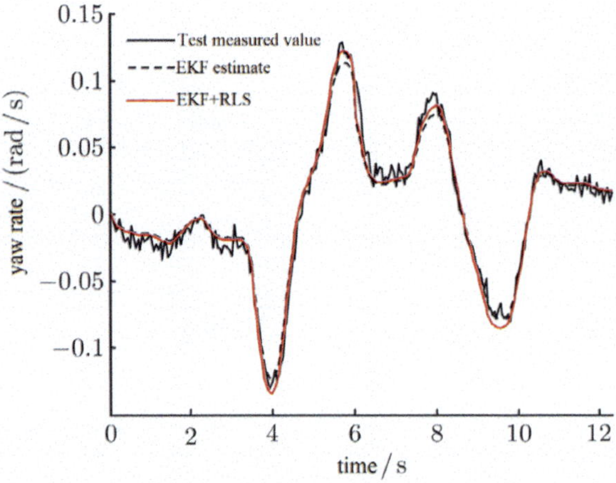

Fig. 7.67 Estimation of the yaw rate (double lane change)

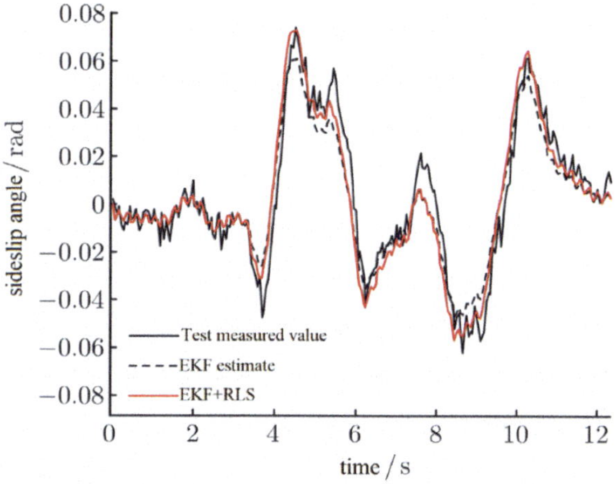

Fig. 7.68 Estimation of the sideslip angle (double lane change)

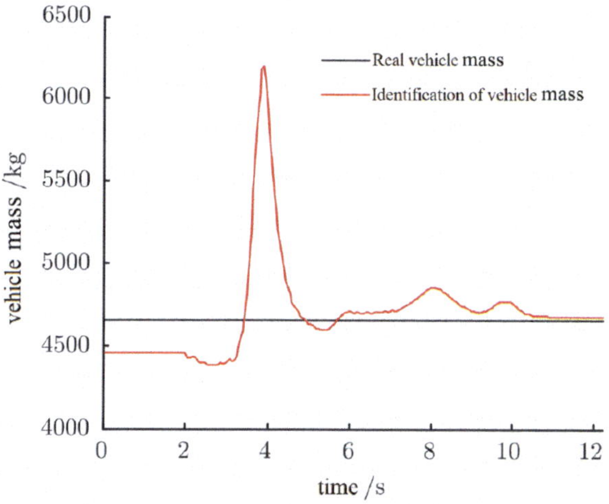

Fig. 7.69 Identification process of vehicle mass (double lane change)

References

1. Guo KH. Vehicle handling dynamics. Changchun: Jilin Science and Technology Press; 1991.
2. ZONG CF. Research on the improvement of fidelity of development driving simulator and the comprehensive evaluation of vehicle handling stability. Changchun: Jilin University; 1998.
3. Matusko J, Petrovic I, Peric N (2003) Application of the RBF neural networks for tire-road friction force estimation. In: IEEE international symposium on industrial electronics, Rio de Janeiro, Brazil, pp 701–706
4. Wang WC. Neural network and its application in automobile engineering. Beijing: Beijing Institute of Technology Press; 1998.
5. Lin F, Zhao YQ. Comparison of vehicle sideslip angle estimation methods. J Nanjing Univ Sci Technol (Nat Sci). 2009;33(1):122–6.
6. Fang KT. Uniform design and uniform design table. Beijing: Science Press; 1994.
7. Song WY, Zang Y. Kalman filter. Beijing: Science Press; 1988.

8. Julier SJ, Uhlmann JK, Durrant-Whyten HF (1995) A new approach for filtering nonlinear system. In: Proceedings of the American control conference, Washington, pp 1628–1632
9. Julier SJ (2002) The scaled unscented transformation. In: Proceedings of American control conference, Anchorage, pp 4555–4559
10. Wan EA, Nelson AT. Kalman filtering and neural networks. New York: John Wiley & Sons; 2001.
11. Zhao YQ, Lin F. Vehicle state estimation based on unscented Kalman filter algorithm. China Mech Eng. 2010;21(5):615–9.
12. Bakker E, Pacejka HB (1989) A new tire model with an application in vehicle dynamics studies. SAE Paper 890087
13. Lin F, Zhao YQ. Vehicle state and parameter estimation based on dual extended adaptive Kalman filter. China Mech Eng. 2009;20(6):750–5.
14. Deng ZL. Self-tuning filtering theory and its application. Harbin: Harbin Institute of Technology Press; 2003.
15. Huang C, Lin F. Vehicle states estimation algorithm based on combination of S-correction adaptive Kalman filter and fuzzy Kalman filter. China Mech Eng. 2013;24(20):2831–5.
16. Guo KH. Principles of vehicle handling dynamics. Nanjing: Phoenix Science Press; 2011.
17. Huang C. Research on vehicle states estimation based on adaptive filter. Nanjing: Nanjing University of Aeronautics and Astronautics; 2013.

Chapter 8
Vehicle Handling Inverse Dynamics

8.1 Optimal Control Theory and Solution Methods

8.1.1 Optimal Control Theory

Optimal control theory, as an extension of the variational method, is a mathematical optimization method for derived control strategies. The method is largely the work of Pontryagin in the USSR and his American collaborator Bellman.

Optimal control theory is a well-established mathematical discipline with numerous applications in science and engineering. Optimal control deals with the class of problems in which a control law is found for a given system to obtain a certain optimality criterion. Optimal control problems include functions of the objective function, state variables, and control variables.

Optimal control can be obtained by using Pontryagin's principle of maximal values (necessary condition, also known as Pontryagin's principle of minimal values) or by solving the Hamilton–Jacobi-Bellman equation (sufficient condition).

Let the system equation of state is $\dot{X} = f(x(t), u(t), t)$, $x(t_0) = x_0$, the objective set is $S = \left\{x(t_f) : x(t_f) \in \mathbf{R}^n, \psi(x(t_f), t_f = 0\right\}$, $x(t)$ is n dimensional state vector, $u(t)$ is p dimensional control vector, $u(t)$ is segmentally continuous on $[t_0, t_f]$, $J = \theta(x(t_f), t_f) + \int f(x(t), u(t), t)\mathrm{d}t$. And subject to certain constraints, seek a permissible control from all available controls $u(t) \in u$, $t \in [t_0, t_f]$, let the system state $x(t)$ starts from a given initial state x_0, moves to the terminal state at $x(t_f) \in S$, $t_f > t_0$, and the performance index J reaches an extreme value. The solution of the problem $u^*(t)$ is the optimal control, $x^*(t)$ is the optimal trajectory, and the performance index The optimal value of $J^* = J(u^*(t))$ is called the optimal performance indicator [1–3].

In general, the optimal control problem can be described as follows, i.e., finding control variables such that Bolza-type performance indicators with generality are minimized.

Y. Zhao and F. Lin, *Vehicle System Dynamics*,
https://doi.org/10.1007/978-981-97-2019-4_8

$$J = \Phi(x(t_0), t_0, x(t_f), t_f) + \int_{t_0}^{t_f} L(x(t), u(t), t)\mathrm{d}t \tag{8.1}$$

where the state variables $x(t) \in \mathbf{R}^n$, initial time t_0 and terminal time t_f (free or fixed) satisfy the kinetic differential equation constraint:

$$\dot{x}(t) = f(x(t), u(t), t),\ t \in [t_0, t_f] \tag{8.2}$$

Boundary Conditions:

$$\Phi(x(t_0), t_0, x(t_f), t_f) = 0 \tag{8.3}$$

And equation and/or inequality constraints:

$$C(x(t), u(t), t) \leq 0 \tag{8.4}$$

Also, consider the control constraints $u \in U \subset \mathbf{R}^m, U = \{u \in \mathbf{R}^m : u_L \leq u \leq u_R\}$, u_L and u_R as the lower and upper bounds of the control variables, respectively.

8.1.2 *Optimal Control Solution Method*

1. Indirect method

The indirect method is a method for converting the optimal control Eqs. (8.1)–(8.4) into a Hamiltonian side value problem (HBVP) based on the Pontryagin principle of minimal values [4].

Introducing the Hamiltonian function

$$H(x(t), u(t), \lambda(t), t) = L(x(t), u(t), t) + \lambda^{\mathrm{T}}(t) f(x(t), u(t), t) \tag{8.5}$$

where,

$$\lambda^{\mathrm{T}}(t) = [\lambda_1(t), \lambda_2(t), \cdots, \lambda_n(t)] \tag{8.6}$$

The accompanying vector or coordination vector called the state $x(t)$. To apply the principle of minimal values, the decomposition boundary condition (8.3) is the initial condition $x(t_0) = x_0$ and the terminal constraint $N(x(t_f), t_f) = 0$. If u^* is a necessary condition for optimal control is the existence of a non-zero vector function $\lambda^*(t)(t \in [t_0, t_f])$, the $u^*(t), \lambda^*(t), x^*(t), t_f^*$ should satisfy the following conditions:

(1) Hamilton system equations

$$x^* = \frac{\partial H(x^*(t), u^*(t), \lambda^*(t), t)}{\partial \lambda} \tag{8.7}$$

$$\lambda^* = \frac{\partial H(x^*(t), u^*(t), \lambda^*(t))}{\partial x} \tag{8.8}$$

$$H(x^*(t), u^*(t), \lambda^*(t), t) = \min_{u \in U} H(x^*(t), u^*(t), \lambda^*(t), t) \tag{8.9}$$

(2) Terminal transect conditions

$$\lambda^*(t_f^*) = \left(\frac{\partial \Phi}{\partial x} + \frac{\partial N^{\mathrm{T}}}{\partial x} v \right)_{t=t_f} \tag{8.10}$$

(3) Terminal constraints

$$N(x(t_f^*), t_f^*) = 0 \tag{8.11}$$

$$\left(H + \frac{\partial \Phi}{\partial t_f} + \frac{\partial N^{\mathrm{T}}}{\partial t_f} v \right)_{t=t_f^*} = 0 \tag{8.12}$$

The indirect method has the following advantages:

(1) High solution accuracy.
(2) The optimal solution satisfies the necessary conditions for first-order optimality.

The indirect method has the following shortcomings:

(1) The process of deriving the optimal solution by the indirect method is very tedious.
(2) The convergence domain is small when solving the two-point edge value problem.

2. Direct method

[1] Parameterization method for control receptors

(1) Converting Lagrangian problems to Mayer problems [5]

Expand the state space into $n+1$ dimensions and introduce a new state variable $\overline{X}_{n+1}$ to satisfy

$$\dot{\overline{X}}_{n+1} = f(\overline{X}(\overline{x}), Z(\overline{x})) \tag{8.13}$$

$$\overline{X}_{n+1}(t_0) = 0 \tag{8.14}$$

Then the integral expression for the performance index of this problem can be reduced to

$$J = \overline{X}_{n+1}(t_f) \tag{8.15}$$

Therefore, by incorporating Eq. (8.13) into the system state equation and Eq. (8.14) into the constraints, it can be reduced to the standard Mayer problem.

(2) Converting the Mayer problem to a finite-dimensional nonlinear programming problem.

(a) Equalizes the interval $\overline{x} \in [t_0, t_f]$ by n to obtain $n + 1$ nodes.
(b) A set of vectors s_j is introduced as an estimate of the control variables at the node $\overline{x}_j (j = 0, 1, \cdots, n - 1)$. The values of the control variables between nodes are obtained by linear interpolation of the values between two adjacent nodes. When the control variables at the nodes $s_j (j = 0, 1, \cdots, n-1)$ are known, the state variables at each node can be obtained by sequential integration when the initial values of the state variables are known, and thus the values of $\overline{X}_{n+1}(t_f)$ and hence the performance index can be obtained. Therefore, it can be considered that the solution of the differential equation and the performance index are only functions of the control variables at each node.

The above discrete process can be mathematically described as

$$t_0 = t_1 < t_2 < \cdots < t_N = t_f \tag{8.16}$$

$$y = (u_1, u_2, \cdots, u_N) \tag{8.17}$$

$$u_k(t) = \phi(t, u_{k1}, u_{k2}, \cdots, u_{kp}),\ k = 1, 2, \cdots, m \tag{8.18}$$

$$u_k(t) = u_{k(i)} + \frac{u_{k(i)} - u_{k(i-1)}}{t_i - t_{i-1}}(t - t_{i-1}),\ t_{i-1} \le t \le t_i \tag{8.19}$$

(3) Direct method with simultaneous discrete state variables and control variables.

The principle of the direct configuration nonlinear programming method is as follows (see literature [6–9] for details). Given a "time" division.

$$0 = \overline{x}_0 < \overline{x}_1 < \cdots < \overline{x}_n = \overline{x}_e \tag{8.20}$$

The corresponding parameter Y is taken as the value of the control variable as well as the state variable at the node.

$$Y = (Z(\overline{x}_0), \cdots, Z(\overline{x}_n), \overline{X}(\overline{x}_0), \cdots, \overline{X}(\overline{x}_n))$$

In the interval $\overline{x}_j \leq \overline{x} < \overline{x}_{j+1}(j = 0, 1, \cdots, n-1)$, the approximate control variable $Z_a(\overline{x}_n)$ is represented by a segmented linear interpolation function.

$$Z_a(\overline{x}) = Z(\overline{x}_j) + \frac{\overline{x} - \overline{x}_j}{\overline{x}_{j+1} - \overline{x}_j}(Z(\overline{x}_{j+1}) - Z(\overline{x}_j)) \tag{8.21}$$

The approximate state variable $\overline{X}_a(\overline{x})$ is represented by a continuously differentiable segmented Ermitian cubic polynomial interpolation function.

$$\overline{X}_a(\overline{x}) = \sum_{k=0}^{3} c_{kj}\left(\frac{\overline{x} - \overline{x}_j}{s_j}\right)^k \tag{8.22}$$

where, $c_{0j} = \overline{X}(\overline{x}_j)$, $c_{1j} = s_j f_j$, $c_{2j} = -3\overline{X}(\overline{x}_j) - 2s_j f_j + 3\overline{X}(\overline{x}_{j+1}) - s_j f_{j+1}$, $s_j = \overline{x}_{j+1} - \overline{x}_j$, $c_{3j} = 2\overline{X}(\overline{x}_j) + s_j f_j - 2\overline{X}(\overline{x}_{j+1}) + s_j f_{j+1}$, $f_j = f(\overline{X}(\overline{x}_j), Z(\overline{x}_j))$.

The derivative of the state variable at the midpoint of each segment is

$$\dot{\overline{X}}_a(\overline{x}_{cj}) = \frac{1}{4s_j}(-6\overline{X}(\overline{x}_j) + 6\overline{X}(\overline{x}_{j+1}) - s_j f_j - s_j f_{j+1}) \tag{8.23}$$

where, $\overline{x}_{cj} = \overline{x}_{j+1/2} = (\overline{x}_j + \overline{x}_{j+1})/2$, $j = 0, 1, \cdots, n-1$.

The approximate state function must satisfy the state equation at the node $\overline{x}_j(j = 0, 1, \cdots, n)$ and the midpoint $\overline{x}_{cj}(j = 0, 1, \cdots, n-1)$, which is satisfied at the node $\overline{x}_j$ as seen by the above configuration.

[2] Matching point method

The method of applying the matching point method is as follows.

(1) Discrete time.

$$t_0 = t_1 < t_2 < \cdots < t_N = t_f \tag{8.24}$$

The state and control variables corresponding to the time points (nodes) are discretized as $(x_1, x_2, \cdots, x_N)$ and $(u_1, u_2, \cdots, u_N)$ respectively.

(2) Approximate state variables and control variables.

The state variables on the subinterval are approximated by a cubic Hermite interpolation polynomial as

$$\begin{cases} x = c_0 + c_1 s + c_2 s^2 + c_3 s^3 \\ h_i = t_i - t_{i-1},\ i = 2, 3, \cdots, N \\ s = (t - t_i)/h_i,\ s \in [0, 1] \end{cases} \tag{8.25}$$

The subinterval $[t_{i-1}, t_i]$ boundary condition is expressed as

$$\begin{cases} x(0) = x_{i-1},\ x(1) = x_i \\ \left.\dfrac{dx}{ds}\right|_{s=0} = \dot{X}_{i-1},\ \left.\dfrac{dx}{ds}\right|_{s=1} = \dot{X}_i \end{cases} \tag{8.26}$$

The segmented linear interpolation function is still used to approximate the control variables, and the k-th value of the control variable at a given moment is denoted by u_k

$$u_k(t) = u_{ki} + \frac{u_{ki} - u_{k(i-1)}}{t_i - t_{i-1}}(t - t_{i-1}),\ t_{i-1} \le t \le t_i \tag{8.27}$$

(3) Defect vector calculation at the mating point.

To ensure that the state variables are better fitted by the polynomial (8.18), the derivatives of the state variables at the interval matching points $\dot{\boldsymbol{X}}_{\mathrm{EOM}}$ and the derivatives of the state variables derived from Eq. (8.18) $\dot{\boldsymbol{X}}_{\mathrm{H}}$ should be equal, i.e., the vector $\boldsymbol{d} = \dot{\boldsymbol{X}}_{\mathrm{EOM}} - \dot{\boldsymbol{X}}_{\mathrm{H}}$ is expected to converge to zero. where, $\boldsymbol{d}$ is the Defect vector.

(4) Solving nonlinear programming (NLP) problems

The design variables for the NLP problem include the state variable x_i at the node, the control variable u_i and the control variable u_{ci} at the mating point, with $\boldsymbol{d} = 0$ and the conditions (8.3), (8.4) as constraints such that a certain performance metric J is minimized.

Figure 8.1 depicts the basic principles of the two direct methods.

$$\begin{cases} \min\limits_{y \in \mathbf{R}^M} F(y) \\ \text{s.t. } g_j(y) \ge 0,\ j = 1, 2, \cdots p \\ h_j(y) = 0,\ j = 1, 2, \cdots l \end{cases} \tag{8.28}$$

where y denotes the design variables, which can be state variables at discrete time points, terminal moments, and control variable F denotes the objective function, $g_j(y)$ and $h_j(y)$ denote the inequality constraint and equation constraint of the nonlinear programming problem, respectively.

Advantages of the direct method.

(1) No need to derive first-order optimality conditions.
(2) The direct method has a wider convergence domain and therefore does not require high accuracy for the initial value estimation.

Disadvantages of the direct method.

(1) There is no guarantee that the obtained NLP solution is the solution of the original optimal control problem.

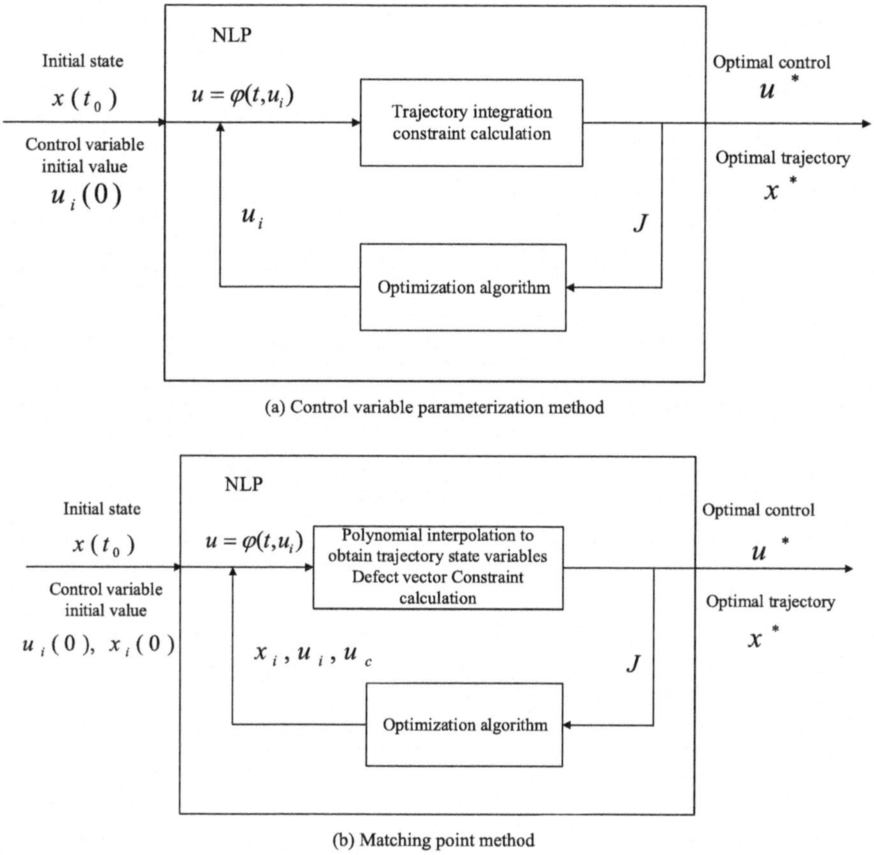

Fig. 8.1 Schematic diagram of the two direct methods

(2) It is easier to converge to a locally optimal solution, i.e., there is a large dependence on the initial guesses. For this reason, some scholars use genetic algorithms for solving vehicle maneuvering inverse dynamics problems.

3. Dynamic planning

In the 1950s, Bellman et al. proposed a relatively novel mathematical method for solving the optimization of multi-stage decision processes, the dynamic programming method.

In mathematics, computer science, economics and bioinformatics, dynamic programming solves complex problems by decomposing them into simpler subproblems. The advantage of dynamic programming methods over traditional methods is that a depth-first search takes less time for problems with overlapping subproblems.

The idea of dynamic programming is relatively simple. In general, to solve a given problem, different parts of the problem (subproblems) need to be solved, and then the solutions of the subproblems are combined to reach a comprehensive solution method. Usually, when using traditional methods, many subproblems are generated and solved many times. The dynamic programming approach seeks to solve each subproblem only once, thus significantly reducing the number of computations: once a given subproblem is solved, it is stored or backed up, and the next time the same problem is encountered, it is solved using the same method and procedure and then stored and backed up. This approach is particularly useful when the input of repeated subproblems grows exponentially.

Dynamic programming methods use all possible ways and select the best solution when studying the problem to be solved. Therefore, we can roughly assume that dynamic programming methods are guaranteed to find the best solution if the scope of the problem to be solved is such that all possible solutions are guaranteed to be feasible and the solution is fast enough.

[1] Algorithm description

If the optimal control problem solved by the first type of direct method is considered as a subproblem consisting of individual time periods $[t_{i-1}, t_i]$, i.e.

$$\begin{cases} \min\limits_{u_1} \min\limits_{u_2} \cdots \min\limits_{u_N} J(u_1, u_2, \cdots, u_N) \\ \text{s.t.} \dot{x} = f(x, u, t) \\ x(t_0) = x_0 \end{cases} \tag{8.29}$$

Then the optimal control problem described by Eqs. (8.13)–(8.16) can be solved directly by using the dynamic programming optimality principle.

[2] Features

(1) Dynamic programming methods have the advantage of simple computational principles, relatively high accuracy and rigorous theoretical support.
(2) The disadvantage of dynamic planning methods is the phenomenon of "dimensional catastrophe".
(3) Variational method

Let the system equations of state be $\dot{X} = f(x(t), u(t), t)$, $J = \theta(x(t_f), t_f) + \int F(x(t), u(t), t)\mathrm{d}t$, and if the initial moments t_0 and $x(t_0)$ are given, $x(t_0) = x_0$, its canonical equations are $\dot{X} = \partial H/\partial\lambda$, $\dot{\lambda} = -\partial H/\partial x$, the control equations are $\partial H/\partial u = 0$, and the boundary conditions and cross-sectional conditions vary depending on the terminal state.

(1) t_f fixed, $x(t_f)$ free time, $x(t_0) = x_0$, $\lambda(t_f) = \partial\theta/\partial x(t_f)$.
(2) When t_f is fixed, $x(t_f)$ is constrained, $x(t_0) = x_0$, $\Phi(x(t_f), t_f) = 0$, $\lambda(t_f) = \partial\theta/\partial x(t_f) + v^{\mathrm{T}}(\partial\psi/\partial x(t_f))$.
(3) When t_f is fixed and $x(t_f)$ is also fixed, $x(t_0) = x_0$ and $x(t_f) = x_f$.
(4) When t_f change, $x(t_f)$ is free, $x(t_0) = x_0$, $\lambda(t_f) = \partial\theta/\partial x(t_f)$.

Solving the canonical equation, the control equation and the boundary conditions jointly, we can find the optimal control $u^*(t)$ and the optimal state trajectory $x^*(t)$ when the performance index reaches the extreme value.

5. Maximum value principle

Let the equations of state of the system be $\dot{X} = f(x(t), u(t), t)$, $J = \theta(x(t_f), t_f) + \int F(x(t), u(t), t)\mathrm{d}t$, $x(t_0) = x_0$, $u(t)$ belongs to the control vectors in some bounded closed set Ω in the vector space of dimension p, and the inequality constraints are $G(x(t), u(t), t) \geq 0$,G are continuous differentiable vector functions of dimension m, $m \leq p$, and the necessary conditions that the system should satisfy for the optimal control to move from the initial state x_0 to the terminal state satisfying the boundary condition $\Phi(x(t_f), t_f) = 0$ and minimizing J are as follows.

(1) The optimal trajectory $x^*(t)$ and the covariance vector satisfy the canonical equations $\dot{X} = \partial H/\partial\lambda$, $\dot{\lambda} = -\partial H/\partial x$.
(2) The expression for the minimum value of the Hamiltonian function corresponding to the optimal control $u^*(t)$ on the optimal trajectory curve $H(x^*, u^*,\lambda^*, t) = \min H(x, u, \lambda, t)$.
(3) Boundary conditions.

   (a) When $x(t_f)$ is unrestricted, $x(t_0) = x_0$, $\lambda(t_f) = \partial\theta/\partial x(t_f)$.
   (b) When the terminal is restricted, i.e.,$\Phi(x(t_f), t_f) = 0$, $x(t_0) = x_0$, $\lambda(t_f) = \partial\theta/\partial x(t_f) + v^{\mathrm{T}}(\partial\Phi/\partial x(t_f))$.
   (c) When $x(t_f)$ is fixed, there are $x(t_0) = x_0$, $x(t_f) = x_f$.
   (d) t_f fixed for constant system H function is constant and t_f not fixed for constant system H function holds zero value.

6. Differential methods

The vehicle system dynamics equations and the endpoint constraint equations are the same as Eqs. (8.3) and (8.4), and the algebraic constraint equation for the desired output $p(t)$ is defined to be satisfied as

$$\omega(x(t), p(t), t) = 0 \tag{8.30}$$

where, $\omega : \mathbf{R}^n \times \mathbf{R}^b \times \mathbf{R} \rightarrow \mathbf{R}^b$ is completely differentiable; expect the outputs $p(t) \in \mathbf{R}^b$, $t_0 \leq t \leq t_f$. Continuously differentiating Eq. (8.23) until an explicit control variable appears, one obtains

$$\Omega(x(t), u(t), p(t), \cdots, t) = 0 \tag{8.31}$$

The basic principle of the differential method is shown in Fig. 8.2.

7. Pseudospectrum method [10–21]

The pseudospectral method, also known as the orthogonal matching point method, is an optimal control method derived from the spectral method for solving fluid dynamics problems. The pioneering work on the orthogonal matching point method for optimal control problems originated in 1979 with the work of Reddien. Some of the first studies in engineering using the orthogonal matching point method can be found in the chemical engineering literature, and more studies in the chemical

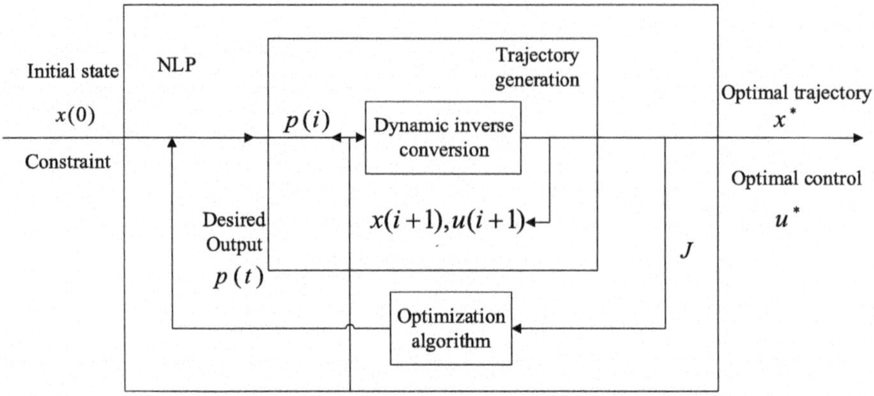

Fig. 8.2 Basic principles of differential methods

Table 8.1 Various pseudospectral method polynomials and types of matching points

Methods	Polynomial	Matching points
CMP	Chebyshev	Chebyshev-Gauss–Lobatto
LPM	Legendre	Legendre–Gauss-Lobatto
GPM	Lagrange	Legendre–Gauss-Lobatto

and aerospace fields have used the matching point method at Lejeune de Gauss points. In aerospace engineering, many well-known pseudospectral methods have been developed to solve optimal control problems, such as the Chebyshev pseudospectral method (CPM), the Legendre pseudospectral method (LPM), and the Gauss pseudospectral method (GPM). pseudospectral method (GPM), see Table 8.1.

The Lejeune pseudospectral method is a method for solving optimal control problems based on Lejeune polynomials. It is a very important part of the pseudospectral method of optimal control problems. The Legendre pseudospectral method was originally proposed by Elnagar and his colleagues in 1995. Since then, Ross and Fahroo have extended and applied it in solving some important problems, the most important application of which is to obtain real-time trajectories for the International Space Station.

The Chebyshev pseudospectral method is an optimal control problem based on the first class of Chebyshev polynomials. It is a very important part of the pseudospectral method of optimal control problems. Unlike the Legendre pseudospectral method, the Chebyshev pseudospectral method does not immediately provide orthogonal solution results with high accuracy. Therefore, two different versions of the method have been proposed: one through Elnagar et al. and the other by Fahroo and Ross. The biggest difference between the two versions lies in their different orthogonal techniques. The Fahroo-Ross method is more commonly used today due to the ease of implementing the Clenshaw-Curtis orthogonal technique (as opposed to

Elnagar-Kazemi's cell averaging method.) In 2008, numerical linear algebraic solution characteristics showed that the Clenshaw-Curtis method is almost as accurate as the Gaussian pseudospectral method. This breakthrough result opened the door to a convection mapping theorem for the Chebyshev pseudospectral method, and in 2009, a complete mathematical theory of the Chebyshev pseudospectral method was developed. The Chebyshev pseudospectral method is often confused with other Chebyshev methods.

The Gaussian pseudospectral method is a method that directly transforms a discretized continuous optimal control problem into a nonlinear programming problem. The Gaussian pseudospectral method differs from several other pseudospectral methods in that instead of placing the collocation points at the two endpoints of the time interval, this method derives a set of KKT conditions by combining the placement of the collocation points with the approximate co-state, that is, the KKT conditions are in the same discrete form as the first order optimality conditions.

The Gaussian pseudospectral method is based on the theory of orthogonal configurations, where the configuration point (i.e., the point at which this optimal control problem is discretized) is the Lejeune Gaussian (LG) point. The method used by the GPM approximates the state variables using Lagrangian polynomials, which include the initial state containing the coefficients and the state at the LG point values. At the same time, the approximation is made with Lagrangian polynomials at the collocation points, where the Lagrangian polynomials include the final values of the co-states and the values of the co-states at the LG points. These two approximations together improve the ability to map from the KKT multiplier of the NLP to the point of the optimal control problem at LG and the boundary point co-state.

The pseudospectral method has very obvious advantages in terms of solution accuracy and computational efficiency, and does not require high initial values. Due to these advantages, the pseudospectral method is considered to have the potential to be applied to real-time optimal control problems of practical systems.

8.1.3 Sequential Quadratic Programming Algorithm

The classical NLP problem is represented in the following form.

The objective function is $\min f(x)$.

where $x = [x_1, x_2, x_3, \cdots, x_n]^{\mathrm{T}}$ is the vector; $f(x)$ is the scalar form.

The constraint is $G(x) \leq 0$.

where $G(x) = [g_1(x), g_2(x), \cdots, g_m(x)]^{\mathrm{T}}$ is a function vector, either equatorially or unequally constrained, $f(x)$ and $g(x)$ are both nonlinear functions.

The optimization method used in this chapter is one of the sequential quadratic programming (SQP) algorithms: the Wilson-Han-Powell method. This method is based on the general Lagrange-Newton method. A modified Hessian array approximation B_k is used in each iteration instead of $W(x_k, \lambda_k)$ [22].

The basic idea of the SQP algorithm is that the nonlinear planning problem is transformed into solving a series of quadratic planning problems, hence the name sequential quadratic programming algorithm. In general, if the approximate solution x_k and the approximate Lagrange multiplier vector λ_k are known at the beginning of the k-th iteration, then based on them, we can obtain the k-th quadratic programming subproblem P_k and we can obtain the new approximate solution x_{k+1} by solving P_k and the corresponding Lagrange multiplier vector λ_{k+1} by iterating until we obtain the approximate optimal solution for the NLP problem. If we make $d_k = x_{k+1} - x_k$, then we can obtain x_{k+1} by solving the problem P_k, which translates into solving the quadratic programming subproblem for P_k and d_k.

The linearized nonlinear objective function is followed by the construction of the quadratic programming subproblem P_k.

The objective function is

$$\min \frac{1}{2} d^{\mathrm{T}} B_k d + \nabla f(x_k)^{\mathrm{T}} d \tag{8.32}$$

where, d is the search direction, B_k is the k-th approximation of the Hessian matrix, and $\nabla f(x_k) = \left[\frac{\partial f(x)}{\partial x_1} \ \frac{\partial f(x)}{\partial x_2} \ \cdots \ \frac{\partial f(x)}{\partial x_n} \right]^{\mathrm{T}}$.

Constraints are

$$\begin{cases} \nabla g_i(x)^{\mathrm{T}} d + g_i(x) = 0, \ i = 1, 2, \cdots, m_{\mathrm{e}} \\ \nabla g_i(x)^{\mathrm{T}} d + g_i(x) \le 0, \ i = m_{\mathrm{e}} + 1, \cdots, m \end{cases} \tag{8.33}$$

where, $\nabla g_i(x)^{\mathrm{T}} = \left[\frac{\partial g_i(x)}{\partial x_1} \ \frac{\partial g_i(x)}{\partial x_2} \ \cdots \ \frac{\partial g_i(x)}{\partial x_n} \right]$.

The solution of Eqs. (8.22) and (8.23) is d_k and the Wilson-Han-Powell method uses d_k as the search direction for the k-th iteration.

The stepwise quadratic programming algorithm proposed by Han is given below.

Step 1: Give $x_1 \in \mathbf{R}^n, \sigma > 0, \delta > 0, B_1 \in \mathbf{R}^{n \times n}, \varepsilon \ge 0, k = 1$.

Step 2: Give the search direction d_k, if $\|d_k\| \le \varepsilon$ then stop. Find $\alpha_k \in [0, \delta]$ so that

$$P(x_k + \alpha_k d_k, \sigma) \le \min_{0 \le \alpha \le \delta} P(x_k + \alpha_k d_k, \sigma) + \varepsilon_k \tag{8.34}$$

Step 3: $x_{k+1} = x_k + \alpha_k d_k$; Compute the new Hessian matrix approximation array B_{k+1}, $k = k + 1$. Go to step 2.

In Eq. (8.34), the penalty function $P(x, \sigma)$ is L_1 exact penalty function and ε_k is a non-negative series and satisfies

$$\sum_{k=1}^{\infty} \varepsilon_k < +\infty \tag{8.35}$$

Regarding the calculation of B_{k+1} in the third step, it is generated by stepwise iteration using a modified proposed Newton formula. Taking

$$s_k = x_{k+1} - x_k \tag{8.36}$$

$$y_k = \nabla f(x_{k+1}) - \nabla f(x_k) - \sum_{i=1}^{m} (\lambda_k)_i (\nabla c_i(x_{k+1}) - \nabla c_i(x_k)) \tag{8.37}$$

Calculation B_{k+1} using the BFGS method

$$B_{k+1} = B_k - \frac{B_k s_k s_k^{\mathrm{T}} B_k}{s_k^{\mathrm{T}} B_k s_k} + \frac{y_k y_k^{\mathrm{T}}}{s_k^{\mathrm{T}} y_k} \tag{8.38}$$

8.2 Research on Linear Vehicle Angle Input Recognition

8.2.1 Vehicle Angular Input Steering Motion Model

Assuming that the tire lateral deflection characteristics are in the linear range, it is simplified to a 3-degree-of-freedom linear vehicle model with lateral, yaw and longitudinal motions as shown in Fig. 8.3. Then the differential equations of motion are expressed as

$$\begin{cases} \dot{v} = -u\omega_r + \dfrac{F_{yf}\cos\delta + F_{yr} + F_{xf}\sin\delta}{m} \\ \dot{\omega}_r = \dfrac{aF_{yf}\cos\delta - bF_{yr} + aF_{xf}\sin\delta}{I_z} \\ \dot{u} = v\omega_r + \dfrac{F_{xf}\cos\delta - F_{yf}\sin\delta + F_{xr} - F_f - F_w}{m} \end{cases} \tag{8.39}$$

where v is the lateral velocity of the vehicle, u is the longitudinal velocity of the vehicle, ω_r is the yaw rate of the vehicle, m is the mass of the vehicle, I_z is the inertia of the vehicle around the vertical axis, a and b are the distances from the center of mass to the front and rear axles, respectively, δ is the steering angle of the front wheel, F_{yf} is the front wheel lateral deflection force, F_{yr} is the rear wheel lateral deflection force, F_{xf} is the front wheel drive/braking force ($F_{xf} \geq 0$ is the drive force, $F_{xf} < 0$ is the braking force), F_{xr} is the rear wheel drive/braking force, F_f is the rolling resistance ($F_f = mgf, f$ is the rolling F_w is air resistance ($F_w = C_D A(3.6u)^2/21.15$, C_D is air resistance coefficient, A is windward area). The vehicle is driven on a flat road with no grade resistance. To simplify the problem, only air resistance and rolling resistance are considered, and no acceleration resistance is considered.

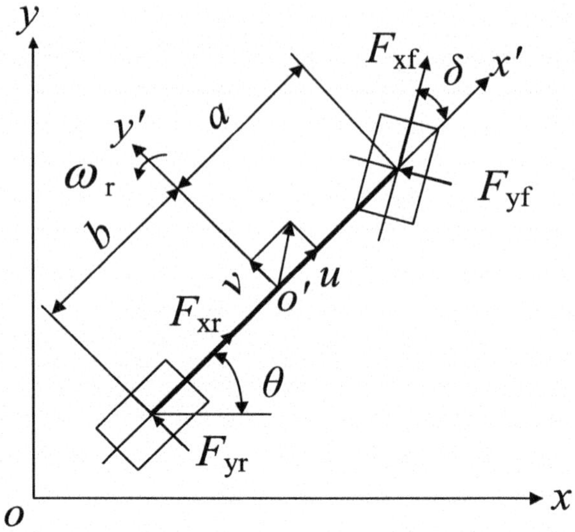

Fig. 8.3 3-degree-of-freedom vehicle model

If the effect of drive/braking force on the lateral deflection force is considered, then

$$\begin{cases} F_{yf} = k_1\left(\dfrac{v + a\omega_r}{u} - \delta\right)\sqrt{1 - \left(\dfrac{F_{xf}}{\phi F_{zf}}\right)^2 + \left(\dfrac{F_{xf}}{k_1}\right)^2} \\ F_{yr} = k_2\left(\dfrac{v - b\omega_r}{u}\right)\sqrt{1 - \left(\dfrac{F_{xr}}{\phi F_{zr}}\right)^2 + \left(\dfrac{F_{xr}}{k_2}\right)^2} \end{cases} \tag{8.40}$$

where, k_1 and k_2 are the integrated lateral cornering stiffness of the front and rear wheels respectively, ϕ is the road friction coefficient, F_{zf} is the vertical force of the front wheel, F_{zr} is the vertical force of the rear wheel. Considering the longitudinal load transfer, we have

$$\begin{cases} F_{zf} = \dfrac{mgb - (F_{xf} + F_{xr})h_g}{a + b} \\ F_{zr} = \dfrac{mga + (F_{xf} + F_{xr})h_g}{a + b} \end{cases} \tag{8.41}$$

where h_g is the height of the center of mass of the vehicle.

The coordinates of the vehicle's center of mass in the *xoy* coordinate system *x*, *y* are related as follows

$$\begin{cases} \dot{x} = u\cos\theta - v\sin\theta \\ \dot{y} = v\cos\theta + u\sin\theta \end{cases} \tag{8.42}$$

The control variables $Z(t)$ are steering wheel angle δ_{sw} and front wheel drive/braking force F_{xf} and the control objective is to pass the given path in the shortest possible time. Therefore, the minimum time performance indicator is [23]

$$J(Z) = \int_{t_0}^{t_e} \mathrm{d}t \tag{8.43}$$

where t_0 and t_e are the initial and terminal moments of the motion, respectively.

According to Eqs. (8.39) and (8.42), the equation of state can be expressed as

$$\dot{X} = f(X(t), Z(t)) \tag{8.44}$$

where the state variable $X(t) = [v(t), \gamma(t), u(t), x(t), y(t), \theta(t)]^{\mathrm{T}}$.

The longitudinal speed u is constrained by the maximum speed of the vehicle; the lateral displacement y is constrained by the left and right boundaries of the trajectory; the steering wheel angle δ_{sw} is constrained by the physiological limits of the driver, and the control variable F_{xf} is constrained by the road adhesion. When the vehicle is subject to the driving force and is front-wheel drive, there is

$$\begin{cases} F_{xf} \le \dfrac{\phi mgb}{a + b + \phi h_g} \\ F_{xr} = 0 \end{cases} \tag{8.45}$$

If the vehicle is subjected to the braking force, assuming that both front and rear wheels are in a holding condition, then

$$\begin{cases} F_{xf} \ge -\dfrac{\phi mg(b + \phi h_g)}{a + b} \\ F_{xr} = \dfrac{a - \phi h_g}{b + \phi h_g} F_{xf} \end{cases} \tag{8.46}$$

The control variable F_{xf} is also constrained by the maximum driving force provided by the powertrain. According to the relationship between engine speed and vehicle driving speed, as well as the relationship between engine output torque and driving force, the relationship between the maximum driving force and vehicle driving speed can be obtained from the engine external characteristic curve.

According to the relevant literature, the constraint to prevent the vehicle from rolling over during driving is

$$\frac{u^2 \delta}{(a + b)(1 + Ku^2)g} \le \frac{B}{2h_g} \tag{8.47}$$

where B is the wheelbase, K is the stability factor.

All constraints are expressed by the following equation

$$\psi(X(t), Z(t)) = 0 \tag{8.48}$$

The transformation method of the state variables is similar to that of the first type of manipulation of the inverse dynamics steering wheel angle input and will not be described in detail here.

The transformed optimal control problem is solved by a direct configuration nonlinear method.

8.2.2 Simulation Results

Two models of vehicles are used as examples for the study of double-shift linear energy. The values of the specific parameters of the vehicles are shown in Table 8.2. To simplify the problem, the width of the wheels is neglected. The steering wheel turning angle is controlled between ±360°.

1. $u = 108$ km/h

After 12 iterations, Model 1 is optimized to obtain the minimum time of 9.3 s for the double lane change; after 15 iterations, Model 2 obtains the minimum time of 9.9 s for the double lane change. Therefore, Model 1 takes less time than Model 2 to complete the double lane change. Figure 8.4a–d show the simulation results of some state variables and control variables in the double lane change process at $u = 108$ km/h.

Figure 8.4a shows the simulation results of the lateral displacement of the two models. It can be seen that the lateral displacements of the two models almost overlap,

Table 8.2 List of vehicle parameters

Parameters	Vehicle type I	Vehicle type II
m/kg	1500	1265
$I_z/(\text{kg} \cdot \text{m}^2)$	2080	1800
a/m	1.185	1.170
b/m	1.283	1.195
k_1/(N/rad)	−60,533	−60,042
k_2/(N/rad)	−110,185	−109,295
i	20	20
Φ	0.8	0.8
h_σ/m	0.53	0.53
i_σ	0.97	0.914
i_0	4.12	4.38
r/m	0.289	0.280
$u_{max}/(km/h)$	190	170

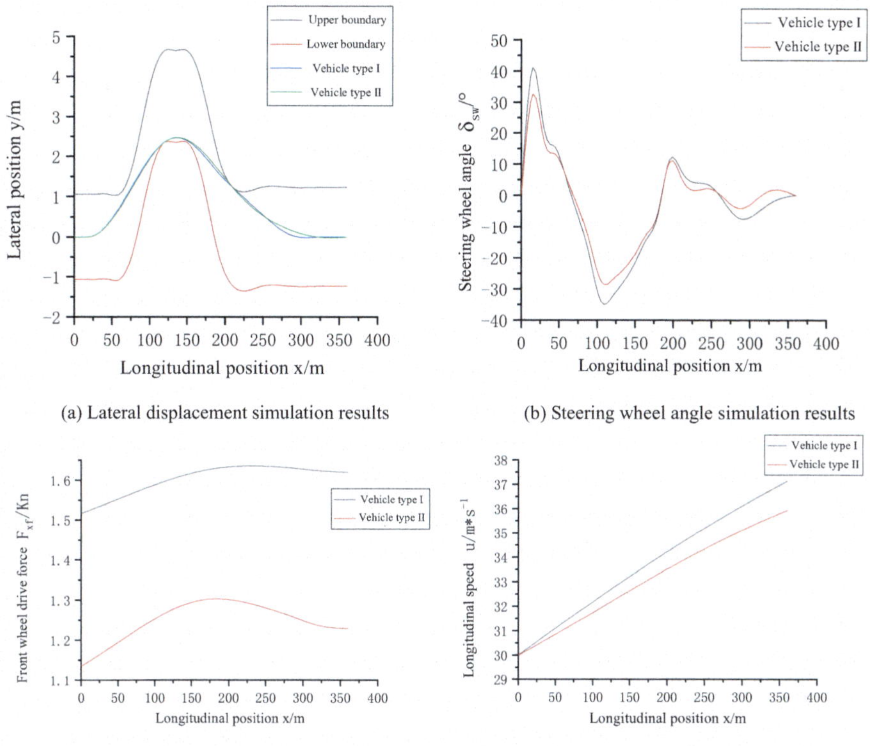

(a) Lateral displacement simulation results

(b) Steering wheel angle simulation results

(c) Front wheel drive simulation results

(d) Longitudinal speed Simulation results

Fig. 8.4 Double lane change simulation results (u = 108 km/h)

and their motion laws are as follows: the first drive in a straight line for a distance, then turn left to change lanes, and in the process of the first and second lane change, the vehicle is close to driving in a straight line between road boundaries. Figure 8.4b shows the simulation results of the steering wheel turning angle for both models. It can be seen that the steering wheel corner has two magnitudes, which are located at the longitudinal displacement of 15 m and 105 m; model one has a slightly larger steering wheel corner magnitude than model two. Figure 8.4c shows the simulation results of the front-wheel drive for both models. It can be seen that the driving force first increases and then decreases during the double lane change; the driving force on model one is larger than that on model two. Figure 8.4d shows the simulation results of longitudinal speed for both models. It can be seen that during the double-shift linear performance process, type I increases rapidly from 30 m/s to 37 m/s; type II increases rapidly from 30 m/s to 36 m/s. Therefore, type I increases faster than type II.

2. $u = 90$ km/h

After 28 iterations, the minimum time to complete the double lane change process is 10.5 s for type 1 and 10.8 s for type 2 after 11 iterations. Therefore, when the speed of u decreases, the minimum time to complete the double lane change process increases, and model 1 still takes less time than model 2 to complete the double lane change process. Figure 8.5a–d show the simulation results of some state and control variables in the double lane change process at $u = 90$ km/h. The motion trend is similar to that of $u = 108$ km/h.

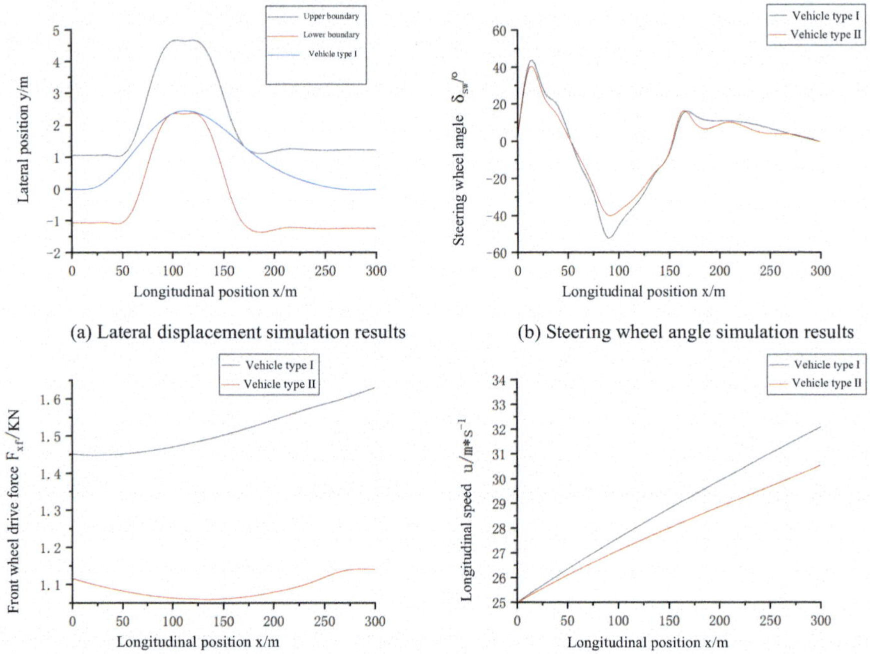

(a) Lateral displacement simulation results

(b) Steering wheel angle simulation results

(c) Front-wheel drive simulation results

(d) Longitudinal speed simulation results

Fig. 8.5 Double-shift simulation results ($u = 90$ km/h)

8.3 Research on Linear Vehicle Force Input Recognition

8.3.1 Vehicle Force Input Steering Motion Model

1. Mathematical model of steering wheel torque input

Assuming that the tire lateral deflection characteristics are in the linear range and considering the steering system rotation inertia, it is simplified to a 4-degree of freedom linear vehicle model with lateral motion, yawing motion, longitudinal motion and steering system rotation shown in Fig. 8.6. Then the differential equation of motion is expressed as

$$\begin{cases} \dot{v} = -u\omega_r + \dfrac{F_{yf}\cos\delta + F_{yr} + F_{xf}\sin\delta}{m} \\ \dot{\omega}_r = \dfrac{aF_{yf}\cos\delta - bF_{yr} + aF_{xf}\sin\delta}{I_z} \\ \dot{u} = v\omega_r + \dfrac{F_{xf}\cos\delta - F_{yf}\sin\delta + F_{xr} - F_f - F_w}{m} \\ \dot{\delta} = p \\ \dot{p} = -\dfrac{k_1\xi_1}{I_w u}v - \dfrac{k_1\xi_1 a}{I_w u}\omega_r + \dfrac{(k_1\xi_1 - k_w)}{I_w}\delta - \dfrac{c_w}{I_w}p + \dfrac{T_{sw}i}{I_w} \end{cases} \tag{8.49}$$

where I_w is the steering inertia, ξ_1 is the front wheel return force arm, p is the state variable, c_w is the steering drag coefficient, k_w is the overall steering stiffness, i is

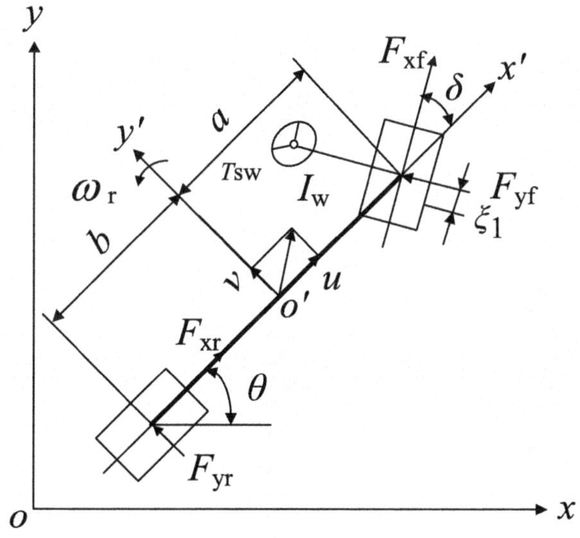

Fig. 8.6 4-degree-of-freedom vehicle model

the steering gear ratio, T_{sw} is the steering wheel torque, the other parameters are the same as the mathematical model parameters for steering wheel angle input.

2. Steering wheel torque input optimal control model

The control variables $Z(t)$ are steering wheel torque $T_{sw}(t)$, front wheel drive/braking force $F_{xf}(t)$. The control objective is to pass the given path in the shortest possible time. Therefore, the minimum time performance indicator is

$$J(Z) = \int_{t_0}^{t_e} \mathrm{d}t \tag{8.50}$$

where t_0 and t_e are the initial and terminal moments of the motion, respectively.

According to Eq. (8.49), the equation of state can be expressed as

$$\dot{X} = f(X(t), Z(t)) \tag{8.51}$$

where the state variable is $X(t) = \left[v(t)\ \omega_r(t)\ u(t)\ \delta(t)\ p(t)\ x(t)\ y(t)\ \theta(t) \right]^{\mathrm{T}}$.

The constraint is similar to the constraint on the steering wheel corner input.

The conversion method of the state variables is similar to that of the steering wheel angle input and will not be described in detail here.

The transformed optimal control problem is solved using a direct configuration nonlinear method.

8.3.2 Simulation Results

Example of two models of sedan for the study of snaking linear energy. The specific parameter values of the sedan are shown in Table 8.2. The steering wheel torque is controlled between $\pm 8\,\mathrm{N \cdot m}$.

1. $u = 90\,\mathrm{km/h}$

The minimum time for the snake line process is 14.9 s for model 1 after 15 iterations and 15.2 s for model 2 after 18 iterations. Therefore, the minimum time used for model 1 is less than that for model 2. Figure 8.7a–d shows the simulation results of some state variables and control variables in the snaking line process at $u = 90\,\mathrm{km/h}$.

Figure 8.7a shows the simulation results of the lateral displacement of the two types. It can be seen that the lateral displacements of the two models almost overlap, and their motion laws are such that the vehicle is close to driving in a straight line between the road boundaries. Figure 8.7b, c show the simulation results of steering wheel torque and front wheel drive for the two types. Again, it can be seen that type I has a larger steering wheel torque amplitude and is subjected to a larger driving force than type II. Figure 8.7d shows the longitudinal speed simulation results for

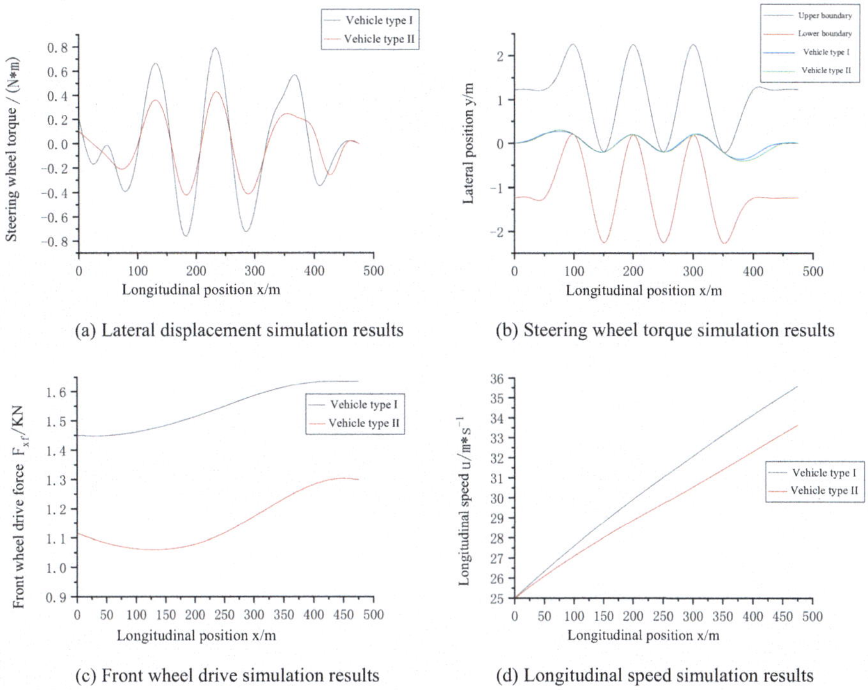

(a) Lateral displacement simulation results

(b) Steering wheel torque simulation results

(c) Front wheel drive simulation results

(d) Longitudinal speed simulation results

Fig. 8.7 Snake line simulation results ($u = 90$km/h)

both models. It can be seen that during the snake line, type I increases rapidly from 25 m/s to 35.6 m/s; Model II increases rapidly from 25 m/s to 33.7 m/s. Therefore, type I increases faster than type II.

2. $u = 72$ km/h

The minimum time for the snake line process is 15.7 s for type I after 14 iterations and 16.2 s for type II after 16 iterations. Therefore, when the speed u decreases, the minimum time to complete the snake line process increases, and type I still takes less time than type II to complete the snake line process. Figure 8.8a–d shows the simulation results of some state and control variables in the snake line process at $u = 72$ km/h. The motion trend is similar to that of $u = 90$ km/h.

8.4 ADAMS/Car Model Building and Test Verification

The Vehicle module in ADAMS software is a complete vehicle virtual design package developed by MDI in cooperation with Audi, BMW, Renault and Volvo, integrating their experience in vehicle design and other aspects. Using this module, engineers can quickly build and simulate high-precision virtual prototypes of the entire vehicle

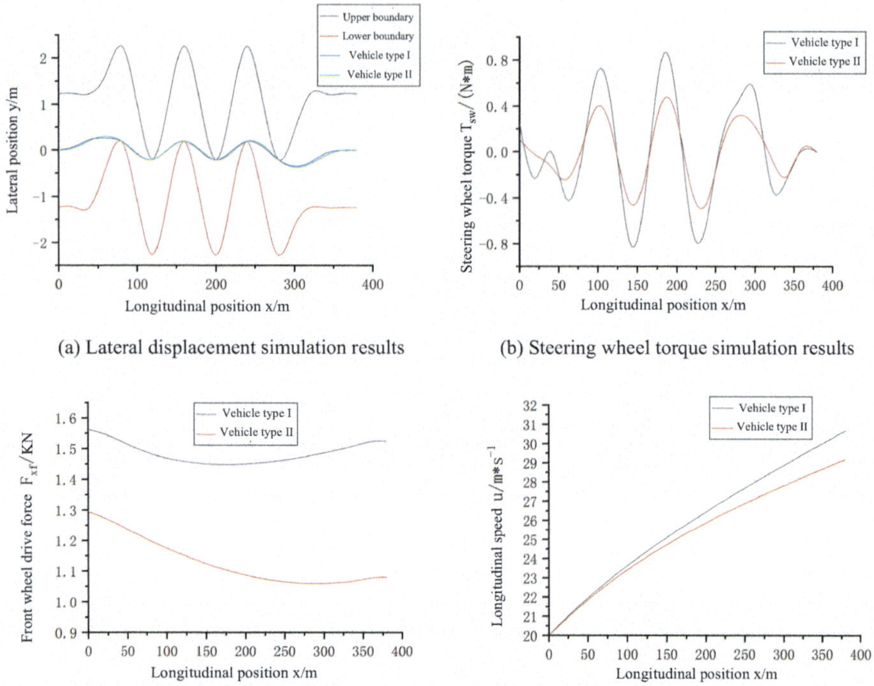

(a) Lateral displacement simulation results

(b) Steering wheel torque simulation results

(c) Front-wheel drive simulation results

(d) Longitudinal speed simulation results

Fig. 8.8 Snake line simulation results ($u = 72$ km/h)

(including body, suspension, drivetrain, engine, steering mechanism, braking system, etc.), and visualize the dynamic response of the entire vehicle under various test conditions (e.g. weather, road conditions, driver experience) through high-speed animations, and output parameters representing handling stability, braking, ride comfort and safety.

The maximum speed maneuvering problem is more dangerous than the real vehicle test, so this section uses the ADAMS/Vehicle virtual simulation test instead of the real vehicle test to verify the correctness of the optimal control model.

8.4.1 ADAMS/Car Vehicle Modelling

Since the vehicle is an extremely complex mechanical system, the workload and difficulty are very large if the modeling is done according to the real structure of the vehicle. Therefore, according to the purpose of the study, the vehicle modeling according to the structure of the vehicle can be appropriately abstracted and simplified: in the suspension parts, except for the elastic parts, the rest of the parts are all regarded as rigid bodies; for the convenience of analyzing the problem, the body is

Fig. 8.9 Simulation model of the whole vehicle

regarded as a rigid body; because the structure of the damper is too complex, the Spring-damper force element provided by ADAMS is applied and used to model the damper and spring The spring-damper force element provided by ADAMS is used to simulate the action of the damper and the spring.

The vehicle studied in this section mainly consists of MacPherson's independent front suspension structure, multi-link independent rear suspension structure, rack and pinion steering structure and rigid chassis structure. Firstly, the virtual prototype model of the whole vehicle is divided into subsystems such as suspension, body, steering system and braking system, and the corresponding subsystem models are established according to the vehicle parameters shown in Table 8.2. Then, the corresponding tire characteristics file and experimental road characteristics file are prepared in ADAMS/Vehicle according to the tire parameters of the real vehicle and the road surface characteristics parameters specified in the national standard "Mechanical Vibration Road Surface Spectrum Measurement Data Report" (GB/T 7031-2005); finally, the whole vehicle virtual prototype test model is assembled according to the system as shown in Fig. 8.9 [24, 25].

8.4.2 Closed-Loop Control Analysis

ADAMS/Vehicle provides open-loop analysis, quasi-static analysis and closed-loop analysis, where open-loop analysis and quasi-static analysis only require inputting the required parameters on the user interface, and closed-loop analysis is more complex than open-loop analysis and quasi-static analysis, and requires providing the control files required for the analysis. According to the experimental verification requirements, this section needs to discuss the closed-loop analysis.

For closed-loop analysis, ADAMS/Vehicle provides dual shift analysis and understeer analysis. The designer can design the driver control file (*.dcf) needed to drive the vehicle according to the specific test requirements, just like a real vehicle test. The speed and trajectory of the vehicle can be set in the control file, and the driver model information can be added to make the vehicle drive intelligently [26].

Documents related to closed-loop control are as follows:

(1) Driver control file (*.dcf). The driver control file describes a series of operations that the vehicle model is prepared to perform, such as steering, braking, etc. In this file, you can set the initial speed of the vehicle, the trajectory file, the control method, and the conditions for the end of the test (e.g., distance traveled). There are two types of control methods, machine control and driver control. In this test, the machine control method is used, and the driver model is used to automatically control the operation of the vehicle, and the input excitation is the steering wheel angle change, and the end of the test condition is distance control.
(2) Driver control data file (*.dcd). The driver control data file contains the data required by the Driver Control File. If the vehicle movement is controlled by the trajectory in *.dcf, the main content of the *.dcd file is to set the coordinate data of the vehicle trajectory. In this test, the vehicle trajectory is given in the data of x and y coordinates.
(3) Driver input file (*.din). The driver input file specifies the values of various control parameters required by the ADAMS/Driver module, such as radial dynamic characteristics, lateral dynamic characteristics, vehicle structural characteristics, and control parameters of the learning file, which are used to guide the whole vehicle analysis.

8.4.3 ADAMS/Car Verification

Using the built model with an initial speed of 90 km/h, the virtual test was conducted using the relationship between lateral displacement and longitudinal displacement shown in Fig. 8.5 as the data file for driver control. Figure 8.10 shows the comparison of the ADAMS/Vehicle virtual prototype test results with the MATLAB simulation results. It can be seen from the Fig. that the two match well and the changing trend is the same, thus verifying that the optimal control model is correct.

8.5 Summary

This chapter first introduces the inverse dynamics solving method based on optimal control, mainly including direct and indirect methods, and introduces several common methods of the direct method. After that, the linear 3-degree-of-freedom vehicle angular input model and 4-degree-of-freedom vehicle force input model are used as the basis, respectively, and the angular input and driving force/braking force applied by the driver to the vehicle is used as the control variables, and the minimum time to complete the double- lane change and snake contour process is used as the

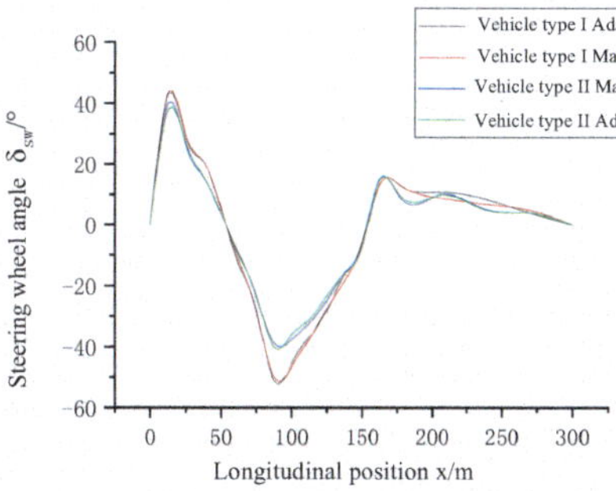

(a) Steering wheel turning angle

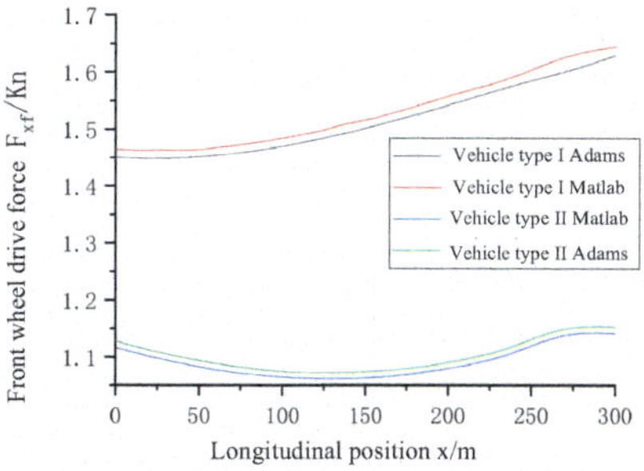

(b) Front wheel drive force

Fig. 8.10 Comparison of MATLAB simulation values and ADAMS simulation values

control objective to transform the optimal control problem into the optimal control problem. The optimal control problem is transformed into a nonlinear programming problem using a direct configuration method, and the transformed nonlinear programming problem is solved by a sequential quadratic programming algorithm. Finally, the correctness of the optimal control model is verified by ADAMS/Vehicle virtual prototype tests. The main conclusions of this chapter are as follows:

(1) Type I has a larger steering wheel turning angle amplitude and front wheel drive amplitude than type II.
(2) Type I takes less time than type II in terms of minimum time.
(3) When the initial speed of the vehicle increases, the minimum time to complete the double lane change process decreases. As the speed of the vehicle increases, the driving burden increases and safety decreases. Therefore, drivers generally have to reduce the vehicle speed to a certain speed before turning to ensure safe driving.
(4) The derived minimum time motion law can provide some guidance to the actual driver or racer in the maneuvering process, and also has some reference value to the automatic driving vehicle.

References

1. Zhang G. Optimal control calculation methods. Chengdu: Chengdu University of Science and Technology Press; 1991. p. 2–22.
2. Ye QK, Wang ZM. Computational methods in optimization and optimal control. Beijing: Science Press; 1986. p. 1–17.
3. Gong X. Computational methods for optimal control problems. Beijing: Science Press; 1979. p. 13–25.
4. Yong EM, Chen L, Tang GJA. Review of numerical methods for flight vehicle trajectory optimization. J Astronaut. 2008;29(2):397–406.

5. Wu C. Theory and methods of optimal control. 2nd ed. Beijing: National Defense Industry Press; 2000. p. 73–102.
6. Von Stryk O. Numerical solution of optimal control problems by direct collocation. In: Proceedings of the optimal control: calculus of variations. Basel: Birkhäuser Verlag; 1993. p. 129–43.
7. Von Stryk O, Bulirsch R. Direct and indirect methods for trajectory optimization. Ann Oper Res. 1992;37:357–73.
8. Enright PJ, Conway BA. Discrete approximations to optimal trajectories using direct transcription and nonlinear programming. In: AIAA Paper 90-2963-CP, 1992. p. 994–1002.
9. Hu GS, Ong CJ, Teo CL. Direct collocation and nonlinear programming for optimal control problem using an enhanced transcribing scheme. In: Proceedings of the 1999 IEEE international symposium on computer aided control system design, Hawaii; 1999. p. 22–27.
10. Elnagar JG, Kazemi MA. Pseudospectral Chebyshev optimal control of constrained nonlinear dynamical systems. Comput Optim Appl. 1998;11:195–217.
11. Fahroo JF, Ross IM. Direct trajectory optimization by a Chebyshev pseudospectral method. J Guid Control Dyn. 2002;25(1):160–6.
12. Gong JQ, Trefethen LN. Is Gauss quadrature better than Clenshaw-Curtis? SIAM Rev. 2008;50(1):67–87.
13. Gong JQ, Ross IM. Costate computation by a Chebyshev pseudospectral method. J Guid Control Dyn. 2010;33(2):623–8.
14. Gong JQ, Ross IM. A Chebyshev pseudospectral method for nonlinear constrained optimal control problems. In: Joint 48th IEEE conference on decision and control and 28th Chinese control conference, Shanghai; 2009. p. 16–18
15. Zhang XL, Zhao YQ, Zhang WX, et al. Study on the optimal lane changing steering input under different road adhesion coefficients. Proc Inst Mech Eng Part I J Syst Control Eng. 2020;234(3):384–95.
16. Zhang XL, Zhao YQ, Zhang WX, et al. Minimum time lane changing problem of handling inverse dynamics considering the driver's intention. Int J Automot Technol. 2019;20(1):109–18.
17. Zhao YQ, Zhang XL, Zhang WX, et al. Minimum time overtaking problem of vehicle handling inverse dynamics based on two kinds of safe distances. Chin J Mech Eng. 2018;31(1):1–10.
18. Zhao YQ, Zhang WX, Zhang XL, et al. Minimum time overtaking problem of vehicle handling inverse dynamics based on interval mathematics. Proc Inst Mech Eng Part D J Automob Eng. 2019;233(6):1534–45.
19. Gong JQ, Huntington GT, Rao AV. Optimal reconfiguration of spacecraft formations using the gauss pseudospectral method. J Guid Control Dyn. 2008;31(3):689–98.
20. Elnagar JG, Kazemi MA. The pseudospectral legendre method for discretizing optimal control problems. IEEE Trans Autom Control. 1995;40:1793–6.
21. Kang JW, Gong Q. On the convergence of nonlinear optimal control using pseudospectral methods for feedback linearizable systems. Int J Robust Nonlinear Control. 2007;17:1251–77.
22. Gong JQ, Ross IM, Fahroo F. Pseudospectral knotting methods for solving nonsmooth optimal control problems. J Guid Control Dyn. 2004;27:397–405.
23. Yuan Y, Sun W. Optimization theory and methods. Beijing: Science Press; 1997. p. 521–53.
24. Zhang L, Zhao R, Song G, et al. An inverse dynamics study of the automobile maximum speed maneuvering problem. China Mech Eng. 2007;18(21):2628–32.
25. Shi P-C, Chen L-Q. Experimental study of virtual prototype of multifunctional vehicle handling stability. Autom Technol. 2005;3:45–8.
26. Wang S. Research on the virtual test system of automobile handling stability. Beijing: China Agricultural University; 2002.

Chapter 9
Road Roughness, Evolution Random Response and Ride Comfort Based on Vehicle Vertical Dynamic

There are many kinds of vibration excitation sources for vehicles, such as inertia force during driving (centrifugal force during turning, inertia force during braking, etc.), aerodynamic force and excitation of uneven road surface. This chapter mainly introduces the random vibration caused by uneven excitation of road surface, because it is a kind of vibration that occurs a lot and often. The research on such problems originated in the 1960s. Due to the continuous development of random vibration theory, the testing and calculation technology has been improved significantly. It has been able to analyze the random vibration of the multi-degree-of-freedom vehicle model that is more in line with the actual situation, and give the relationship between the random vibration response and the structural parameters, which provides a useful basis for vehicle design. The vehicle vibration system is shown in Fig. 9.1.

Vehicle ride comfort is the performance of maintaining a certain level of comfort in the vibration environment of the passengers when the vehicle is driving. For the truck, it also includes the performance of keeping the goods intact. Due to the micro unevenness of the road, it is inevitable to arouse the vibration and dynamic load of the vehicle. How to reduce this vibration and dynamic load has always been one of the efforts of vehicle design [1].

Because of the randomness of the micro roughness of the road surface, the vibration and dynamic load of the vehicle also have randomness.

To reduce the dynamic load transmitted by the car to passengers or goods (equipment), it is necessary to reduce the body vibration acceleration; to improve the service life of the bearing parts and reduce the vehicle's own mass, it is necessary to reduce the dynamic load of the corresponding transmission link; to ensure the good handling stability at high speed, it is necessary to make the wheel and the road have a good adhesion effect, that is, to reduce the fluctuation of dynamic load between the car wheel and the road, because the fluctuation of adhesion reduces the "dynamic lateral rigidity of the tire" [2]; when the phenomenon of "suspension breakdown" occurs on uneven roads (hitting the limit block), the vehicle speed is often greatly reduced. To

Y. Zhao and F. Lin, *Vehicle System Dynamics*,
https://doi.org/10.1007/978-981-97-2019-4_9

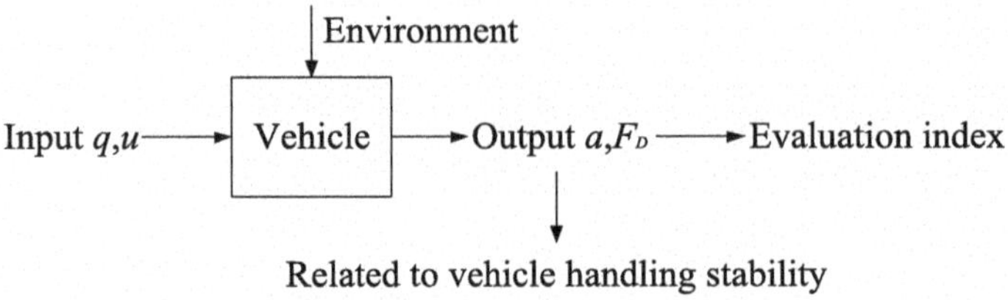

Fig. 9.1 Vehicle vibration system

minimize the probability of "breakdown", the limit travel must be selected reasonably, and try to reduce the dynamic travel of the relative vibration between the sprung mass and unsprung mass. In addition, in order to improve the service life of the spring, it is also desirable to reduce the suspension working space.

Therefore, vehicle ride comfort belongs to the sports performance in the basic mobility of vehicles, the riding performance and loading performance in convenience and comfort, and the safety performance and environmental adaptability (countermeasures to prevent public hazards) in sociality and economy.

A lot of research has been done on the vibration acceleration, dynamic load, suspension working space on the actual road and the relationship between them and the design parameters, and some calculation methods convenient for engineering application have been proposed. However, the design of suspension and seat uses empirical data and inaccurate calculation to a certain extent.

9.1 Development Status and Trends

Since the 1960s until now, with the gradual popularization of the application of electronic computers, the development of random vibration theory and the in-depth study of the basic performance of vehicles. Abroad, extensive experimental verification and theoretical discussion have been carried out on the ride comfort prediction and optimization design of vehicles, and gradually applied to practice production.

As early as 1976, Guo Konghui, a scholar who introduced the theory of system dynamics and random vibration into the study of vehicle vibration and dynamic load and now an academician of the Chinese Academy of Engineering [3], qualitatively discussed how the single input two-degree-of-freedom vehicle system can improve vehicle ride comfort and how to select suspension system parameters.

In 1982, Professor Zhang Yiming of Beijing Institute of Technology and others first applied optimization design to vehicle ride comfort research, quantified the above problems, and obtained some laws and quantitative results of suspension parameters affecting vehicle ride comfort. In 1987, Professor Gao Guihua of Tsinghua University and others discussed the influence of suspension parameters and seat parameters on the ride comfort of BJ-130 vehicle body-wheel system and the five-degree-of-freedom system with "model man" on the basis of the above research, and obtained some conclusions. In 1987, Professor Zhang Hongxin and others of the vehicle ride comfort research group in Jilin University of Technology carried out a series of tests and studies on the DD-680 bus of Dandong Automobile Factory, and put forward the optimization design scheme of seat parameters and suspension parameters to improve the automobile ride comfort, and achieved certain results. In 1993, Zheng Haozhe of Shenyang Glider Factory popularized the fast virtual excitation algorithm of stationary random vibration response proposed by Professor Lin Jiahao of Dalian University of Technology for a single random process, making it suitable for the random vibration of moving vehicles with joint input of multiple processes [4].

The research of Guo Konghui, Zhang Yiming, Gao Guihua, Zhang Hongxin and Zheng Haozhe were based on the theory of stationary random vibration, which are undoubtedly correct when the vehicle is driving at a constant speed. When the vehicle is driving at a variable speed, although the road roughness is a stationary random process in the space domain, the road excitation in the time domain is a non-stationary random process. In 1990, Professor Luo Minglian of South China University of Technology discussed the method of "covariance equivalence" modeling and evolutionary spectrum analysis to study the response of five-degree-of-freedom vehicle model to variable speed driving on uneven roads. This method can basically meet certain accuracy requirements in the low-frequency region, but the deviation is large in the high-frequency region. In 1997, Professor Fang Tong from Northwestern Polytechnic University proposed a solution to the evolutionary random response of variable-speed vehicles excited by the road surface based on complex modal analysis and evolutionary spectrum analysis, which was very similar to the conventional vibration analysis, and illustrated it with the two-degree-of freedom-model of a two-axle vehicle as an example [5].

The above researches on vehicle ride comfort didn't consider the tire containment characteristics, which is the basis of road simulation. Therefore, the research on vehicle ride comfort considering tire containment characteristics is the main development trend of vehicle ride comfort research.

9.2 Study on Frequency Domain Inverse Dynamics of Road Roughness

9.2.1 Identification of Four-Degree-of-Freedom Vehicle Road Roughness in Frequency Domain

Generally, the height q of the pavement relative to the reference plane and the change $q(I)$ of the length I along the road are called the pavement profile curve or roughness function. The input of road roughness has a large amount of data in space–time domain. Thus, the frequency domain analysis method can be applied to the linear analysis of the system. According to random vibration theory [6], the Fourier transform of the sample function usually does not satisfy the existence condition of the integral, so the Fourier spectrum cannot be obtained. However, the autocorrelation function of the stationary process satisfies the integral convergence condition, hence, its Fourier transform power spectral density function is used to establish the frequency domain model.

The power spectral density of road roughness comes from the editing and processing of a large number of sample measurement data. The standard road power spectral density is obtained by processing the measured discrete data with the classical nonparametric spectrum estimation method, that is periodic graph method, and then the mathematical expression of the power spectrum is obtained by fitting the rational function or other functions such as power function. Reference [7] gives the power spectrum simulation formula of different levels of standard pavement. They describe the frequency structure of random process, which can clearly give the frequency components involved in the vibration and the role of each frequency component in the vibration in the automobile vibration analysis.

1. Establishment of four-degree-of-freedom vehicle vibration model

Vehicle is a complex vibration system with many particles, which is generally considered to have 18 degrees of freedom. Therefore, it is quite intricate to study it completely according to the actual vibration of vehicle, and it is difficult to achieve the desired purpose. Therefore, it should be simplified according to the problems analyzed. The following assumptions are made in this section:

① The vehicle body is regarded as a rigid body with concentrated mass, and only the vertical and pitching vibration of the vehicle body are considered to affect the vehicle ride comfort, ignoring the front and rear vibration and lateral vibration of the vehicle body;
② Assuming that the body vibrates slightly near the equilibrium position, the stiffness and damping of the suspension are regarded as constants, that is, the body is treated as a linear system;
③ The vehicle is symmetrical along the longitudinal centerline and moves uniformly in a straight line;
④ Road surface is a stationary, ergodic, normal stochastic process with isotropic properties;
⑤ Ignore other vibration sources except the road surface.

In this way, the vehicle is simplified into a four-degree-of-freedom plane model as shown in Fig. 9.2.

From the Lagrange equation, the vibration equation of the vehicle can be obtained as follows:

$$M\ddot{Z} + C\dot{Z} + KZ = K_t Q + C_t \dot{Q} \tag{9.1}$$

where, M is the mass matrix, C is the damping matrix, K is the stiffness matrix of the vehicle, Z is the displacement vector corresponding to each degree of freedom of the vehicle, Q is the road input vector, K_t is the tire stiffness matrix, C_t is the tire damping matrix. The specific expression is as follows.

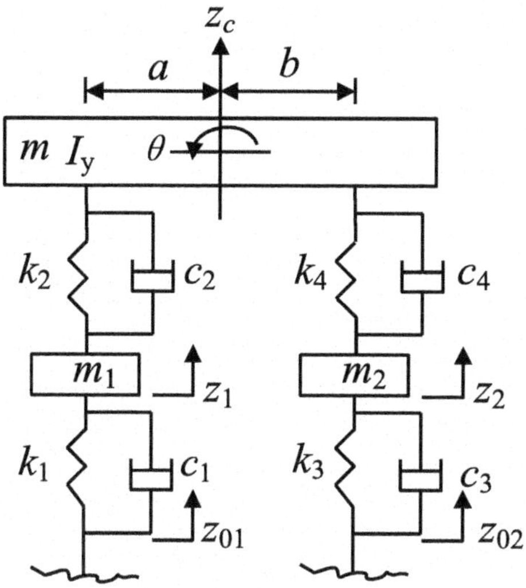

Fig. 9.2 Four-degree-of-freedom vehicle vibration model

$$M = \mathrm{diag}[\, m\ I_y\ m_1\ m_2\,]$$

$$Q = \begin{bmatrix} z_{01}\ z_{02} \end{bmatrix}^{\mathrm{T}}, Z = \begin{bmatrix} z_c\ \theta\ z_1\ z_2 \end{bmatrix}^{\mathrm{T}}$$

$$K_t = \begin{bmatrix} 0 & 0 \\ 0 & 0 \\ k_1 & 0 \\ 0 & k_3 \end{bmatrix}, C_t = \begin{bmatrix} 0 & 0 \\ 0 & 0 \\ c_1 & 0 \\ 0 & c_3 \end{bmatrix}$$

$$K = \begin{bmatrix} k_2 + k_4 & -ak_2 + bk_4 & -k_2 & -k_4 \\ -ak_2 + bk_4 & a^2k_2 + b^2k_4 & ak_2 & -bk_4 \\ -k_2 & ak_2 & k_2 + k_1 & 0 \\ -k_4 & -bk_4 & 0 & k_3 + k_4 \end{bmatrix}$$

$$C = \begin{bmatrix} c_2 + c_4 & -ac_2 + bc_4 & -c_2 & -c_4 \\ -ac_2 + bc_4 & a^2c_2 + b^2c_4 & ac_2 & -bc_4 \\ -c_2 & ac_2 & c_2 + c_1 & 0 \\ -c_4 & -bc_4 & 0 & c_4 + c_3 \end{bmatrix}$$

2. Power spectral density function of road surface input to two-wheeled vehicle

As the input of vehicle vibration, the statistical characteristics of road roughness are mainly described by road power spectral density. This is reflected in the "Draft Method for Expressing Road Roughness" proposed by the international organization

Table 9.1 Eight-level classification standard of road roughness

Road level	$G_q(n_0)/(10^{-6}\text{m}^3)$		
	Lower limit	Geometrical mean	Upper limit
A	8	16	32
B	32	64	128
C	128	256	512
D	512	1024	2048
E	2048	4096	8192
F	8192	16,384	32,768
G	32,768	65,536	131,072
H	131,072	262,144	524,288

for standardization in document ISO/TC108/SC2N67 in 1984 and the domestic standard "Vehicle Vibration Input - Road Roughness Representation Method" drafted by Changchun Automobile Research Institute [8]. Both of the two documents suggest that the following formula can be used as the fitting expression of pavement power spectral density $G_q(n)$.

$$G_q(n) = G_q(n_0)/\left(\frac{n}{n_0}\right)^{-W} \tag{9.2}$$

where n is the space frequency (m^{-1}), n_0 is the reference space frequency, $n_0 = 0.1\text{m}^{-1}$, $G_q(n_0)$ is the road spectral value at the reference spatial frequency n_0, which is called road roughness coefficient (m^3), W is the frequency index, generally $W = 2$.

The above two documents also propose that the road surface is divided into 8 levels according to the road surface power spectral density, as shown in Table 9.1.

Table 9.1 sets out the range and geometric mean of road roughness coefficient $G_q(n_0)$ at all levels.

In addition to the road roughness, the vehicle speed is also considered as the input of the vehicle vibration system. According to the vehicle speed u, the spatial frequency spectral density $G_q(n)$ is converted into the temporal frequency spectral density $G_q(f)$.

When the vehicle is driving on the road with spatial frequency n at a certain speed u, the input time frequency f is the product of n and u, that is:

$$f = un \tag{9.3}$$

Therefore, the conversion formula of $G_q(n)$ and $G_q(f)$ can be obtained.

$$G_q(f) = \frac{1}{u} G_q(n) \tag{9.4}$$

The expression of time frequency road spectrum $G_q(f)$ obtained by substituting Eqs. (9.2) and (9.3) into Eq. (9.4) is

$$G_q(f) = \frac{1}{u} G_q(n_0) \left(\frac{n}{n_0} \right)^{-2} = G_q(n_0) n_0^2 \frac{u}{f^2} \tag{9.5}$$

The spectral matrix of road surface input of two wheeled vehicle is expressed as

$$G_{ik}(f) = G_q(f) \begin{bmatrix} 1 & e^{\frac{-j2\pi f(a+b)}{u}} \\ e^{\frac{j2\pi f(a+b)}{u}} & 1 \end{bmatrix} \tag{9.6}$$

3. Solution of frequency response of vehicle vibration

For Eq. (9.1), in order to obtain the vibration response spectrum under random input, the differential equation is transformed into a frequency equation by Fourier transform at first. Then, the frequency response function of the system is obtained. Finally, the vibration response spectrum is obtained according to the road spectrum. This section uses matrix inversion method to calculate frequency response function.

Fourier transform is performed on both sides of Eq. (9.1) to obtain:

$$Z(\omega)(-M\omega^2 + jC\omega + K) = Q(\omega)(K_t + jC_t\omega) \tag{9.7}$$

Consequently, the frequency response function is

$$H(j\omega) = \frac{Z(\omega)}{Q(\omega)} = \frac{K_t + jC_t\omega}{-M\omega^2 + jC\omega + K} \tag{9.8}$$

where $H(j\omega)$ is a complex matrix, which is the frequency response matrix of the displacement response output to the input of road roughness. The frequency response matrix $H(j\omega)$ corresponding to the plane four-degree-of-freedom vehicle vibration model is a 4-row and 2-column matrix, which takes the vibration circumferential frequency ω as the independent variable. However, in general, the time frequency f needs to be used in the calculation. According to the relationship between ω and f, $\omega = 2\pi f$, the matrix elements ω can be replaced with f. Then the power spectrum matrix of the vibration displacement response is:

$$G_z(\omega) = H(j\omega)^* G_{ik} H(j\omega)^{\mathrm{T}} \omega = 2\pi f \tag{9.9}$$

where "*" means to conjugate the matrix, "T" means transposing the matrix.

The relationship between acceleration spectrum $G_{\ddot{z}}(\omega)$ and displacement spectrum $G_z(\omega)$ of a certain point is

$$G_{\ddot{z}}(\omega) = \omega^4 G_z(\omega) \tag{9.10}$$

4. Establishment of RBF neural network model

Due to the function of the vehicle vibration isolation system, the displacement or acceleration response of the vehicle to the road excitation of certain frequencies is very small, so the influence of these frequency components can be ignored when calculating the road roughness. Assuming that the upper and lower limits of the required pavement time frequency component (effective time frequency) are f_u and f_l, respectively, then [9]

$$G_q(f) = \begin{cases} G_q(n_0)n_0^2\frac{u}{f^2}, & f_l \leq f \leq f_u \\ 0\,, & \text{others} \end{cases} \tag{9.11}$$

In the training of RBF neural network in this section, the vertical acceleration power spectral density of the body centroid and the pitch angular acceleration power spectral density are taken as the ideal input samples of the network, and the corresponding road power spectral density is taken as the ideal output samples of the network. When the network training is completed, the nonlinear mapping relationship between the vertical acceleration power spectral density of the body centroid, the pitch angular acceleration power spectral density and the road power spectral density is established [10, 11]. At this time, the model for identifying the road surface spectrum is completed.

9.2.2 *Frequency Domain Identification of Seven-Degree-of-Freedom Vehicle Road Surface Roughness*

1. Establishment of seven-degree-of-freedom vehicle vibration model

When the vehicle is driving on the uneven road, although the mass distribution of the vehicle in the longitudinal plane is approximately symmetrical, the different changes of the left and right wheel rut heights in the actual driving process will inevitably cause the vehicle body to roll. Therefore, the plane four-degree-of-freedom vehicle vibration model mentioned above has certain errors with the actual situation. In order to take the vehicle roll motion into account, a spatial seven-degree-of-freedom vehicle vibration model is introduced here, which is closer to the actual situation than the plane four-degree-of-freedom model. The 7-DOF vehicle vibration model takes into account the vertical, roll and pitch motions of the vehicle body.

Similarly, to establish the seven-degree-of-freedom vibration model of the vehicle, the following assumptions must be made first:

① The vehicle body is regarded as a rigid body with concentrated mass, and only the influence of vertical, roll and pitch vibration of the vehicle body on the ride comfort is considered, while the influence of torsional vibration of the vehicle body on the ride comfort is ignored;

② Assuming that the body vibrates slightly near the equilibrium position, the stiffness and damping of the suspension are regarded as constants, that is, the body is treated as a linear system;

③ The vehicle is symmetrical left and right along the longitudinal centerline and moves in a straight line at a constant speed;

④ The road surface is a stationary ergodic normal random process with isotropy;

⑤ Ignore vibration sources other than road surface.

Therefore, the vehicle with four-wheel independent suspension can be simplified into a spatial seven-degree-of-freedom vehicle vibration model as shown in Fig. 9.3.

From the Lagrange equation, the vibration equation of the vehicle can be obtained as follows

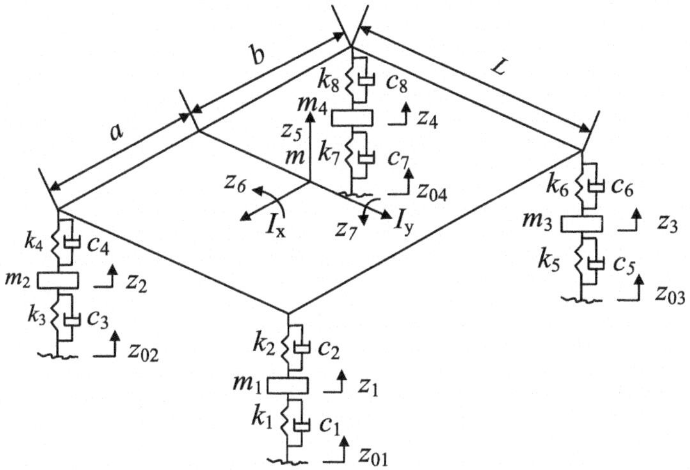

Fig. 9.3 Seven-degree-of-freedom vehicle vibration model. m_1, m_2 are the mass of two front wheels (kg); m_3, m_4 are the mass of two rear wheels (kg); m is the mass of vehicle body (kg); I_x is the moment of inertia of the vehicle body about the longitudinal axis of the center of mass (kg·m^2); I_y is the moment of inertia of the vehicle body about the transverse axis of the center of mass (kg·m^2);k_2, k_4 are the stiffness of two front suspensions (N/m);k_6, k_8 are the stiffness of two rear suspensions (N/m);c_2, c_4 are equivalent damping coefficient of two front suspensions (N·s/m); c_6, c_8 are equivalent damping coefficient of two rear suspensions (N·s/m); k_1, k_3 are the stiffness of two front tires (N/m); k_5, k_6 are the stiffness of two rear tires (N/m); c_1, c_3 are equivalent damping coefficient of two front tires (N·s/m); c_5, c_7 are equivalent damping coefficient of two rear tires (N·/m); z_{01}, z_{02}, z_{03}, z_{04} are the displacement excitation of the front left wheel, front right wheel, rear left wheel and rear right wheel in the vertical direction of the road surface (m); z_1, z_2, z_3, z_4 are the displacement of the front left wheel, front right wheel, rear left wheel and rear right wheel in the vertical direction (m); z_5 the vertical displacement of the body centroid (m); z_6 is the roll angle of the body (rad); z_7 is the pitch angle of the vehicle body (rad); a is the distance from the front axle to the centroid (m); b is the distance from the rear axle to the centroid (m); L is the track width of the left and right wheels (m)

$$M\ddot{Z} + C\dot{Z} + KZ = K_t Q + C_t \dot{Q} \tag{9.12}$$

where, M is the mass matrix of the vehicle, C is the damping matrix of the vehicle, K is the stiffness matrix of the vehicle, Z is a displacement vector corresponding to each degree of freedom of the vehicle, Q is the input vector of the road surface, K_t is the tire stiffness matrix, C_t is the tire damping matrix. The specific expression is as follows.

$$M = \text{diag}[\, m_1 \; m_2 \; m_3 \; m_4 \; m \; I_{\text{x}} \; I_{\text{y}} \,]$$

$$Z = \begin{bmatrix} z_1 & z_2 & z_3 & z_4 & z_5 & z_6 & z_7 \end{bmatrix}^{\text{T}}$$

$$Q = \begin{bmatrix} z_{01} & z_{02} & z_{03} & z_{04} \end{bmatrix}^{\text{T}}$$

$$K_t = \begin{bmatrix} k_1 & 0 & 0 & 0 \\ 0 & k_3 & 0 & 0 \\ 0 & 0 & k_5 & 0 \\ 0 & 0 & 0 & k_7 \\ 0 & 0 & 0 & 0 \\ 0 & 0 & 0 & 0 \\ 0 & 0 & 0 & 0 \end{bmatrix} \quad C_t = \begin{bmatrix} c_1 & 0 & 0 & 0 \\ 0 & c_3 & 0 & 0 \\ 0 & 0 & c_5 & 0 \\ 0 & 0 & 0 & c_7 \\ 0 & 0 & 0 & 0 \\ 0 & 0 & 0 & 0 \\ 0 & 0 & 0 & 0 \end{bmatrix}$$

$$K = \begin{bmatrix} k_1 + k_2 & 0 & 0 & 0 & -k_2 & -\frac{L}{2}k_2 & ak_2 \\ 0 & k_3 + k_4 & 0 & 0 & -k_4 & \frac{L}{2}k_4 & ak_4 \\ 0 & 0 & k_5 + k_6 & 0 & -k_6 & -\frac{L}{2}k_6 & -bk_6 \\ 0 & 0 & 0 & k_7 + k_8 & -k_8 & \frac{L}{2}k_8 & -bk_8 \\ -k_2 & -k_4 & -k_6 & -k_8 & \begin{matrix} k_2 + k_4 \\ +k_6 + k_8 \end{matrix} & \begin{matrix} \frac{L}{2}(k_2 - k_4 \\ +k_6 - k_8) \end{matrix} & \begin{matrix} -ak_2 - ak_4 \\ +bk_6 + bk_8 \end{matrix} \\ -\frac{L}{2}k_2 & \frac{L}{2}k_4 & -\frac{L}{2}k_6 & \frac{L}{2}k_8 & \begin{matrix} \frac{L}{2}(k_2 - k_4 \\ +k_6 - k_8) \end{matrix} & \begin{matrix} \frac{L^2}{4}(k_2 + k_4 \\ +k_6 + k_8) \end{matrix} & \begin{matrix} \frac{L}{2}(-ak_2 + ak_4 \\ +bk_6 - bk_8) \end{matrix} \\ ak_2 & ak_4 & -bk_6 & -bk_8 & \begin{matrix} -ak_2 - ak_4 \\ +bk_6 + bk_8 \end{matrix} & \begin{matrix} \frac{L}{2}(-ak_2 + ak_4 \\ +bk_6 - bk_8) \end{matrix} & \begin{matrix} a^2(k_2 + k_4) \\ +b^2(k_6 + k_8) \end{matrix} \end{bmatrix}$$

$$C = \begin{bmatrix} c_1+c_2 & 0 & 0 & 0 & -c_2 & -\frac{L}{2}c_2 & ac_2 \\ 0 & c_3+c_4 & 0 & 0 & -c_4 & \frac{L}{2}c_4 & ac_4 \\ 0 & 0 & c_5+c_6 & 0 & -c_6 & -\frac{L}{2}c_6 & -bc_6 \\ 0 & 0 & 0 & c_7+c_8 & -c_8 & \frac{L}{2}c_8 & -bc_8 \\ -c_2 & -c_4 & -c_6 & -c_8 & \begin{matrix} c_2+c_4 \\ +c_6+c_8 \end{matrix} & \begin{matrix} \frac{L}{2}(c_2-c_4 \\ +c_6-c_8) \end{matrix} & \begin{matrix} -ac_2-ac_4 \\ +bc_6+bc_8 \end{matrix} \\ -\frac{L}{2}c_2 & \frac{L}{2}c_4 & -\frac{L}{2}c_6 & \frac{L}{2}c_8 & \begin{matrix} \frac{L}{2}(c_2-c_4 \\ +c_6-c_8) \end{matrix} & \begin{matrix} \frac{L^2}{4}(c_2+c_4 \\ +c_6+c_8) \end{matrix} & \begin{matrix} \frac{L}{2}(-ac_2+ac_4 \\ +bc_6-bc_8) \end{matrix} \\ ac_2 & ac_4 & -bc_6 & -bc_8 & \begin{matrix} -ac_2-ac_4 \\ +bc_6+bc_8 \end{matrix} & \begin{matrix} \frac{L}{2}(-ac_2+ac_4 \\ +bc_6-bc_8) \end{matrix} & \begin{matrix} a^2(c_2+c_4) \\ +b^2(c_6+c_8) \end{matrix} \end{bmatrix}$$

2. Power spectral density function of road surface input to four-wheel vehicle

The fitting expression of road power spectral density is the same as that of four-degree-of-freedom.

When the statistical characteristics of the left and right wheel ruts are the same and the phase spectrum is zero, the spectrum matrix of the road input to the four-wheel vehicle can be expressed as [12]

$$G_{ik}(f) = G_q(f)\begin{bmatrix} 1 & \mathrm{coh}(f) & A^{-1} & \mathrm{coh}(f)A^{-1} \\ \mathrm{coh}(f) & 1 & \mathrm{coh}(f)A^{-1} & A^{-1} \\ A & \mathrm{coh}(f)A & 1 & \mathrm{coh}(f) \\ \mathrm{coh}(f)A & A & \mathrm{coh}(f) & 1 \end{bmatrix} \tag{9.13}$$

where $A = e^{\frac{j2\pi f(a+b)}{u}}$, $\mathrm{coh}(\cdot)$ denotes the coherence function of the two wheel trajectories.

Coherence function is generally used to describe the correlation degree between left and right wheel ruts. After data fitting, the expression is [13]

$$\mathrm{coh}(f)^2 = \begin{cases} 1-0.45f, & f \le 2\mathrm{Hz} \\ 0.1, & f > 2\mathrm{Hz} \end{cases} \tag{9.14}$$

The method of solving the vibration response of 7-DOF vehicle and the method of establishing RBF neural network model are the same as that of 4-DOF vehicle.

9.2.3 Virtual Test Verification

1. Four-degree-of-freedom simulation case

In this case, the displacement power spectral density of class B and class C road surface is simulated. The lower limit of time and frequency of the ground acting on the tire excitation is $f_l = 0.5$ Hz and the upper limit is $f_u = 30$ Hz.

Refer to the four-degree-of-freedom vehicle vibration model in Fig. 9.2, and use the data of a mini-car for simulation calculation. The basic data are [14]: $m = 920$ kg, $I_y = 948\,\text{kg}\cdot\text{m}^2$, $m_1 = m_2 = 50$ kg, $a = 1.109$ m, $u = 80$ km/h, $c_2 = 3340\,\text{N}\cdot\text{s/m}$, $k_2 = 33.4$ kN/m,.

$k_4 = 42.2$kN/m, $b = 1.09$m, $c_4 = 2845\text{N}\cdot\text{s/m}$, $k_1 = k_3 = 294$kN/m, $c_1 = c_3 = 80\text{N}\cdot\text{s/m}$.

(1) Power spectral density simulation results of B-class road

The power spectral density of various roads obtained by Eq. (9.6) are taken as the ideal output samples of the network. According to the above data, the MATLAB software is used to program, and the power spectral density of vertical acceleration of vehicle body centroid and pitch angular acceleration corresponding to various road surfaces are obtained by simulation, which are used as ideal input samples of the network. The function Newrb in MATLAB is adopted to establish the network sample training process.

After the network training, in order to test the accuracy of the network model, the simulated power spectral density of vertical acceleration and pitching angle acceleration of the vehicle body centroid on the B-class road are substituted into the trained network for calculation, as shown in Figs. 9.4 and 9.5.

Figure 9.6a shows the results of RBF neural network identifying B-class road spectrum with double logarithmic coordinates. It can be seen that the accuracy of

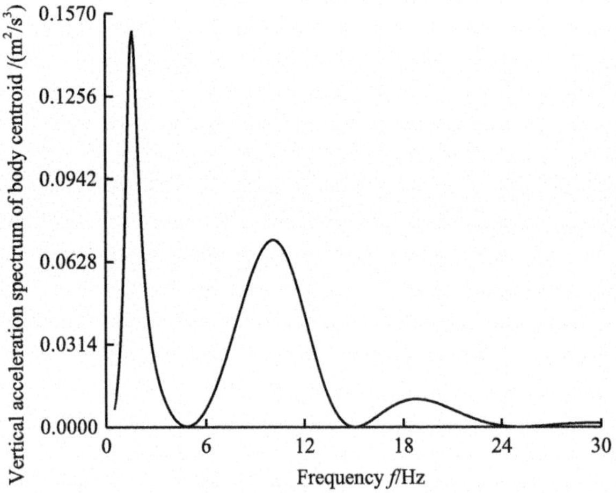

Fig. 9.4 Vertical acceleration spectrum curve of body centroid on B-class road (4-DOF)

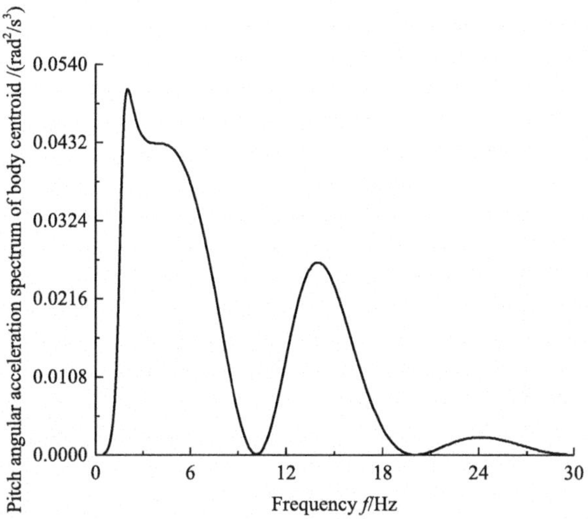

Fig. 9.5 Pitch angular acceleration spectrum curve of body centroid on B-class road (4-DOF)

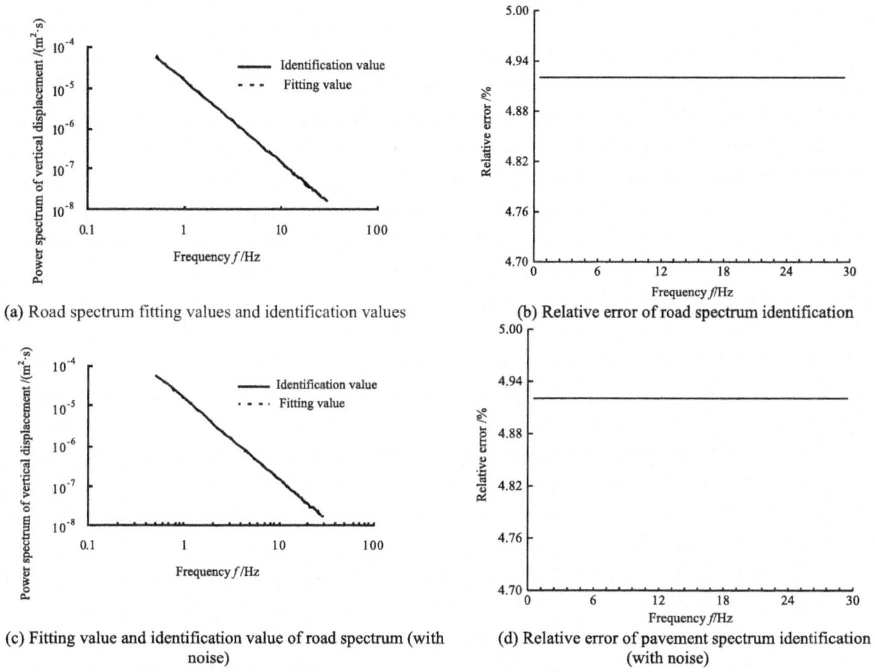

Fig. 9.6 Simulation results on B-class road (4-DOF)

RBF network in identifying the road spectrum is very high, and the identification value and the fitting value have a good agreement.

Figure 9.6b shows the relative error between the fitting value and the identification value. As can be seen, the relative error is very small, almost equal to zero. However, in reference [15], the error of road surface spectrum fitted by AR model method is 10.22%, and the error of road surface spectrum fitted by pseudo white noise method is 9.29%. Therefore, using RBF neural network to identify the road spectrum is simple, convenient, clear and accurate.

In order to examine the anti-interference ability of the RBF network after training, the uniformly distributed random noise with the maximum amplitude of 10% is artificially added to the vehicle body centroid acceleration response data and the pitch angular acceleration response data, and the body centroid acceleration and pitch angular acceleration after interference are input into the trained RBF network. The identification results are shown in Fig. 9.6c, d. It can be seen that the relative error of the network recognition is slightly increased, but the accuracy of the road spectrum recognition is higher than that of other methods, which indicates that the RBF network has better anti-noise ability.

(2) Simulation results of power spectral density on C-class road

For C-class road, the power spectral density of vertical acceleration and pitch angular acceleration of the body centroid are obtained by simulations in MATLAB software, as shown in Figs. 9.7 and 9.8 respectively, and they are substituted into the trained network for calculation.

Figure 9.9a shows the results of RBF neural network identifying the road surface spectrum with double logarithmic coordinates. Figure 9.9b shows the relative error between the fitting value and the identification value. It can be seen that RBF neural network has high accuracy in identifying road spectrum. However, the relative error is larger than that of identifying B-class road spectrum. Figure 9.9c, d are simulation

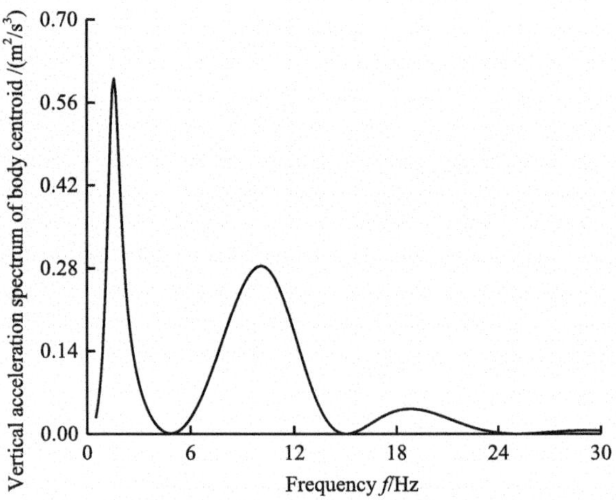

Fig. 9.7 Vertical acceleration spectrum curve of body centroid on C-class road (4-DOF)

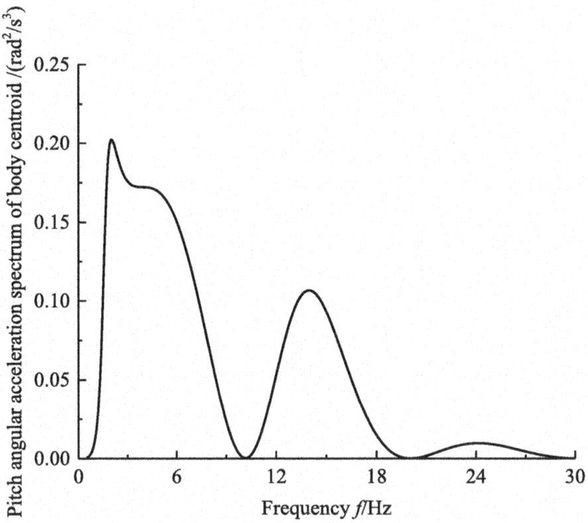

Fig. 9.8 Pitching angular acceleration spectrum curve of body centroid on C-class road (4-DOF)

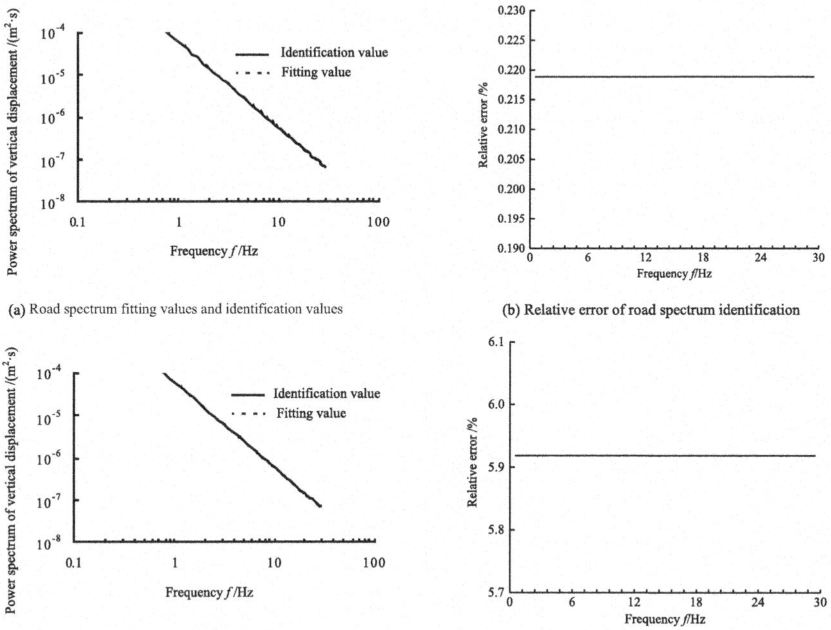

(a) Road spectrum fitting values and identification values

(b) Relative error of road spectrum identification

(c) Fitting value and identification value of road spectrum (with noise)

(d) Relative error of pavement spectrum identification (with noise)

Fig. 9.9 Simulation results on C-class road (4-DOF)

results after adding noise. It can be seen that the RBF neural network has strong anti-noise ability, and the relative error of identifying C-class road spectrum is larger than that of identifying grade B-class road spectrum.

2. Seven-degree-of-freedom simulation case

Refer to the seven-degree-of-freedom vehicle vibration model in Fig. 9.3, and use the data of a mini-car for simulation calculation. The basic data are:

$$I_x = 239\,\text{kg}\cdot\text{m}^2,\ a = 1.109\,\text{m},\ m_1 = m_2 = m_3 = m_4 = 25\,\text{kg},\ L = 1.22\,\text{m},\ u = 80\,\text{km/h}$$
$$c_6 = c_8 = 3340\,\text{N}\cdot\text{s/m},\ k_2 = k_4 = 16.7\,\text{kN/m},\ k_6 = k_8 = 21.1\,\text{kN/m},\ b = 1.09\,\text{m}$$
$$c_2 = c_4 = 2845\,\text{N}\cdot\text{s/m},\ k_1 = k_3 = k_5 = k_7 = 147\,\text{kN/m},\ c_1 = c_3 = c_5 = c_7 = 40\,\text{N}\cdot\text{s/m}$$

(1) Power spectral density simulation results of B-class road

The MATLAB software is applied to program, and the power spectral density of vertical acceleration and pitching angular acceleration of the body centroid on the B-class road are obtained by simulation, as shown in Figs. 9.9, 9.10 and 9.11, respectively. These are substituted into the trained network for calculation.

Figure 9.12a shows the results of RBF network identifying the road surface spectrum with double logarithmic coordinates. Figure 9.12b gives the relative error between the fitting value and the identification value. It can be seen from the figure that the fitting value and the identification value are in good agreement. The relative error is smaller than that of the four-degree-of-freedom model, which indicates that the accuracy obtained by using the seven-degree-of-freedom model is high. Therefore, the seven-degree-of-freedom model can better describe the actual motion of the vehicle. Figure 9.12c, d are the simulation results with noise. It can be seen that RBF neural network has strong anti-noise ability. The relative error is smaller than that of four-degree-of freedom model.

Fig. 9.10 Vertical acceleration spectrum curve of body centroid on B-class road (7-DOF)

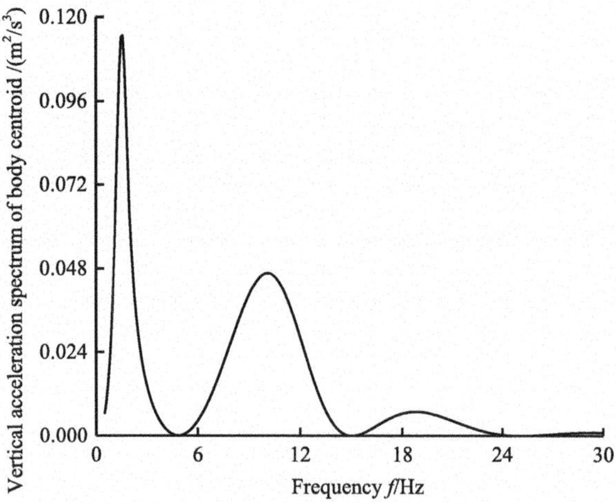

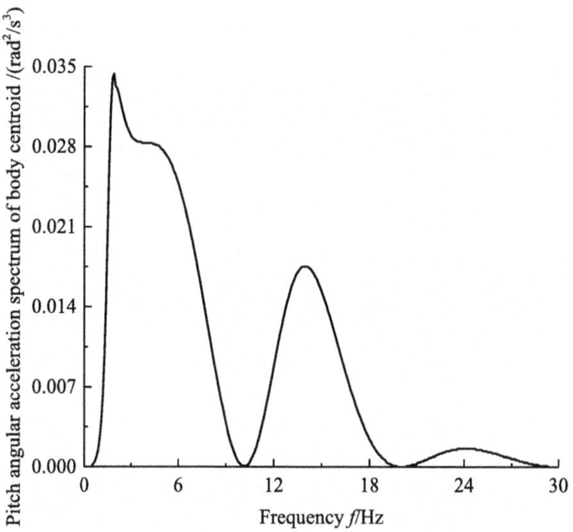

Fig. 9.11 Pitch angular acceleration spectrum curve of body centroid on B-class road (7-DOF)

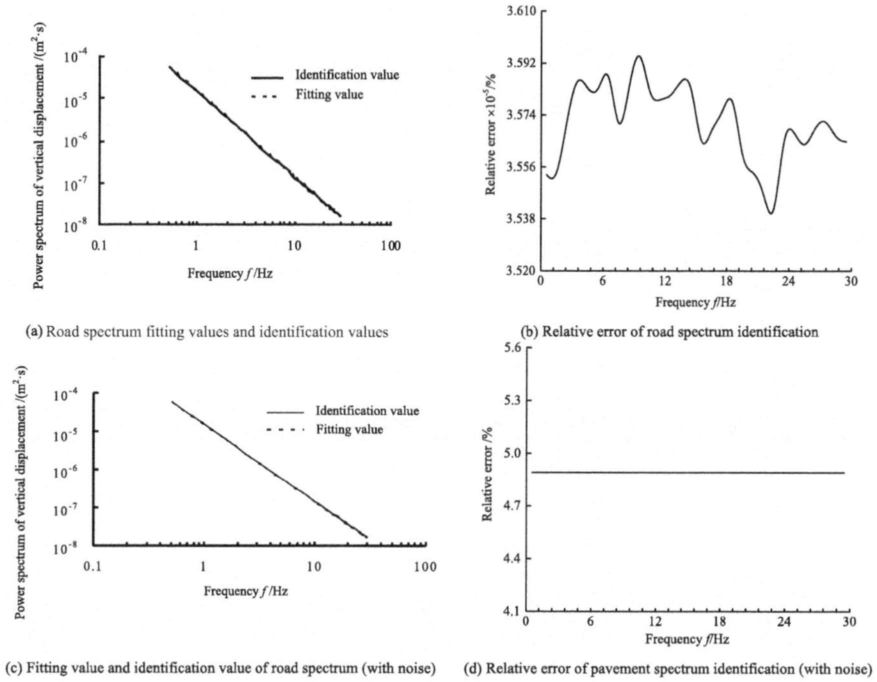

Fig. 9.12 Simulation results on B-class road (7-DOF)

(2) Simulation results of power spectral density on C-class road

The vertical acceleration power spectrum density and pitch angular acceleration power spectrum density of the body centroid on the C-class road are obtained by simulation, as shown in Figs. 9.13 and 9.14, which are substituted into the trained network for calculation.

Figure 9.15a shows the results of RBF neural network identifying the road surface spectrum with double logarithmic coordinates. Figure 9.15b illustrates the relative

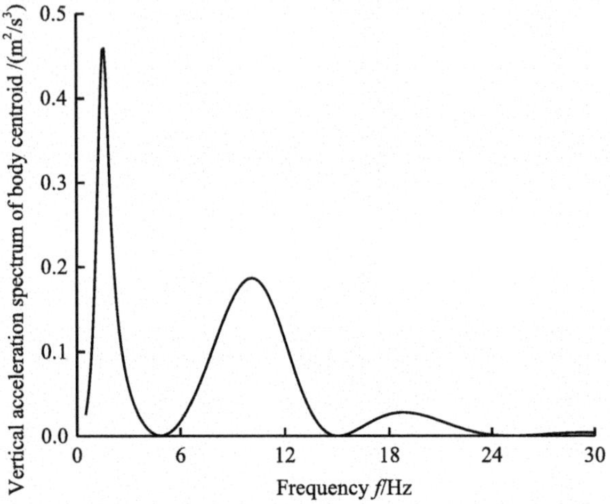

Fig. 9.13 Vertical acceleration spectrum curve of body centroid on C-class road (7-DOF)

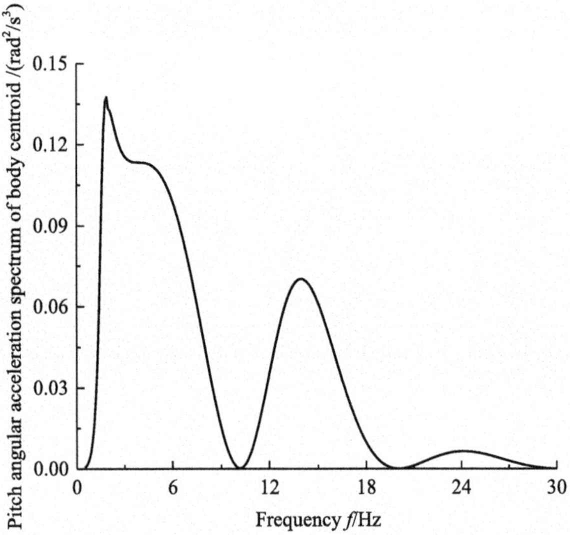

Fig. 9.14 Pitch angular acceleration spectrum curve of body centroid on C-class road (7-DOF)

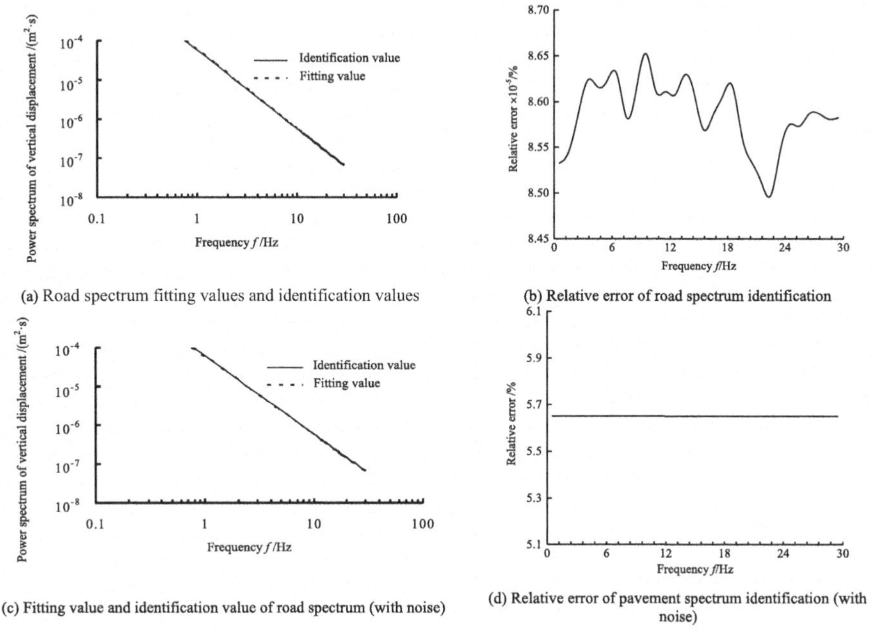

(a) Road spectrum fitting values and identification values

(b) Relative error of road spectrum identification

(c) Fitting value and identification value of road spectrum (with noise)

(d) Relative error of pavement spectrum identification (with noise)

Fig. 9.15 Simulation results on C-class road (7-DOF)

error between the fitting value and the identification value. Figure 9.15c, d are simulation results after adding noise. It can be seen that the relative error is larger than that of identifying B-class road.

9.3 Study on Time Domain Inverse Dynamics of Road Roughness

9.3.1 Identification of Four-Degree-of-Freedom Vehicle Road Roughness in Time Domain

1. Function of road surface input to two-wheeled vehicles

This section uses filtered white noise as a single-point excitation model for road roughness. Suppose the front wheel road roughness model is

$$\dot{q}_1(t) = -auq_1(t) + W(t) \tag{9.15}$$

Suppose the distance between the rear contact point and the front contact point is l_2, the time delay of the rear wheel relative to the front wheel is $\Delta_2 = l_2/u$. This kind of pure time delay systems are approximated by Pade, which takes the first-order algorithm, and the transfer function relationship between the anterior and posterior excitation inputs is [16]

$$G_{21} = \frac{q_2(s)}{q_1(s)} = e^{-\Delta_2 s} = \frac{1 - \Delta_2 \frac{s}{2}}{1 + \Delta_2 \frac{s}{2}} \tag{9.16}$$

where $q_1(s)$, $q_2(s)$ are Laplace transform of $q_1(t)$, $q_2(t)$, respectively, s is Laplace operator. The Laplace inverse transformation is applied and the results are rearranged to obtain

$$\dot{q}_2(t) = -\frac{2}{\Delta_2} q_2(t) + \frac{2}{\Delta_2} q_1(t) - \dot{q}_1(t) \tag{9.17}$$

Consequently, the state equation of the front and rear wheel excitations can be expressed as

$$\begin{bmatrix} \dot{q}_1(t) \\ \dot{q}_2(t) \end{bmatrix} = \begin{bmatrix} -au & 0 \\ au + \frac{2}{\Delta_2} & -\frac{2}{\Delta_2} \end{bmatrix} \begin{bmatrix} q_1(t) \\ q_2(t) \end{bmatrix} + \begin{bmatrix} 1 \\ -1 \end{bmatrix} W(t) \tag{9.18}$$

Take the power spectral density as the standard form, the B-class pavement constant $\alpha = 0.1303\,\boldsymbol{m}^{-1}$, $\sigma = 0.0032\,\boldsymbol{m}$, and the C-class pavement constant $\alpha = 0.12\,\boldsymbol{m}^{-1}$, $\sigma = 0.006\,\boldsymbol{m}$ [17]. Take the speed $u = 80\,\text{km/h}$, using the general numerical simulation software MATLAB, the road roughness of front and rear road wheels is solved by computer as shown in Figs. 9.16 and 9.17. Obviously, the size of the axle distance of the front and rear wheels is reflected as the delay between the simulation curves.

2. Solution of time domain response to vehicle vibration

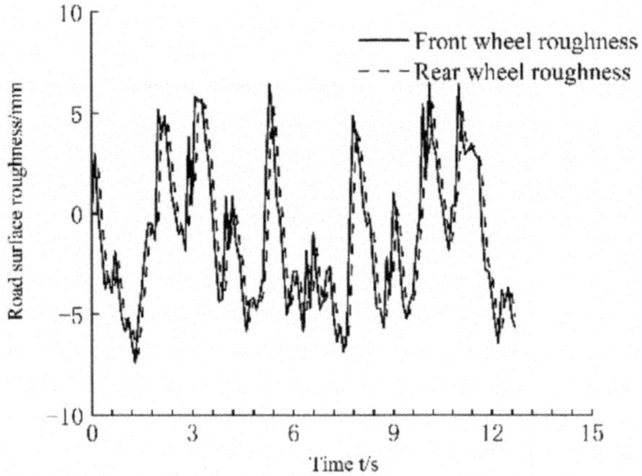

Fig. 9.16 Front and rear wheel unevenness of B-class road

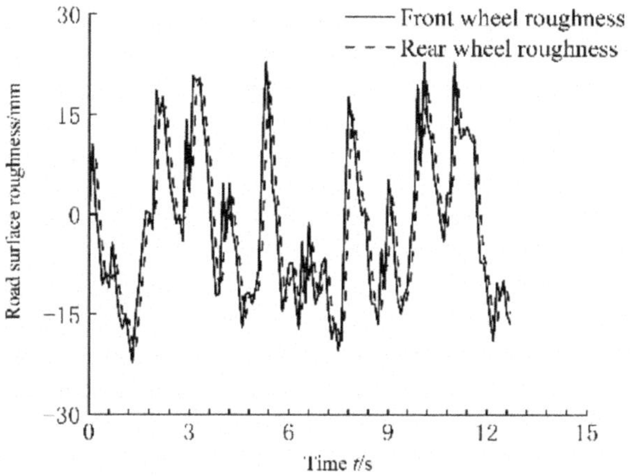

Fig. 9.17 Front and rear wheel unevenness of C-class road

For the convenience of solving, multiply the two sides of Eq. (9.1) by M^{-1}

$$\ddot{Z} = -M^{-1}C\dot{Z} - M^{-1}KZ + M^{-1}K_tQ + M^{-1}C_t\dot{Q} \tag{9.19}$$

Let $A = -M^{-1}C$, $B = -M^{-1}K$, $I_1 = M^{-1}K_t$, $I_2 = M^{-1}C_t$, $Y = [Z \quad \dot{Z}]^{\mathrm{T}}$, the Eq. (9.15) is translated into the state equation as follows

$$\dot{Y}_{8\times 1} = \begin{bmatrix} 0_{4\times 4} & E_{4\times 4} \\ B_{4\times 4} & A_{4\times 4} \end{bmatrix} Y_{8\times 1} + \begin{bmatrix} 0_{4\times 2} \\ I_{14\times 2} \end{bmatrix} Q_{2\times 1} + \begin{bmatrix} 0_{4\times 2} \\ I_{24\times 2} \end{bmatrix} \dot{Q}_{2\times 1} \tag{9.20}$$

The establishment method of RBF neural network model for road roughness time domain recognition is the same as the method for establishing RBF neural network model for road roughness frequency domain recognition.

9.3.2 *Identification of Seven-Degree-of-Freedom Vehicle Road Roughness in Time Domain*

Practice has proved that there is a spatial correlation of outliers between the left and right wheel tracks. Let $W_x(s)$ be the input white noise of the rut x and $W_y(s)$ be the input white noise of the other round of rut y, be first establish the relationship between the two wheels of rut input white noise, and then find excitation elevation input of the front and rear points on the y line. According to the random vibration theory, for linear systems, when the input is a stationary process, its output is also a stationary process. In this way, the problem can be virtualized as follows: The input of one rut is known, and the output of another rut is found, and the input and output are joined by the coherence function $\cos_{xy}(f)$. That is, the transfer relationship between the left and right wheels is established, and the known rut of random processes derives another rut of random processes.

The previous wheel is the reference point, and the transfer function between the left and right wheel road surface white noise inputs $W_x(s)$, $W_y(s)$ is [16]·

$$G_{xy} = \frac{W_y(s)}{W_x(s)} = \frac{a_0 + a_1 s + \cdots + a_n s^n}{b_0 + b_1 s + \cdots + b_n s^n} \tag{9.21}$$

Take the second-order approximation and get

$$G_{xy} = \frac{a_0 + a_1 s + a_2 s^2}{b_0 + b_1 s + b_2 s^2} \tag{9.22}$$

where, the coefficients are obtained by fitting and transforming the coherence function measured on different roads.

Introduce the intermediate variable $M(s)$ and change the Eq. (9.22) into

$$G_{xy} = \frac{W_y(s)}{W_x(s)} = \frac{\left(a_0 s^{-2} + a_1 s^{-1} + a_2\right) M(s)}{\left(b_0 s^{-2} + b_1 s^{-1} + b_2\right) M(s)} \tag{9.23}$$

Furthermore, obtained by the numerator and denominator, respectively

$$M(s) = \frac{1}{b_2} W_x(s) - \frac{b_0}{b_2} s^{-2} M(s) - \frac{b_1}{b_2} s^{-1} M(s) \tag{9.24}$$

$$W_y(s) = a_2 M(s) + a_0 s^{-2} M(s) + a_1 s^{-1} M(s) \tag{9.25}$$

Let the state variable be $x_1 = L^{-1}\left(s^{-1} M(s)\right), \quad x_2 = L^{-1}\left(s^{-2} M(s)\right)$.

Because $L\dot{x}_2 = s \cdot Lx_2 = s \cdot s^{-2} M(s) = s^{-1} M(s) = Lx_1$, there is $\dot{x}_2 = x_1$.

Equation (9.24) is inversely Laplace transformed and combined with $\dot{x}_2 = x_1$ to obtain two-round correlation state equation

$$\begin{bmatrix} \dot{x}_1 \\ \dot{x}_2 \end{bmatrix} = \begin{bmatrix} -\frac{b_1}{b_2} & -\frac{b_0}{b_2} \\ 1 & 0 \end{bmatrix} \begin{bmatrix} x_1 \\ x_2 \end{bmatrix} + \begin{bmatrix} \frac{1}{b_2} \\ 0 \end{bmatrix} W_x \tag{9.26}$$

Similarly, apply the Laplace inverse transformation to Eq. (9.25), and we get the equation for the output W_y

$$W_y = a_2 \dot{x}_1 + a_0 x_2 + a_1 x_1 \tag{9.27}$$

Substituting the equation $\dot{x}_1$ precipitated by the Eq. (9.26) into the Eq. (9.27), then

$$H(\omega) = \mathrm{coh}(\omega) \tag{9.28}$$

When the road surface is generated, the input white noise W_y of another rut can be obtained by the input white noise W_x of one rut, which through intermediate state vibration x_1, x_2 in Eqs. (9.26) and (9.27), then follow the same filter Eq. (9.15) to get time domain model of the second rut associated with the first rut space.

Suppose the average phase difference of the two ruts excitation is zero, according to the relationship between the modulo of the frequency response function and the coherence function $H(\omega) = \text{coh}(\omega)$, and according to the coherence function and processing method cited in Ref. [16], the values of each coefficient in Eq. (9.22) are obtained: $a_0 = 3.1815$, $a_1 = 0.2063$, $a_2 = 0.0108$, $b_0 = 3.223$, $b_1 = 0.59$, $b_2 = 0.0327$. Use MATLAB software to program, the simulation results show that the unevenness excitation of the four-wheeled vehicle on the class-B and class-C roads shown in Figs. 9.18, 9.19, 9.20 and 9.21.

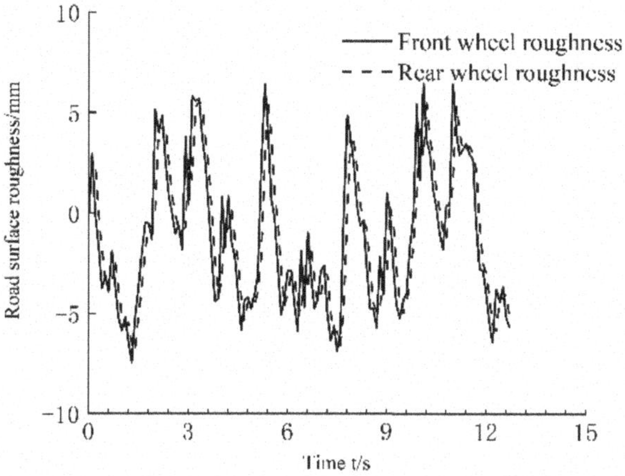

Fig. 9.18 Left front and rear wheel unevenness of B-class road

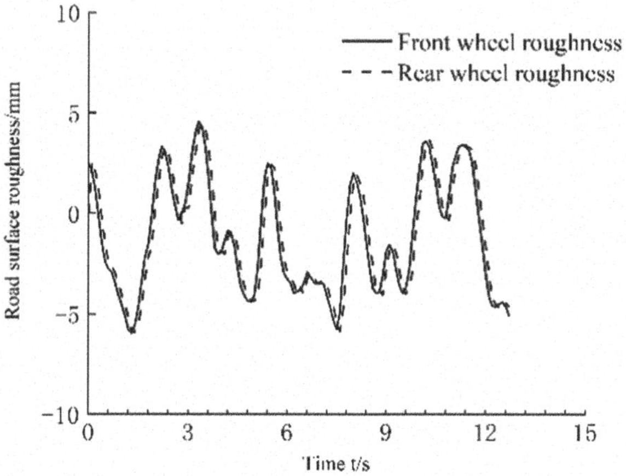

Fig. 9.19 Right front and rear wheel unevenness of B-class road

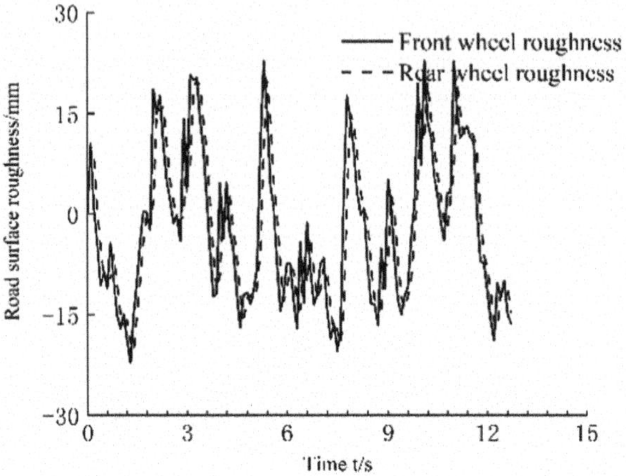

Fig. 9.20 Left front and rear wheel unevenness of C-class road

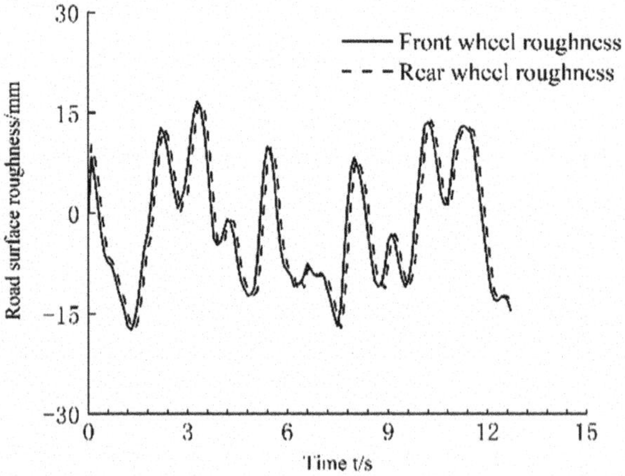

Fig. 9.21 Right front and rear wheel unevenness of C-class road

The establishment method of RBF neural network model for road roughness time domain recognition is the same as the method for establishing RBF neural network model for road roughness frequency domain recognition.

9.3.3 *Virtual Test Verification*

1. Four degrees of freedom simulation example

(1) Simulation results on B-Class road

According to the four-degree-of-freedom vehicle data, programmed using MATLAB software, the simulation obtains the vertical acceleration and pitch angle acceleration of the body centroid on the class-B road, as shown in Figs. 9.22 and 9.23, Substitute the trained network for computation.

Figure 9.24a shows the results of the RBF neural network's identification of the unevenness of the front wheels on the B-class road. Figure 9.24b shows the relative error between the identified value and the filtered white noise fitted value. In the first small period of time, the relative error is large, mainly due to the large vibration generated when the car is just starting. Figure 9.24c, d are simulation results with the addition of noise. It can be seen that the relative error of RBF neural network recognition of road roughness is small, the recognition accuracy is high, and the anti-noise ability is strong.

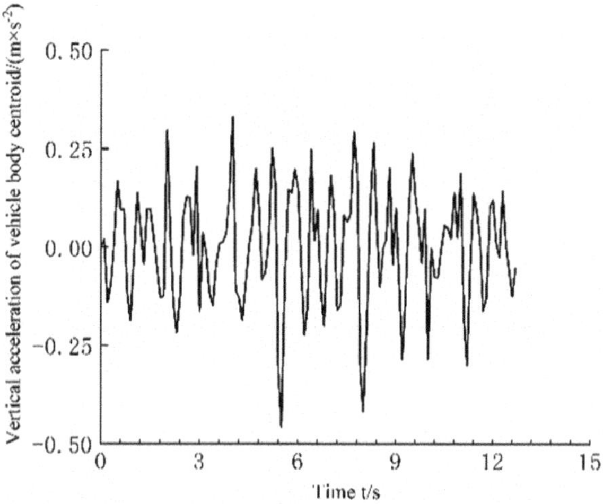

Fig. 9.22 Vertical acceleration curve of the centroid on B-class road (4-DOF)

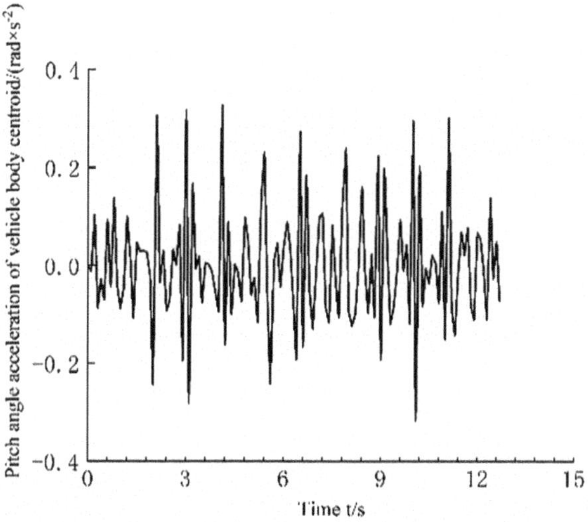

Fig. 9.23 Pitch angular acceleration curve of the centroid on B-class road (4-DOF)

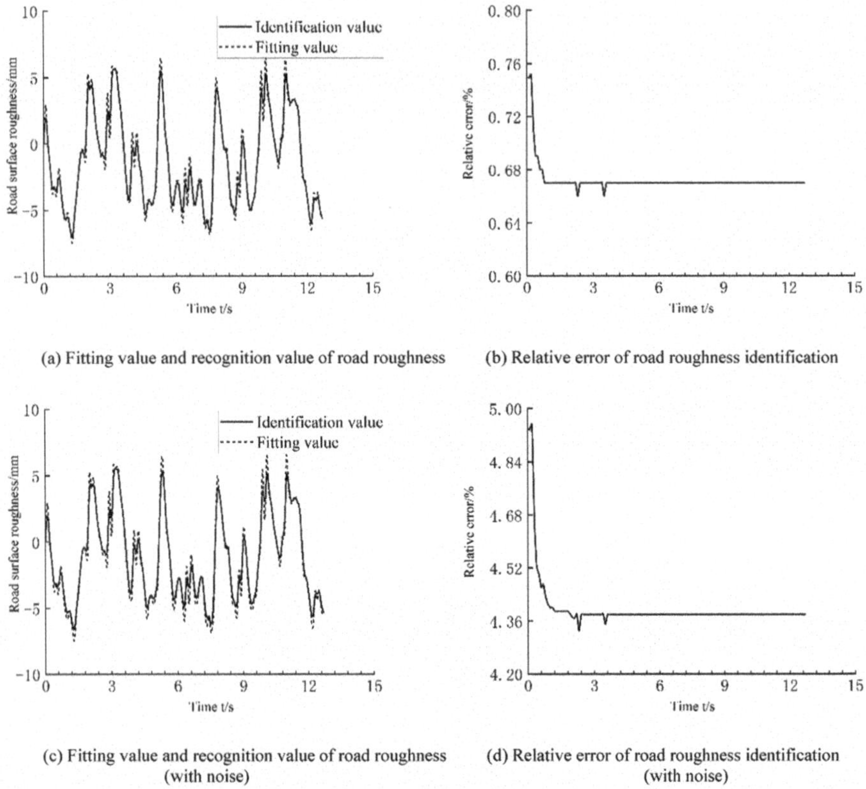

(a) Fitting value and recognition value of road roughness

(b) Relative error of road roughness identification

(c) Fitting value and recognition value of road roughness (with noise)

(d) Relative error of road roughness identification (with noise)

Fig. 9.24 Simulation results on B-class road (4-DOF)

(2) Simulation results on C-class road

Programmed using MATLAB software, the vertical acceleration and pitch angle acceleration of the body centroid on the C-class road surface are simulated, as shown in Figs. 9.25 and 9.26, the trained network is substituted for computation.

Figure 9.27a shows the results of the RBF neural network to identify road roughness. Figure 9.27b shows the relative error between the identification value and the filtered white noise fitting value. Its relative error is larger than the relative error of identifying B-class road roughness. Figure 9.27c, d are simulation results with the addition of noise.

2. Seven degrees of freedom simulation example

(1) Simulation results on B-class road

Based on the seven degrees of freedom vehicle data, the MATLAB software is applied for programming, and the vertical acceleration and pitch angle acceleration of the body centroid on the B-class road surface are simulated, as shown in Figs. 9.28 and 9.29, respectively. The trained network is substituted for computation.

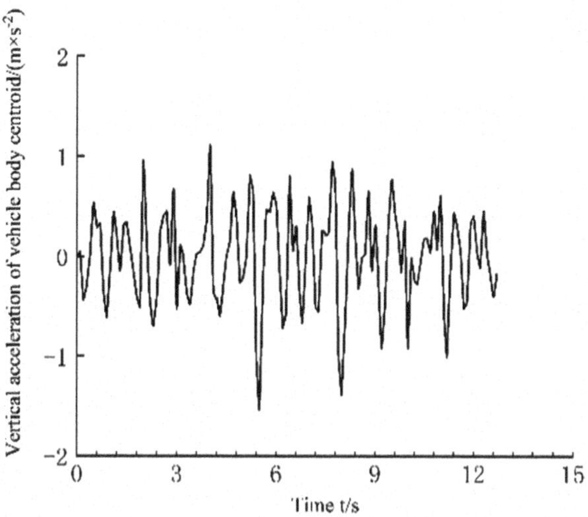

Fig. 9.25 Vertical acceleration curve of the centroid on C-class road (4-DOF)

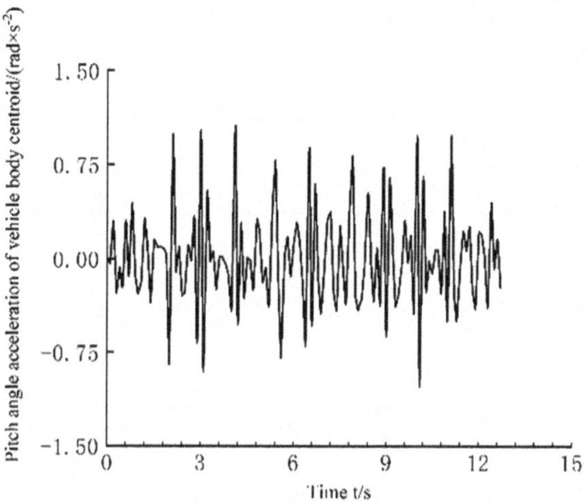

Fig. 9.26 Pitch angle acceleration curve of the centroid on C-class road (4-DOF)

Figure 9.30a gives the results of the RBF neural network identifying the pavement excitation of the left front wheel on the B-class road. Figure 9.30b gives the relative error of the identification value and the filtered white noise fitting value. Figure 9.30c, d is the simulation result after noise is added. It can be seen that its relative error is small, the noise resistance is strong, and the recognition accuracy is higher than that of the four-degree-of-freedom vehicle model.

(2) Simulation results on C-class road

Programmed using MATLAB software, the vertical acceleration and pitch angle acceleration of the body centroid on the C-class road surface are simulated, as shown

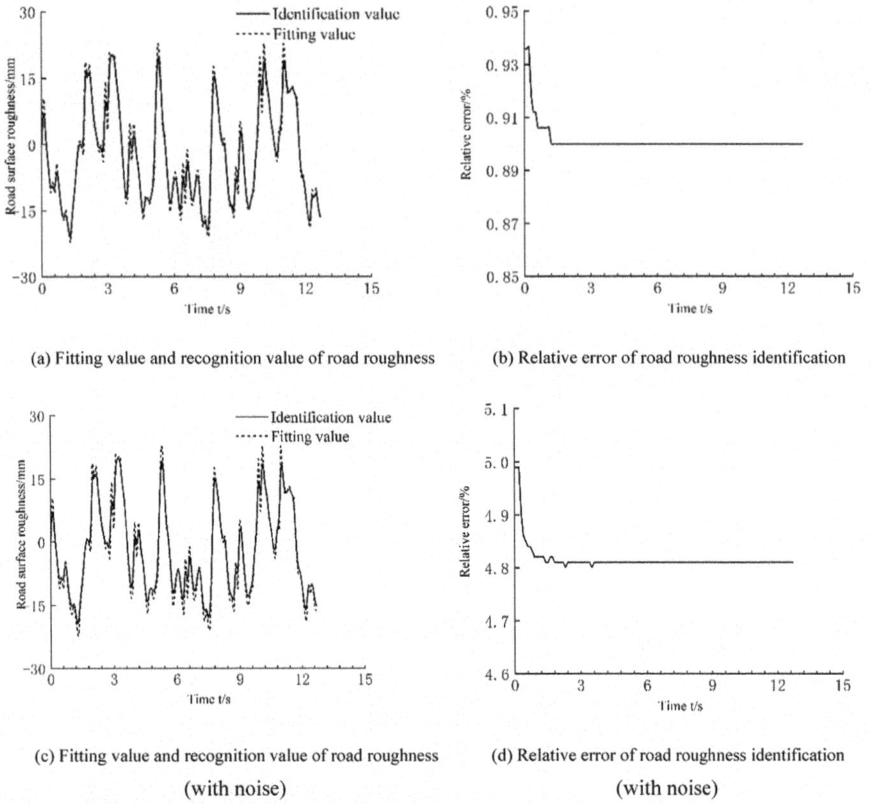

Fig. 9.27 Simulation results on C-class road (4-DOF)

Fig. 9.28 Vertical acceleration curve of the centroid on B-class road (7-DOF)

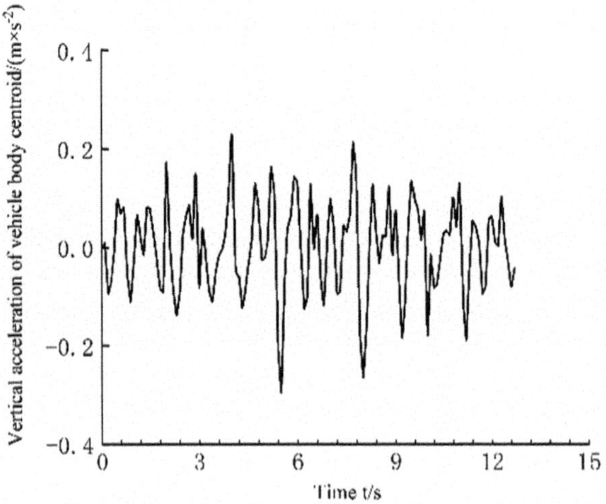

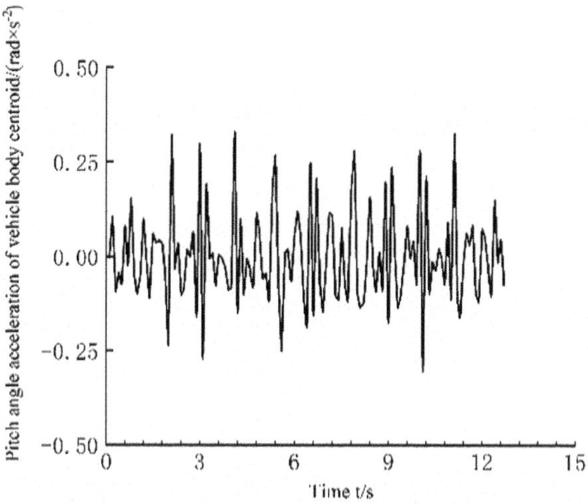

Fig. 9.29 Pitch angular acceleration curve of the centroid on B-class road (7-DOF)

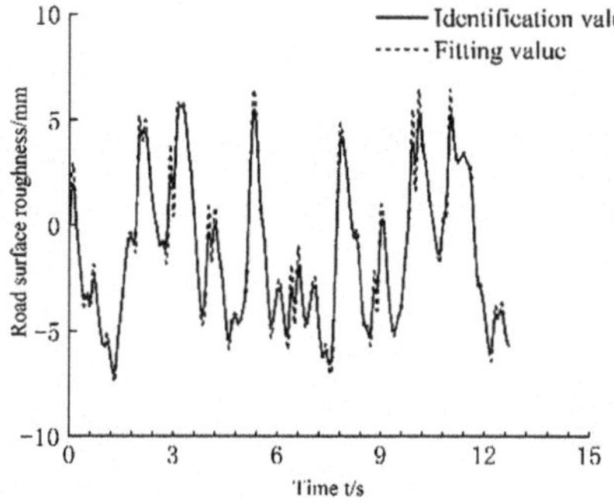

(a) Fitting value and recognition value of road roughness

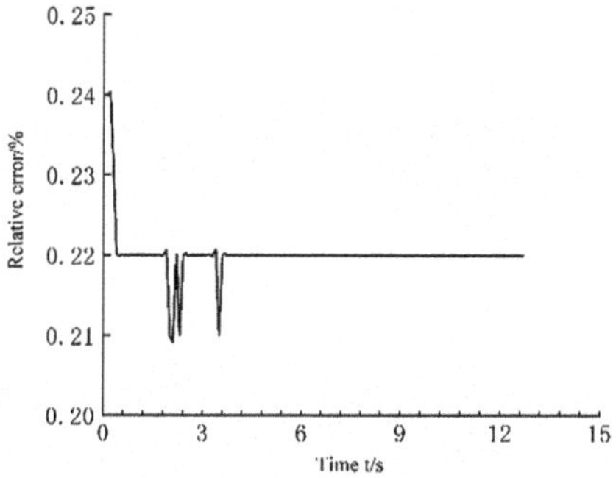

(b) Relative error of road roughness identification

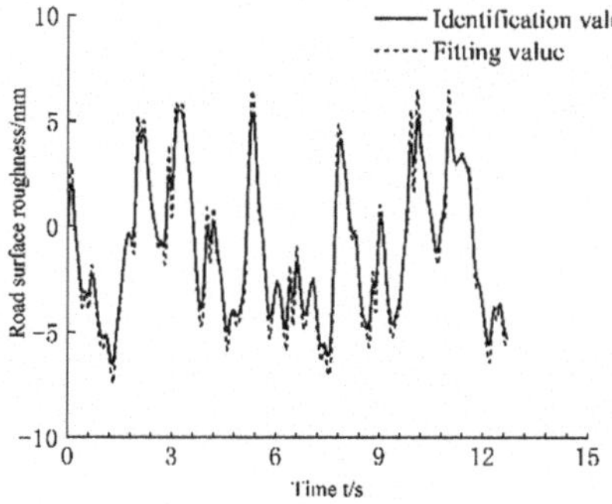

(c) Fitting value and recognition value of road roughness (with noise)

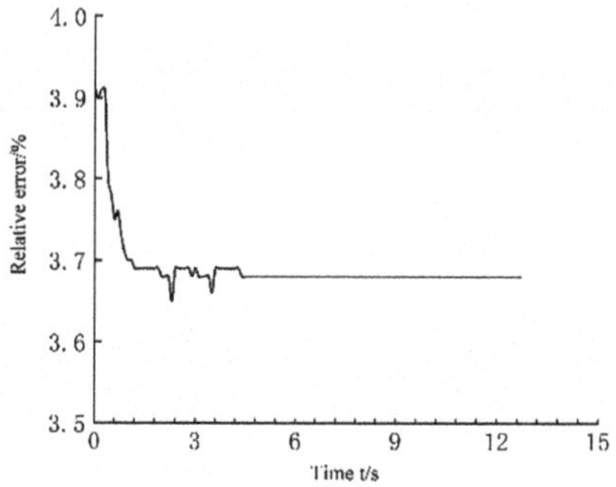

(d) Relative error of road roughness identification (with noise)

Fig. 9.30 Simulation results on B-class road (7-DOF)

in Figs. 9.31 and 9.32, respectively. As shown in Figs. 9.31 and 9.32, the trained network is substituted for computation.

Figure 9.33a gives the results of the RBF neural network identifying the pavement excitation of the left front wheel on a C-class road surface. Figure 9.33b gives the relative error of the identification value and the filtered white noise fitting value. Figure 9.33c, d is the simulation result after noise is added. The result is also more accurate than the four-degree-of-freedom vehicle model recognition.

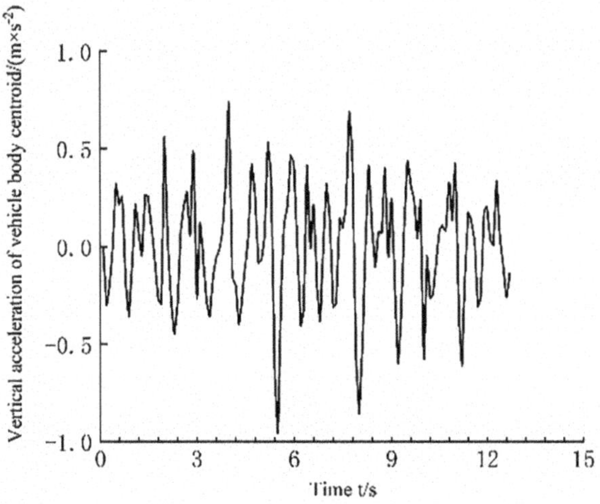

Fig. 9.31 Vertical acceleration curve of the centroid on C-class road (7-DOF)

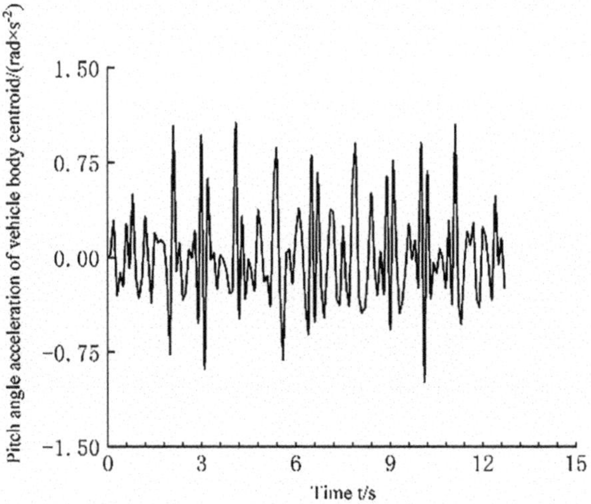

Fig. 9.32 Pitch angular acceleration curve of the centroid on C-class road (7-DOF)

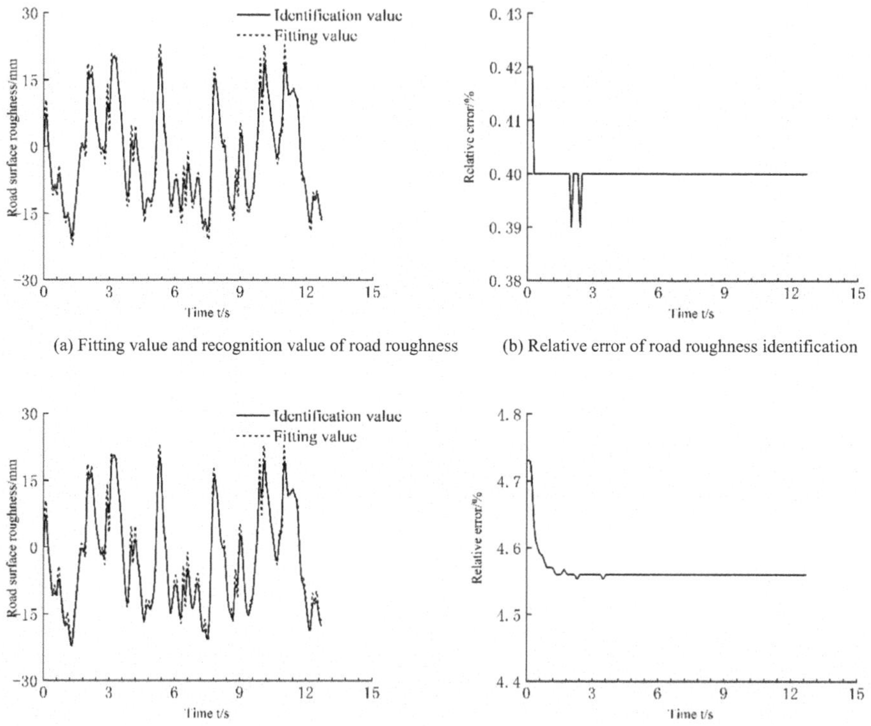

(a) Fitting value and recognition value of road roughness

(b) Relative error of road roughness identification

(c) Fitting value and recognition value of road roughness (with noise)

(d) Relative error of road roughness identification (with noise)

Fig. 9.33 Simulation results on C-class road (7-DOF)

9.4 Prediction of Evolution Random Response of Vehicle Excited by Road Surface [18]

9.4.1 Vehicle Vertical Dynamics Model

1. Two-degree-of-freedom vehicle vibration model

The actual driving vehicle is an extremely complex vibration system, and its vibration belongs to the vibration of an infinite-degree-of-freedom continuum. To simulate the vibration of automobiles, we should first establish a vibration model that basically reflects the actual working conditions without much complexity should be established at first.

This section adopts the vibration model of body and wheel double mass system, as shown in Fig. 9.34. The model can well reflect the vibration of body m_2 and wheel m_1, which is close to the actual situation of automobile system. The differential equation of motion is

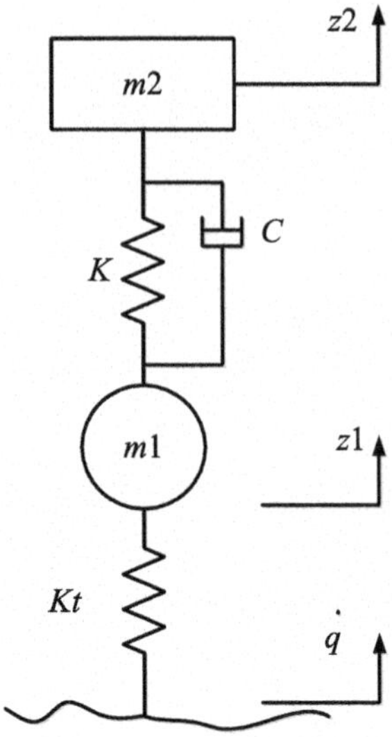

Fig. 9.34 Two-degree-of-freedom vibration system of vehicle body and wheel

$$m\ddot{z} + c\dot{z} + kz = F \tag{9.29}$$

where m is mass matrix, $m = \begin{bmatrix} m_1 & 0 \\ 0 & m_2 \end{bmatrix}$, c is damping matrix, $c = \begin{bmatrix} c & -c \\ -c & c \end{bmatrix}$, k is stiffness matrix, $k = \begin{bmatrix} k + k_t & -k \\ -k & k \end{bmatrix}$, z is displacement vector, F is road roughness input vector.

The first-order differential equation derived from Hamilton principle or transformed from Eq. (9.29) is

$$\dot{x} = Hx + r \tag{9.30}$$

where $H = \begin{bmatrix} 0 & I \\ B & G \end{bmatrix}$, $x = \begin{Bmatrix} z \\ \dot{z} \end{Bmatrix}$, $r = \begin{Bmatrix} 0 \\ m^{-1}F \end{Bmatrix}$, $B = -m^{-1}k$, $G = -m^{-1}c$.

2. Evolutionary spectral density of road roughness

Assuming that the vehicle accelerates uniformly at the speed shown in Eq. (9.31)

$$u(t) = u_0 + at \tag{9.31}$$

where, u_0 is the speed at the initial time, a is acceleration.

When the vehicle passes through the space frequency $n(\mathrm{m}^{-1})$ and road roughness q, the input time frequency $f\,(s^{-1})$ is the product of n and u

$$\Delta f \approx u(t)\Delta n \tag{9.32}$$

$$G_q(n) = \lim_{\Delta n \to 0} \frac{\sigma_{q-\Delta n}^2}{\Delta n} \tag{9.33}$$

where $\sigma_{q-\Delta n}^2$ is the "Power" of q in Δn, $G_q(n)$ is the self-power spectral density of q in spatial frequency domain.

At the speed of u, the harmonic component of uneven vertical displacement contained in Δf corresponding to Δn is the same, and its 'power' is still $\sigma_{q-\Delta n}^2$, then

$$G_q(f) = \lim_{\Delta f \to 0} \frac{\sigma_{q-\Delta n}^2}{\Delta f} = \lim_{\substack{\Delta n \to 0 \\ \Delta u \to 0}} \frac{\sigma_{q-\Delta n}^2}{\Delta n(u_0 + at)} \tag{9.34}$$

$$G_q(f) = \frac{1}{(u_0 + at)} G_q(n) \tag{9.35}$$

Obviously, in the case of variable speed driving, $G_q(f)$ is both a function of time and a function of spatial frequency. Therefore, $G_q(f)$ is the evolving self-power spectral density.

3. Pseudo excitation method

It can be seen from Eq. (9.35) that the input of road roughness during variable speed driving, that is, r in Eq. (9.30) can be written as follows:

$$r(t) = \rho g(t) y(t) \tag{9.36}$$

where, ρ represents the non-time varying vector of road roughness input distributed along each degree of freedom, $g(t)$ is a slowly varying modulation function,$g(t) = \frac{1}{u_0+at}$, $y(t)$ is a stationary random process with the self-power spectral density $G_q(n)$.

Now construct a virtual road input as

$$r(t) = \rho g(t) \sqrt{s_r(\omega)} e^{i\omega t} \tag{9.37}$$

This will transform random road input into deterministic input.

9.4.2 Response of Vehicle Vertical Vibration

The purpose of vehicle vertical dynamics modeling is to predict vehicle ride comfort. Therefore, one of the core problems of ride comfort research is to solve Eq. (9.29) or formula (9.30). Based on the precise time integration method [19] and the virtual excitation method, a simple and practical solution to the random vibration problem of Eq. (9.30) is proposed.

The homogeneous solution of Eq. (9.30) is:

$$x_h(t) = T(\tau)c \tag{9.38}$$

where

$$T(\tau) = e^{H\times\tau} \tag{9.39}$$

In the integral step $t \in [t_k, t_{k+1}]$, $\tau = t - t_k$. Assuming that the special solution $x_p(t)$ of Eq. (9.30) has been obtained. Then, at the initial time of this step, $t = t_k$, namely, $\tau = 0$, $T(\tau) = I$ (unit matrix). Thus, the integral constant vector c can be determined, and the general solution of Eq. (9.30) is

$$x(t) = T(\tau)\big(x(t_k) - x_p(t_k)\big) + x_p(t_k) \tag{9.40}$$

The problem boils down to finding the special solution $x_p(t_k)$ and fine computing $T(\tau)$.

1. Calculation of Matrix $T(\tau)$

Integration step τ, the internal subdivision is $m = 2^N$ equal parts:

$$\Delta t = \tau/m = 2^{-N}\tau \tag{9.41}$$

Generally, $N = 20$, so $\Delta t \approx 10^{-6}\tau$. Δt is generally much smaller than all natural vibration cycles of the system. Substitute Eq. (9.41) into Eq. (9.39) yields:

$$\begin{aligned} T(\tau) &= \left(e^{H\times\Delta t}\right)^m \\ &\approx \left(I + H\times\Delta t + (H\times\Delta t)^2/2! + (H\times\Delta t)^3/3! + (H\times\Delta t)^4/4!\right)^m \\ &\equiv (I + T_{a0})^m \end{aligned} \tag{9.42}$$

Let

$$I + T_{ai} \equiv \left(I + T_{a,i-1}\right)^2 = I + 2\times T_{a,i-1} + T_{a,i-1}\times T_{a,i-1} \tag{9.43}$$

then

$$I + T_{aN} = \left(I + T_{a,N-1}\right)^2 = (I + T_{a0})^m = T(\tau) \tag{9.44}$$

The above two formulas show that $T_{ai}(i = 1, 2, \ldots, N)$ can be calculated according to the following formula to obtain $T(\tau)$.

$$T_{ai} = 2\times T_{a,i-1} + T_{a,i-1}\times T_{a,i-1} \tag{9.45}$$

The recursive calculation according to Eq. (9.45) excludes I from participating in the addition operation, so it avoids the serious loss of significant digits of the elements in T_{ai}, thus ensuring that each T_{ai} and $T(\tau)$ extremely high calculation accuracy.

The only approximation introduced in the above derivation is that Δt^5 and higher order terms are omitted in Taylor expansion of Eq. (9.42); Its order of magnitude is $O(\Delta t^5) = 10^{-30}O(\tau^5)$, which is the rounding error range of general computers. Therefore, $T(\tau)$ calculated according to the above steps can be considered to be accurate within the range of computer accuracy. For the linear-step time-domain integration, $T(\tau)$ only needs to be calculated once, and the calculation amount is small.

2. HPD-S precise integration scheme

According to the characteristics of Eq. (9.37), it is assumed that r changes according to the simple harmonic law in each integration step $t \in [t_k, t_{k+1}]$:

$$r(t) = r_1 \sin(\omega t) + r_2 \cos(\omega t) \tag{9.46}$$

where r_1 and r_2 are time invariant vectors.

Substituting Eq. (9.47) into Eq. (9.47), the special solution of the equation can be obtained as

$$x_p(t) = a \sin(\omega t) + b \cos(\omega t) \tag{9.47}$$

where

$$\begin{cases} a = \left(\omega I + H^2/\omega\right)^{-1}\left(r_2 - H r_1/\omega\right) \\ b = \left(\omega I + H^2/\omega\right)^{-1}\left(-r_1 - H r_2/\omega\right) \end{cases} \tag{9.48}$$

The format of HPD-S (High Precision Direct Integration Method—Sinusoidal) is obtained:

$$\begin{aligned} x(t_{k+1}) &= T(\tau)(x(t_k) - a \sin(\omega t_k) + b \cos(\omega t_k)) \\ &+ a \sin(\omega t_{k+1}) + b \cos(\omega t_{k+1}) \end{aligned} \tag{9.49}$$

where $\tau = t_{k+1} - t_k$.

As long as r is a harmonic change within the time step τ, Eq. (9.49) always gives accurate results; even if τ is large enough to contain multiple r cycles.

For the case of Eq. (9.47), r is generally still a harmonic function in the time step τ, or can be approximately regarded as a harmonic function, $g(t)$ takes the average value of this step length as a constant.

9.4.3 Numerical Simulation Case

Considering the random response of the simplified single-point grounding vehicle model excited by the road in variable speed driving, as shown in Fig. 9.34. The differential equation of motion of this system is shown in Eq. (9.29) or Eq. (9.30). In Fig. 9.8, $m_1 = 60\,\text{kg}$, $m_2 = 500\text{kg}$, $k = 30\text{kN}/m$, $k_t = 300\text{kN}/m$, the power spectrum value of B-class road surface is taken, the acceleration a is taken as $1\text{m}/s^2$,$2\,\text{m}/s^2$,$3\,\text{m}/s^2$, and the damping coefficient c is taken as $0.5\,\text{kN}\cdot s/m$ and $2.0\,\text{kN}\cdot s/m$, respectively.

Figures 9.35, 9.36 and 9.37 show the root mean square of the vertical vibration acceleration of the vehicle body under different driving accelerations and different suspension damping conditions. From the figures, we can not only see the quantitative results, but also get the following qualitative conclusions:

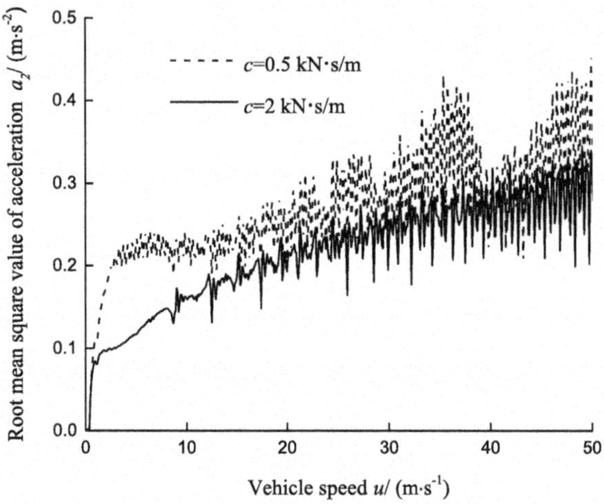

Fig. 9.35 Root mean square of vertical vibration acceleration of vehicle body ($a = 1\,\text{m}/s^2$)

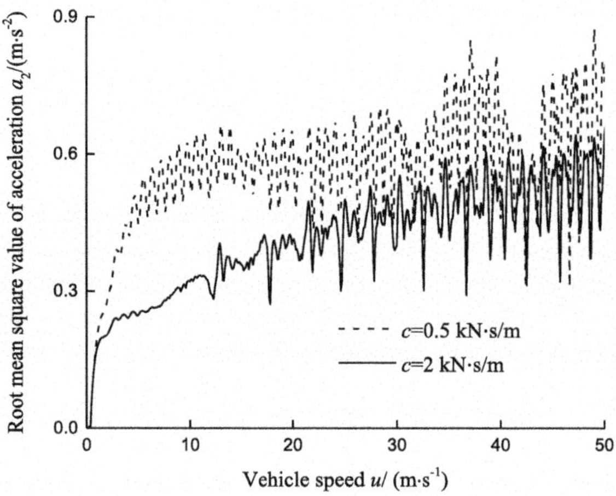

Fig. 9.36 Root mean square of vertical vibration acceleration of vehicle body ($a = 2\,\text{m}/s^2$)

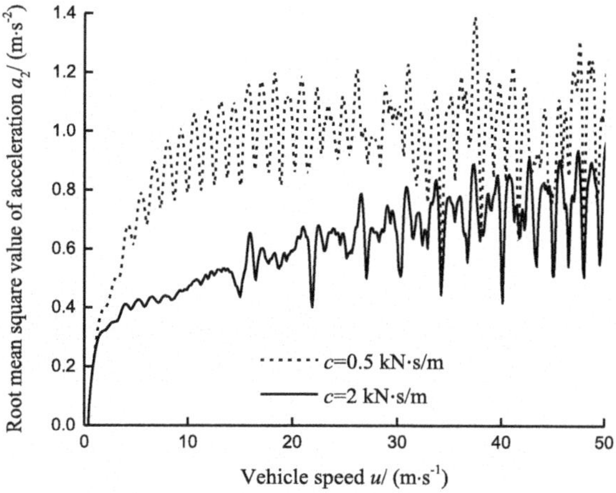

Fig. 9.37 Root mean square of vertical vibration acceleration of vehicle body ($a = 3\,\text{m}/s^2$)

(1) The greater the acceleration of the vehicle, the higher the speed of "passing resonance";
(2) The greater the acceleration of the vehicle is, the higher the peak value of the root mean square of the vertical vibration acceleration of the car body is, and the greater the amplitude of the fluctuation is, that is, the subjective feeling of people tends to be uncomfortable;
(3) With the increase of vehicle driving acceleration, the influence of suspension damper damping on the vertical vibration acceleration of vehicle body further increases. It is appropriate to select active or semi-active suspension to achieve the best damping effect.

9.4.4 Conclusions

The vehicle driving conditions are complex and changeable. Using the up-to-date scientific and technological methods to predict the instantaneous driving condition of vehicle is one of the goals of vehicle design.

The evaluation indexes of vehicle ride comfort include weighted root mean square acceleration, probability of striking suspension limit and driving safety. The requirements of ride comfort on vibration response are sometimes put forward according to probability distribution. For the linear system, the random input of the road basically conforms to the normal distribution, so the vehicle design engineer can use the method and the probability integral table in this section to predict the probability that the vehicle vibration acceleration exceeds a certain threshold value.

The core of this section is the prediction of the vehicle's instantaneous driving conditions. The author is studying how to use the basic theory of this section to make a more comprehensive prediction of vehicle ride comfort according to the data processing method of the "Test Methods for Vehicle Ride Comfort" (GB/T 4970-2009) [20]. The difficulty and risks of this subject ought to be pointed out.

Evolutionary random excitation is close to the road excitation of real vehicles. The method in this section is simple and practical, and does not involve difficult mathematical tools. It is easy to be accepted by vehicle design engineers, and can be used as one of the favorable tools for studying vehicle ride comfort.

9.5 Optimal Design of Vehicle Ride Comfort Based on Bionic Lizard Co-evolution Algorithm

9.5.1 Principles of Bionic Lizard Co-evolution Algorithms

1. Description of the behavior characteristics of bionic lizards

Sinervo and Stephens' study found that the throat of male lateral spotted lizards shous the color of yellow, blue, and orange, as shown in Fig. 9.38; Differences in throat color also lead to differences in behavior patterns during the reproduction of biological species, the yellow throat male lateral spotted lizards are physically weaker than those with blue and orange throats, and their whereabouts are mysterious, often playing the role of stealers when competing for the right to mate with females, the colloquial "little three". Orange throat male lateral spotted lizards are physically strong and aggressive, live in large habitats and have numerous female lateral spotted lizards. Blue throat male lateral spotted lizards can physically be between the yellow and orange throat lizards, and in the event of a foreign invasion, they can form groups with their companions to defend their habitats, and in the process of survival and reproduction, each blue throat lizard has only one female partner and pursues the principle of monogamy. Throughout the evolutionary game cycle of biological species, the orange throat lizard, in order to seek more female lizard mating rights, often develops wars with the blue throat lizard, which will fight selflessly against the enemy for its mate and territory. Due to the wide territory and many mates of the orange throat lizards, it is impossible to take into account the whole territory and mates, which provides a sneak attack opportunity for the yellow throat lizards, and the yellow throat lizards will run to the place where the orange throat lizards cannot reach, and secretly mate with the mates of the orange throat lizards.

Sinervo and other scholars have found that in the evolutionary game of three lizards with different throat colors and personalities, the number of species is mutually restrictive. When the blue throat lizards has the advantage in the number of reproduction, the number of orange throat lizards increases dramatically, thus achieving inhibition of the reproduction rate of the blue throat lizard; When the number of orange throat lizards is dominant, the number of yellow throat lizards increases dramatically, thereby inhibiting the reproduction rate of orange throat lizards; When the yellow throat lizards have the advantage in the number of reproduction, the

Fig. 9.38 Three male lateral spotted lizards

number of blue throat lizards increases dramatically, thereby achieving inhibition of the reproduction rate of the yellow throat lizards; The three biological species cycle back and forth, constraining each other so that the number of lizards of a certain color throat does not multiply too fast. The results show that the performance of these three lizards in the species reproduction game is determined by their respective genetic factors, and this trait is passed on to the offspring during the reproduction process. Although the three lizards with different throat colors behaved in a very different way during the reproduction process, the natural law of "survival of the fittest" in Darwin's theory of biological evolution was finally realized, and the co-evolution of the three lizard species was formed. This section combines the characteristics of lizard reproduction behavior and game theory related knowledge, and constructs a bionic lizard co-evolution algorithm.

2. Principle of bionic lizard co-evolution algorithm

Usually the problems with three optimization objective functions can be described as:

Design variables of the optimization target:

$$X = (x_1, x_2, \cdots, x_n) \in \Omega^n \tag{9.50}$$

Objective function value:

$$F(X) = (f_1(X)\, f_2(X)\, f_3(X)) \to \min \tag{9.51}$$

Corresponding constraint:

$$g_k(X) \le 0 \quad k = 1, 2, \cdots, q \tag{9.52}$$

where n is the number of design variables for optimizing the objective function, q is the number of constraints, Ω^n design the feasible domain of the variable for the entire optimization.

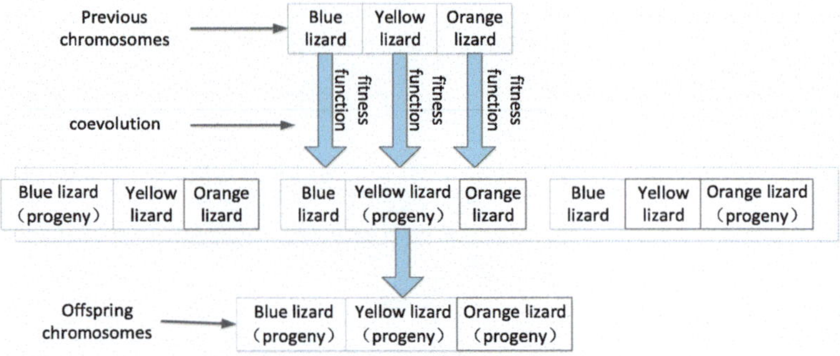

Fig. 9.39 The basic loop of the three-target bionic lizard algorithm

Combined with the reproductive evolutionary phenomena of three male lateral spotted lizard species, a bionic lizard co-evolution algorithm is proposed, and the basic principles of the algorithm are as follows. (1) The survival mating rights of the three male lateral spotted lizards are corresponding to the three optimization design goals. (2) The chromosomes of the three male lateral spotted lizard populations are regarded as design variables in the objective function, and the coding methods in the genetic algorithm are combined to construct a one-to-one correspondence between the design variables and the chromosomes of the lizard population. Combining the three different behavioral characteristics of male lateral spotted lizards, the chromosomes are divided into three parts of genes; Each gene represents the genetic information of three male lateral spotted lizards. (3) According to the behavior characteristics of male lateral spotted lizards, the corresponding adaptation functions are established, and the construction of the three adaptation functions is mainly used to evaluate the individuals' abilities of male lateral spotted lizards, which is a one-to-one correspondence between the three adaptation functions and the three objective functions. (4) Three male lateral spotted lizards co-evolved at target values based on their respective adaptive functions to obtain their own optimal genes. The best genes of the three male lateral spotted lizards recombine into a new chromosome, which undergoes multiple iterations and evolutions according to the conditions of convergence judgment, and finally obtains the optimal chromosome, that is, also the solution of multi-objective optimization. The basic cycle of the three-target bionic lizard co-evolution algorithm is shown in Fig. 9.39.

9.5.2 Key Technologies of Bionic Lizard Co-evolution Algorithms

In the construction of the three-target bionic male lateral spotted lizard co-evolution algorithm, the following three core technologies must be considered: (1) a segmented method for establishing chromosomal genes in male lateral spotted lizard populations, and the mapping relationship with the corresponding design variables and design variables. (2) Establish the adaptation function of three male lateral spotted lizards respectively. (3) Establish the structure of the three-target bionic lizard co-evolution algorithm.

1. Segmentated method of chromosomal genes

Combining the correspondence between the three optimization targets shown in Fig. 9.40 and the three male lateral spotted lizards, the three gene segments of the three male lateral spotted lizards were mapped to the design variable subset of the three goals of the decomposition of the design variable set, this section combines the calculation method of competitive cooperative game; You can get subsets S_1, S_2, S_3 of the design variables for the three objective functions, and satisfy:

$$S_1 \cup S_2 \cup S_3 = X \; S_a \cap S_b = 0, \; a, b = 1, 2, 3; a \neq b$$

2. Construction of adaptive functions of bionic objects

For optimization goals i, it has three bionic objects to choose from, namely orange throat male lateral spotted lizard, blue throat male lateral spotted lizard, and yellow throat male lateral spotted lizard; combined with the evolutionary behavior characteristics of three male lateral spotted lizards, the mutual mapping relationship between the adaptation function and the objective function of the bionic object was established.

(1) The bionic of the orange throat male lateral spotted lizard—egoism

Due to the irritable personality of males with orange throat throats, they are prone to fight and compete with other subpopulations of lizards and show typical self-interest behavior characteristics in the evolutionary reproduction of biological species. As a result, this section considers orange throat male lateral spotted lizards as egoism, and its corresponding adaptation function is constructed as:

$$u_i = f_i \tag{9.53}$$

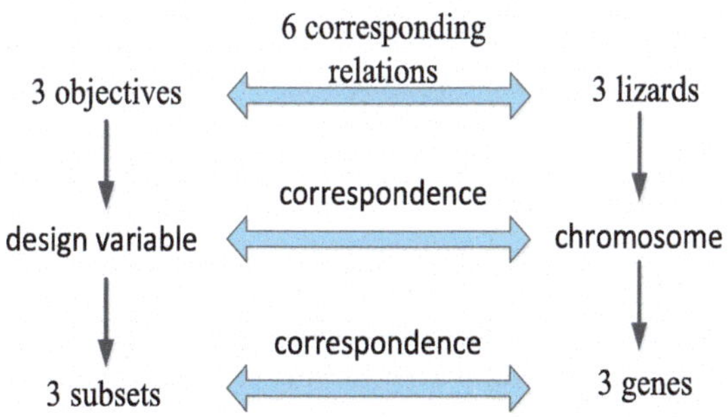

Fig. 9.40 Correspondence of the objective function to three lizard species

(2) The bionic of the blue throat male lateral spotted lizard—collectivism

Because the blue throat male lateral spotted lizards pursues "monogamy", it can combine with other lizards of the same kind to form alliance partnerships when alien species invade, and they have mild personality and do not take the initiative to fight with other subpopulations of lizards. In the evolution and reproduction of biological species they show the typical collectivist behavior, the biological species of the collectivist behavior not only to maximize their own interests, but also to benefit the interests of the other two biological species. Thus, its fitness function embodies the interests of the entire biological population, and for its corresponding fitness function is constructed as:

$$u_i = w_{ii}\frac{f_i}{\overline{f}_i} + \sum_{j=1(j\neq i)}^{3} w_{ij}\frac{f_j}{\overline{f}_j} \tag{9.54}$$

where $\overline{f}$ is the standardized processing reference value for avoiding each objective function at weight level differences, usually the value is taken as the initial state value of the target function, or taken as the objective function value in the loop iteration. $\sum_{j=1}^{3} w_{ij} = 1$, where, w_{ii} is defined as an egoistic factor, $w_{ij}(j \neq i)$ is defined as an altruistic factor, the size of the value w_{ii} is directly reflected in the degree of collectivist behavior of the biological population, and the smaller the value, the higher the characteristics of collectivist behavior; the greater the value, the lower the collectivist behavioral characteristics, that is, the size of its values is inversely proportional to the collectivist behavioral characteristics.

(3) The bionic of the yellow throat male lateral spotted lizard—opportunism (egoism does not harm others)

Yellow throat male lateral spotted lizards act in a mysterious and non-collectivist manner; Even when the orange throat male lateral spotted lizard and the blue throat male lateral spotted lizard fight for mating rights, the yellow throat male lateral spotted lizard does not take advantage of the fight and attack the losing party, but quietly runs to the orange and blue throat male lateral spotted lizards in undefended territory, and secretly mate with female lizards, so the biological species embodies a opportunism in the process of reproduction and evolution, but this opportunism is made without harming other male lateral spotted lizards. This section defines this behavioral characteristic as an opportunism which is egoism does not harm others, and its corresponding fitness function is constructed as

$$u_i = f_i, f_j \leq \overline{f}_j, j = 1, 2, 3; j \neq i \tag{9.55}$$

3. Structure of bionic lizard co-evolution algorithms

In view of the behavioral characteristics of three male lateral spotted lizards, a bionic lizard co-evolution algorithm is proposed in this section [21], which is constructed as follows.

(1) Analyze and calculate the influencing factor index between multiple design variables on the three objective functions respectively, and use the fuzzy clustering method to fuzzy cluster the influencing factor index, and obtain a subset of the design variables belonging to the three objective functions:S_1, S_2, S_3.
(2) Calculate and analyze the correspondence between three male lateral spotted lizards and three objective functions (total $P_3^3 = 6$ species selection) respectively, and each objective function constructs its own adaptability function according to the three male lateral spotted lizards, and the design variable space corresponds to the chromosomal space of three male lateral spotted lizards, and the three design variable subsets correspond to the corresponding gene segments.
(3) In the chromosomal gene space of three male lateral spotted lizards, initial chromosome vectors $s^{(0)} = \left(s_1^{(0)}, s_2^{(0)}, s_3^{(0)}\right)$ are randomly generated.
(4) Let $\overline{s}_1^{(0)}, \overline{s}_2^{(0)}, \overline{s}_3^{(0)}$ be the complements of $s_1^{(0)}, s_2^{(0)}, s_3^{(0)}$ in the initial chromosome vector $s^{(0)}$. The co-evolution of the three lizard species is carried out in a cycle, and the evolution of the target $i(i = 1, 2, 3)$ is carried out as follows.

[1] Set the size of the initial evolutionary population L and the corresponding maximum evolutionary algebra $k_{\max}$.
[2] Set the first generation of evolutionary algebra $k = 0$ and generate any L first generation gene paternal vectors $s_{i1}, s_{i2}, \ldots, s_{iL}$ in its corresponding subset of design variables S_i. Set the initial value of the vector variance ρ as $[1\ 1 \ldots\ 1]_{1\times n_i}^T$, and n_i is the number of design variables in the corresponding subset S_i of design variables.
[3] For each gene vector of the chromosomal gene space, calculate its corresponding adaptation function separately

$$u_i(s_{ij}, \overline{s}_i^{(0)}),\ j = 1, 2, \cdots, L$$

The value of the adaptability function is inversely proportional to the adaptability, that is, the smaller the value, the better the adaptation performance, and the larger the value, the worse the adaptability; Select the gene vector corresponding to the smallest function and note it as s_i^*

[4] Correct the vector variance ρ, let the first k generation of the paternal gene mutate, select the gene individual, and cycle into the first $k + 1$ generation.
[5] Judge the convergence of the cycle, and the basis for judging is

$$\left|\frac{u_{iw} - u_{ib}}{u_{\text{iavg}}}\right| \le \varepsilon \tag{9.56}$$

or

$$k > k_{\max} \tag{9.57}$$

where ε is the decimal number is determined for the initial stage, and there is

$$u_{i\text{avg}} = \frac{1}{L}\sum_{j=1}^{L} u_i(s_{ij}, \overline{s}_i^{(0)})$$

$$u_{iw} = \max_{j\in(1,2,\cdots,L)} (u_i(s_{ij}, \overline{s}_i^{(0)}))$$

$$u_{ib} = \min_{j\in(1,2,\cdots,L)} (u_i(s_{ij}, \overline{s}_i^{(0)}))$$

If the convergence judgment basis is satisfied, the loop is terminated, the algorithm ends, and the calculation result is output; Otherwise turn to step 3).

In the above calculation step, the correction of the variance vector ρ is the most core calculation step in the evolutionary strategy, and its correction strategy is: Let each evolutionary individual contain two aspects of information, that is, gene vectors s_i and vector variances ρ. (s_i, ρ)'s next generation (s_i', ρ') is

$$\begin{cases} \rho_a' = \rho_a \exp(\theta' \cdot N(0, 1) + \theta \cdot N_j(0, 1)) \\ s_{ia}' = s_{ia} + N(0, \rho_a') \end{cases} \tag{9.58}$$

where $a = 1, 2, \cdots, n_i, \theta' = 1/\sqrt{2L}, \theta = 1/\sqrt{2\sqrt{L}}, N(0, 1)$ and $N_j(0, 1)$ is two standard normal random variables that are independent of each other.

(5) After the computational analysis of the evolution of the three targets is completed, define the new chromosome vector as $s^{(1)} = s_1^* \cup s_2^* \cup s_3^*$ and check whether $s^{(1)}$ satisfies the constraints: $g_k(s^{(1)}) \le 0 (k = 1, 2, \cdots, q)$, if the constraint is not met, go to step (3); If the constraint is met, the distance between adjacent chromosome vectors is calculated at this point and check whether the convergence criterion $\left\| s^{(1)} - s^{(0)} \right\| \le \eta$ is met, where η is the initially set value, if the convergence criterion is met, the calculation ends; If not, order $s^{(0)} = s^{(1)}$, return to step (4) to re-cycle the correction variance.

The flowchart of the bionic lizard co-evolution algorithm proposed in this section is shown in Fig. 9.41, if the design variables and design goals are relatively small, the set of design variables X can be directly divided into the corresponding strategy space $S_1, \ldots, S_m$. If there are more design variables and design goals, when performing fuzzy cluster analysis, according to the characteristics of the design variables, the design variables with the same or similar properties can be reduced into one category, which can reduce the computational amount and complexity of fuzzy cluster analysis.

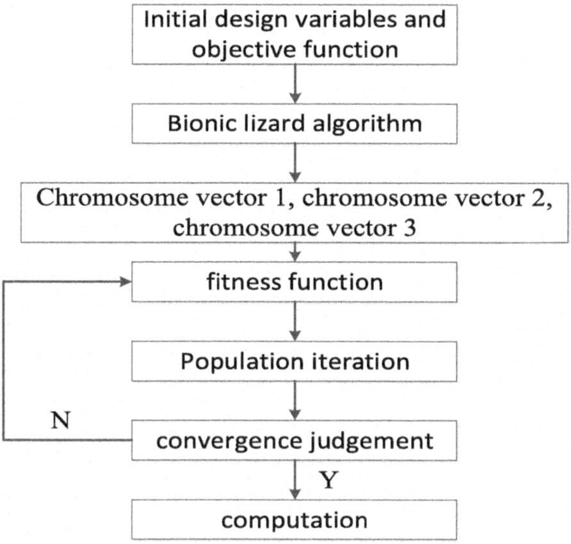

Fig. 9.41 Flowchart of the bionic lizard algorithm

9.5.3 Simulation Results and Analysis

1. Calculation instructions

To prove the correctness of the bionic lizard algorithm, the eight-degree-of-freedom vehicle vertical dynamic model and the four-wheel pavement random excitation time domain simulation data are used as the input data of the model, and the bionic lizard co-evolution algorithm is used to optimize the smoothness of the vehicle model [22–24].

In the bionic lizard co-evolution algorithm, the three male lateral spotted lizards are divided into three game parties according to their corresponding genetic patterns, namely the orange throat male lateral spotted lizards are divided into egoistic game parties, the blue throat male lateral spotted lizards are divided into collectivist game parties, and the yellow throat male lateral spotted lizards are divided into opportunism (egoism does not harm others) game parties, let three male lateral spotted lizards correspond to three optimization design goals, and It can be seen from the arrangement and combination that there are $P_3^3 = 6$ situations, so as not to lose generality, aiming at the three game parties of orange male lateral spotted lizards f_1(Represent: egoism), blue male lateral spotted lizard f_2 (representing collectivism), yellow male lateral spotted lizard f_3 (representing opportunism) to pursue their own maximum benefits, the specific calculation steps of the bionic lizard co-evolution algorithm are as follows.

(1) The initial feasible genes $s_1^{(0)}$, $s_2^{(0)}$, $s_3^{(0)}$ are randomly generated in the gene segments S_1, S_2, S_3 controlled by the three kinds of lizards of orange male lateral spotted lizard, blue male lateral spotted lizard, and yellow male lateral

spotted lizard, respectively, and form their respective corresponding initial chromosomes $s^{(0)} = \left(s_1^{(0)}, s_2^{(0)}, s_3^{(0)}\right)$.

(2) Collaborative bionic evolutionary calculations were performed on three male lateral spotted lizards. In the calculation, the initial population size of the three lizard species is set to L, the maximum evolutionary algebra is set to $k_{\max} = 20$, and the convergence judgment is set to $\varepsilon = 0.001$, the specific calculation steps for the co-evolution of three male lateral spotted lizards are as follows:

① Find the optimal strategy $s_1^* \in S_1$ for orange male lateral spotted lizards and make their corresponding fitness function satisfied:

$$u_1(s_1^*, s_2^{(0)}, s_3^{(0)}) = f_1(s_1^*, s_2^{(0)}, s_3^{(0)}) \to \min$$

② Find the optimal strategy $s_2^* \in S_2$ for the blue male lateral spotted lizards and satisfy its corresponding fitness function:

$$u_2(s_1^{(0)}, s_2^*, s_3^{(0)}) = w_{22} \times \frac{f_2(s_1^{(0)}, s_2^*, s_3^{(0)})}{f_2(s_1^{(0)}, s_2^{(0)}, s_3^{(0)})} + w_{21} \times \frac{f_1(s_1^{(0)}, s_2^*, s_3^{(0)})}{f_1(s_1^{(0)}, s_2^{(0)}, s_3^{(0)})} + w_{23} \times \frac{f_3(s_1^{(0)}, s_2^*, s_3^{(0)})}{f_3(s_1^{(0)}, s_2^{(0)}, s_3^{(0)})} \to \min$$

where $w_{22} = 0.5$, $w_{21} = 0.25$, $w_{23} = 0.25$.

③ Find the optimal strategy $s_3^* \in S_3$ for yellow male lateral spotted lizards, and satisfy the corresponding fitness function when the constraints $f_1(s_1^{(0)}, s_2^{(0)}, s_3^*) \le f_1(s_1^{(0)}, s_2^{(0)}, s_3^{(0)})$ and $f_2(s_1^{(0)}, s_2^{(0)}, s_3^*) \le f_2(s_1^{(0)}, s_2^{(0)}, s_3^{(0)})$ are satisfied:

$$u_3(s_1^{(0)}, s_2^{(0)}, s_3^*) = f_3(s_1^{(0)}, s_2^{(0)}, s_3^*) \to \min$$

(3) Establish a strategy combination of three kinds of lizards:$s^{(1)} = s_1^* \cup s_2^* \cup s_3^*$, test and determine whether the policy group $s^{(1)}$ satisfies the constraints, and if not, go to the step ①. If satisfied, the calculation $\sqrt{\sum_{j=1}^{5}\left(\left((x_j^{(1)} - x_j^{(0)})\Big/x_j^{(0)}\right)^2\Big/5\right)}$ is performed and the settlement result is determined to meet the convergence accuracy η (this section takes 0.0001) and if the convergence condition is met, then the game ends; If not satisfied, order $s^{(0)} = s^{(1)}$, go to step (2) to iterate through the loop.

2. Description of the calculation results

Co-evolutionary calculations were made for orange male lateral spotted lizards f_1 (egoism), blue male lateral spotted lizards f_2 (collectivism), and yellow male lateral spotted lizards f_3 (opportunism). For comparison, this section uses classical genetic algorithm to optimize the design, Table 9.2 gives the three optimization results of the initial design, fuzzy optimization, genetic algorithm and bionic lizard co-evolution algorithm are presented, and the optimization results show the feasibility and effectiveness of the bionic lizard co-evolution algorithm.

In Table 9.2, x_1 is the driver's seat spring rate (kN/m), x_2 is the driver's seat damping coefficient (N · s/m), x_3 is the right front suspension spring rate (kN/m), x_4 is the right front suspension damping coefficient (kN · s/m), x_5 is the left rear suspension spring rate (kN/m), x_6 is the left rear suspension damping coefficient (kN · s/m), F_1 is the rms value of the seat acceleration (m/s^2), F_2 is the rms value of the relative dynamic load of the tire (N), F_3 is the maximum suspension travel in the vehicle model (mm).

Figures 9.42, 9.43 and 9.44 show the optimization calculation results of the bionic lizard co-evolution algorithm with the driver's seat vertical degree of freedom z_1 when simulating the whole vehicle for 120 s, as well as the comparison curve with the fuzzy optimization and initial design. It can be seen from the figure that the calculation results of the bionic lizard co-evolution algorithm proposed in this chapter are better than the fuzzy optimization and initial design.

Table 9.2 Comparison of the parameters of the three design methods with the initial design

Scheme	x_1	x_2	x_3	x_4	x_5	x_6	F_1	F_2	F_3
Initial design	15.000	150.000	15.000	2.500	17.000	2.500	0.0815	0.0264	5.245
Fuzzy optimization	10.328	171.592	16.194	2.495	86.925	2.684	0.0692	0.0265	4.887
Genetic algorithms	10.679	172.645	16.280	2.989	22.489	2.561	0.0679	0.0253	4.989
Bionic lizard co-evolution algorithm	10.102	168.595	22.921	3.498	18.602	3.545	0.0532	0.0261	4.094

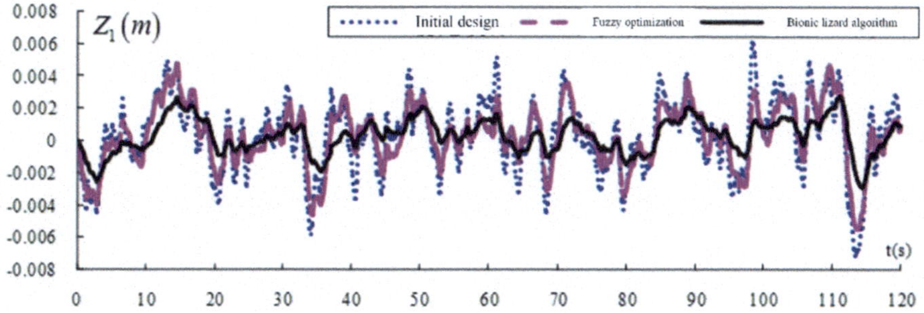

Fig. 9.42 The driver's seat vertical degree of freedom z_1 displacement bionic lizard co-evolution algorithm compares with other algorithms

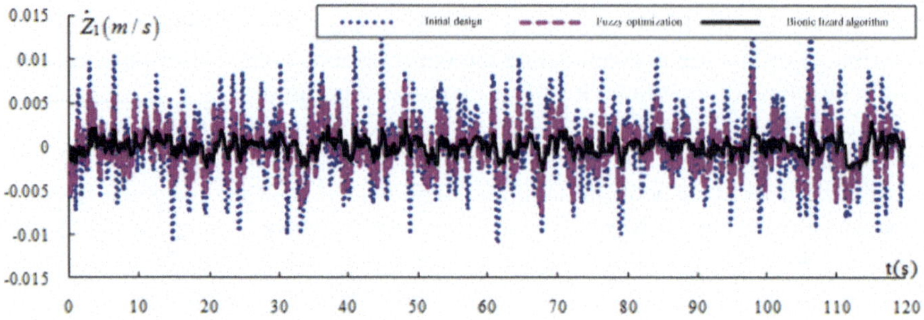

Fig. 9.43 The driver's seat vertical degree of freedom z_1 velocity bionic lizard co-evolution algorithm compares with other algorithms

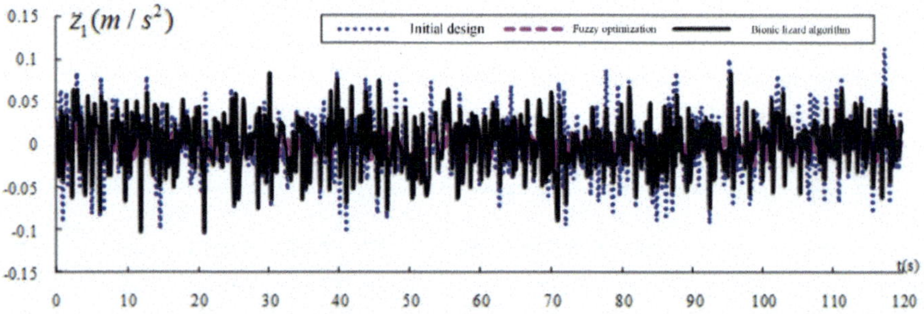

Fig. 9.44 The driver's seat vertical degrees of freedom z_1 acceleration bionic lizard co-evolution algorithm compares with other algorithms

Figures 9.45, 9.46 and 9.47 respectively show the corresponding comparison results of the vertical degree of freedom z_2 vibration of the car body under the B-class road conditions based on different algorithms of the co-evolution algorithm of the bionic lizard and the fuzzy optimization and initial design.

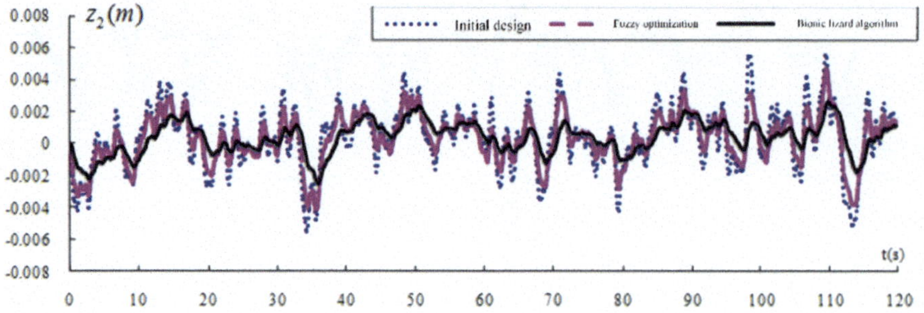

Fig. 9.45 The vertical degree of freedom z_2 displacement bionic lizard co-evolution algorithm of the car body is compared with other design methods

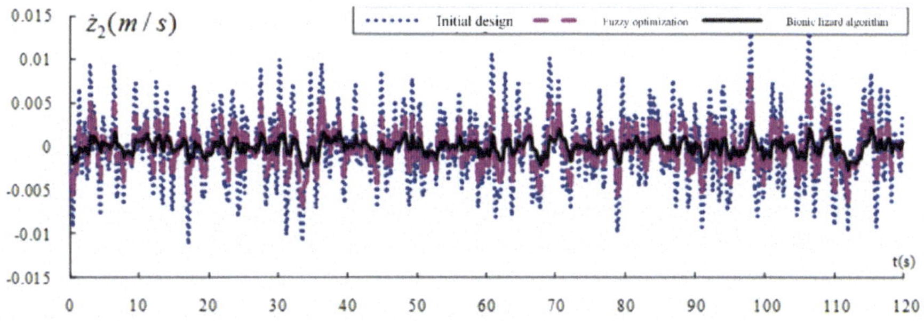

Fig. 9.46 The vertical degree of freedom z_2 velocity bionic lizard co-evolution algorithm of the car body is compared with other design methods

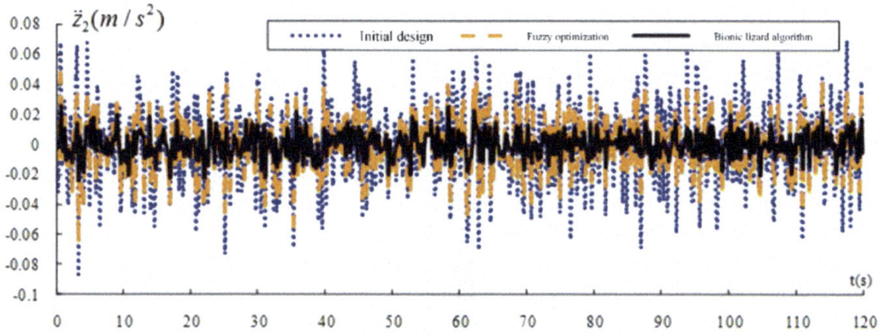

Fig. 9.47 The vertical degree of freedom z_2 acceleration bionic lizard co-evolution algorithm of the car body is compared with other design methods

Figure 9.48 is the comparison of the vibration response of the tire relative dynamic load using the bionic lizard algorithm and the initial design method in the time domain, it can be seen from the figure that the calculation result of the bionic lizard co-evolution algorithm is better than the initial design value, which can reduce the damage of the vehicle to the road surface.

Combining the design results of the researches that have been studied, Fig. 9.49 gives the spectral density of seat acceleration power under different design methods, and it can be seen from the figure that the calculation results of the bionic lizard co-evolution algorithms are better than several other algorithms.

Figure 9.49 gives the results of the comparison of seat acceleration power spectrum density under a variety of design methods, from the figure, it can be seen that the seat acceleration power spectrum density optimized by the bionic lizard co-evolution algorithm is less than that of several other methods, through which the high riding comfort of the whole vehicle can be improved. Table 9.3 lists the increase and decrease percentages in the frequency domain of the three objective functions under different methods.

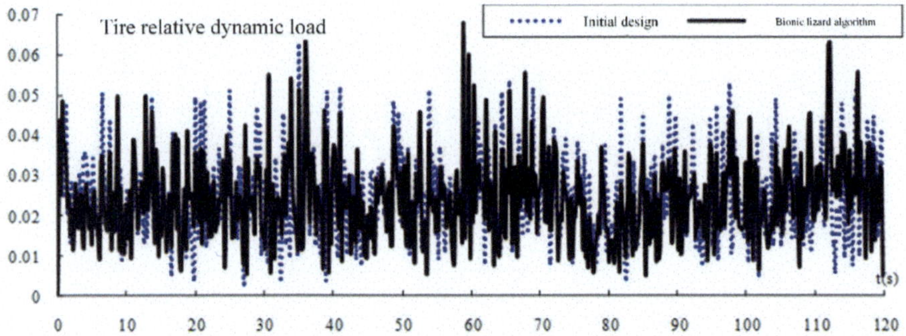

Fig. 9.48 Comparison of calculation results of tire relative dynamic load under two methods

Fig. 9.49 Comparison of seat acceleration power spectrum density under multiple design methods

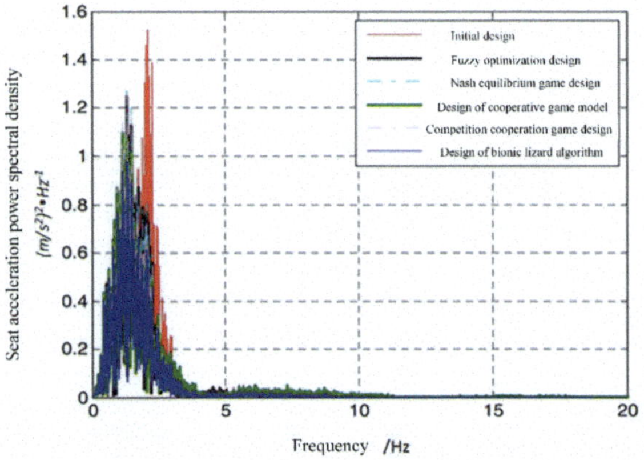

Figures 9.50, 9.51, 9.52 and 9.53 respectively give the comparison results of the right front wheel, left front wheel, left rear wheel, right rear wheel dynamic stroke power spectral density, from the figure can be seen that the use of bionic lizard co-evolution algorithm can be a good improvement of riding comfort.

Figures 9.54 and 9.55 respectively give the number of solving iterations of the objective function F_1, F_2 and in the bionic lizard algorithm, which can be seen from the figure, and it can be seen from the figure that the bionic lizard algorithm only takes 3 to 4 iterations to achieve convergence requirements and complete the calculation. From Table 9.3, the ratio of the maximum increase and decrease of the objective function (compared with the initial design) shows that the bionic lizard algorithm has high computational accuracy and computational efficiency, and the bionic algorithm proposed in this section is feasible.

3. Comparison and analysis of bionic behavior

This section calculates the permutations and combinations of three objective function $P_3^3 = 6$ corresponding to three male lateral spotted lizards, Table 9.4 lists the results of the bionic calculation in six cases, as can be seen from the table, the computational

Table 9.3 The ratio of the maximum increase or decrease of the objective function (compared to the initial design)

Optimization methods	Seat acceleration rms value F_1	Tire relative dynamic load rms value F_{12}	Maximum suspension travel F_3
F_1 single-objective optimization	Down 63.52%	Increase by 23.56%	Down 13.37%
F_2 single-objective optimization	Increase by 6.10%	Down 16.62%	Down 8.58%
Fuzzy optimized design	Down 9.02%	Down 11.38%	Down 9.85%
Nash equilibrium game design	Down 9.13%	Down 12.66%	Down 11.45%
Cooperative game model design	Down 11.15%	Down 13.92%	Down 10.68%
Competitive cooperative game design	Down 11.90%	Down 15.37%	Down 11.64%
Design of bionic lizard co-evolution algorithm	Down 12.10%	Down 18.58%	Down 12.52%

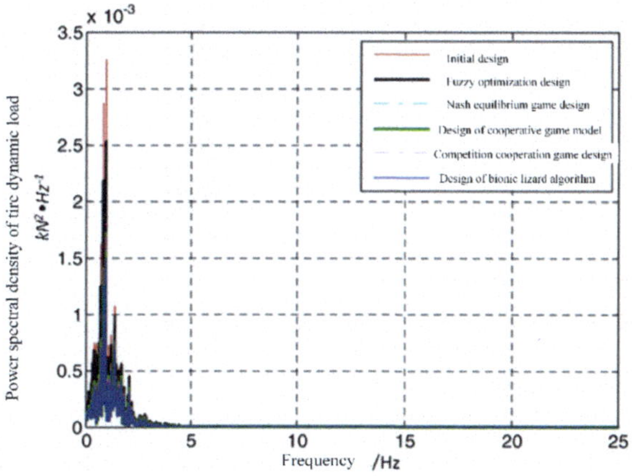

Fig. 9.50 Comparison of right front wheel power spectrum density under multiple design methods

efficiency of the bionic lizard co-evolution algorithm is high, and for different cases, it only takes a small number of iterations to achieve convergence; At the same time, the behavior characteristics of bionic objects (the different behavior patterns exhibited by the evolution of the three male lateral spotted lizards) are assigned different objects, which have a certain impact on the calculation results, and how to make appropriate target assignments for the three male lateral spotted lizards is the work that needs to be studied in the future.

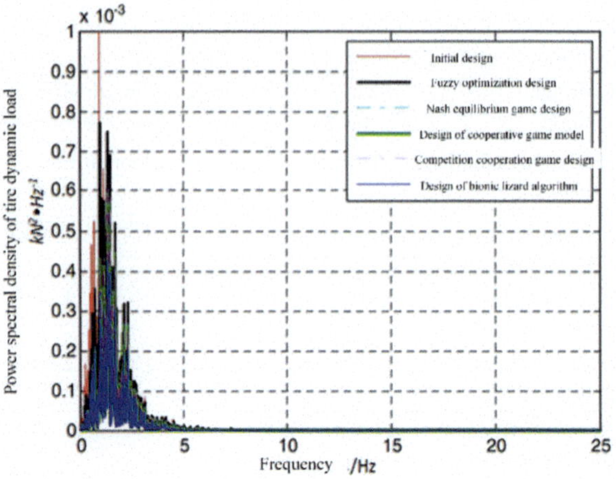

Fig. 9.51 Comparison of left front wheel power spectral density under multiple design methods

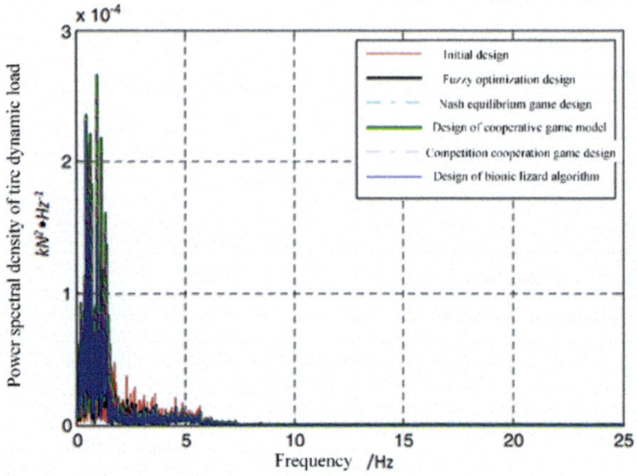

Fig. 9.52 Comparison of power spectral densities of the left rear wheel under various design methods

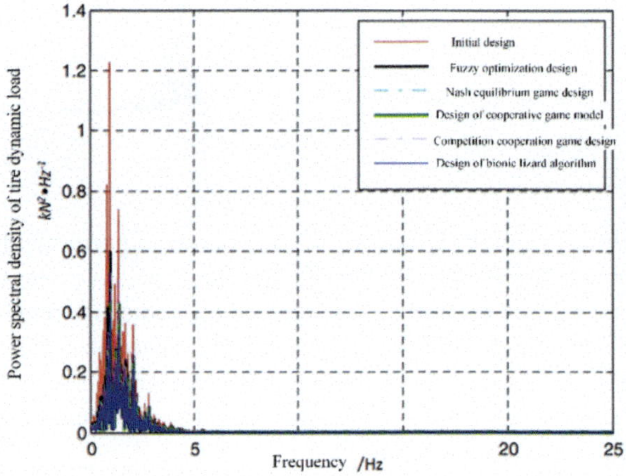

Fig. 9.53 Comparison of right rear wheel power spectrum density under multiple design methods

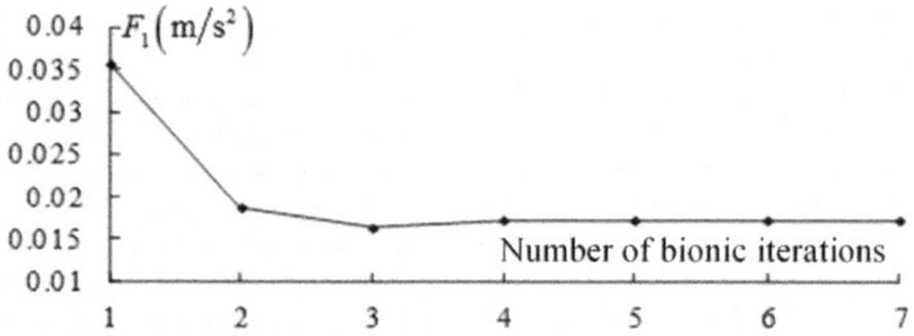

Fig. 9.54 The game side benefits from F_1 game iteration

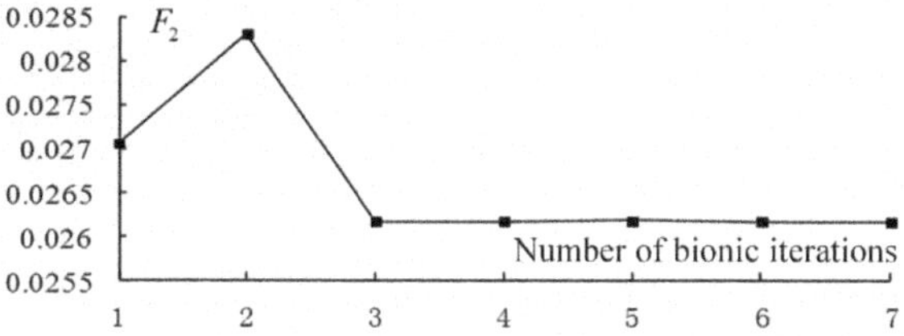

Fig. 9.55 The game side benefits from the F_2 game iteration

Table 9.4 Results of the bionic calculation in six cases

Scheme	Behavioral characteristics			Target response			Iteration times
	F_1	F_2	F_3	F_1/m	F_2/(m/s)	F_3/kJ	
1	Opportunism	Egoism	Collectivism	0.0681	0.0248	4.448	4
2	Opportunism	Collectivism	Egoism	0.0574	0.0266	4.030	3
3	Collectivism	Egoism	Opportunism	0.0756	0.0243	4.398	3
4	Collectivism	Opportunism	Egoism	0.0795	0.0240	4.338	3
5	Egoism	Opportunism	Collectivism	0.0795	0.0240	4.338	5
6	Egoism	Collectivism	Opportunism	0.0530	0.0263	4.075	7

9.6 Hydro-Pneumatic Suspension Control Matched with Mechanical Elastic Wheels

The mechanical elastic wheel structure is shown in Fig. 9.56. The wheel is mainly composed of a hub, a wheel, and a hinge group connecting the two. After simulation analysis and real vehicle test research, it is known that its large cornering stiffness ensures the handling stability of the vehicle, but its larger radial stiffness relative to the radial wheel will adversely affect the ride comfort of the vehicle.

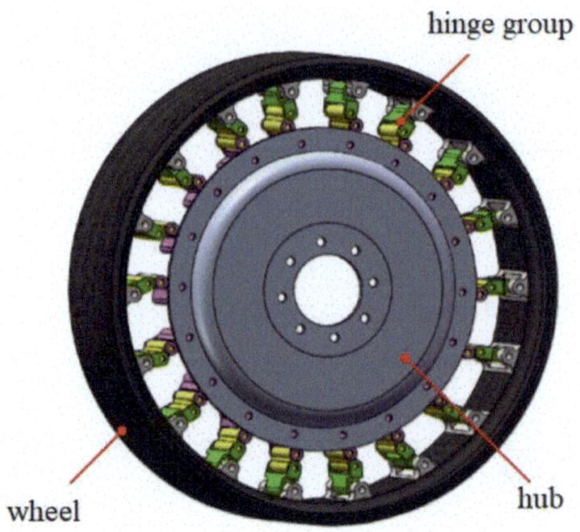

Fig. 9.56 Mechanical elastic wheel structure

9.6.1 *Vehicle Ride Simulation Model*

(1) Single-cylinder oil and gas spring

The oil and gas spring is developed on the basis of the membrane air spring and is the core component of the oil and gas suspension [25–27]. Figure 9.57 shows a schematic diagram of the structure of a single-cylinder oil–gas spring studied in this section, which places the air chamber inside the piston rod, making the structure more compact. Its working principle is that the shock is relieved by the compression and expansion of the inert gas in the gas chamber, and the vibration is damped by the oil flow in the piston and the oil chamber [28].

The elastic force of the oil–gas spring is mainly provided by the state change of the inert gas in the gas chamber [29, 30], and nitrogen is generally used. According to the ideal gas state equation, when the piston displacement is x, the elastic force at this time is

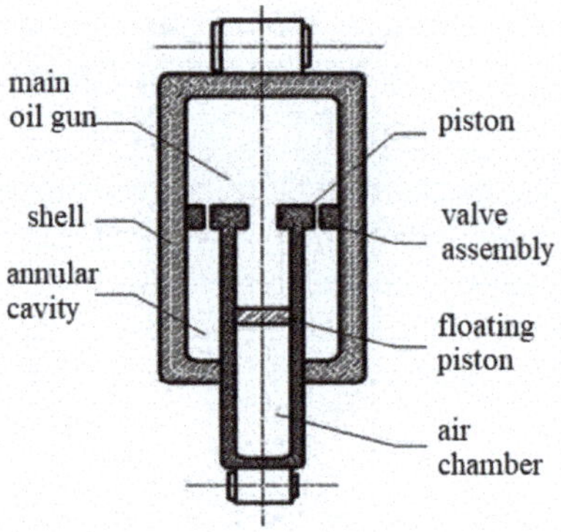

Fig. 9.57 Single-cylinder oil and gas spring structure

$$F_x = pA = m_s g\left(1 - \frac{xA_g}{V_0}\right)^{-n} \tag{9.59}$$

where p is the pressure in the air chamber when the piston is displaced x, A_g is the cross-sectional area of the piston rod cavity, m_s is the static load mass of the front suspension or rear suspension, V_0 is the initial inflation volume, n is the gas variability index, when the oil–gas spring moves fast, and the gas in the gas chamber can be approximately regarded as an adiabatic state, here $n = 1.4$.

The damping force of the oil–gas spring is mainly provided by the oil flowing between the main oil cavity and the annular cavity through the damping hole and the one-way valve, and the one-way valve only opens during the compression stroke [30–32]. It can be obtained from the orifice throttling equation that when the piston speed is v, the damping force at this time is

$$F_s = \frac{\rho A_0^3 v^2}{2C_d^2\left\{A_n + \frac{1}{2}A_d\left[1 + \operatorname{sgn}(v)\right]\right\}^2}\operatorname{sgn}(v) \tag{9.60}$$

where C_d is the flow coefficient of orifice throttling, here $C_d = 0.68$, ρ is the oil density, A_0 is the cross-sectional area of the annular cavity, A_n is the equivalent orifice area, A_d is the equivalent check valve area. The initial parameters of the oil and gas spring studied in this section are shown in Table 9.5.

(2) Four-degree-of-freedom half-vehicle model

In order to improve the efficiency of the optimization calculation and consider the difference between the front and rear suspensions of the vehicle, in this section, a four-degree-of-freedom semi-vehicle model is established to optimize the ride comfort of the hydro-pneumatic suspension. The schematic diagram is shown in Fig. 9.58.

(3) Road white noise input

The power spectral density of the road surface is mainly used to describe the statistical characteristics of the road surface roughness during the study of vehicle vibration. The power spectral density of the road surface is expressed as

Table 9.5 Initial parameters of the oil and gas spring

Parameters	Values
Piston rod inner cavity diameter/m	0.08
Piston rod outer cavity diameter/m	0.09
Inner diameter of main oil cavity/m	0.11
Piston stroke/m	0.266
Initial inflation volume/m^3	5.03×10^{-4}
Equivalent orifice area/m^2	1.20×10^{-3}
Equivalent check valve area/m^2	1.50×10^{-4}
Sprung mass/kg	1.57×10^3

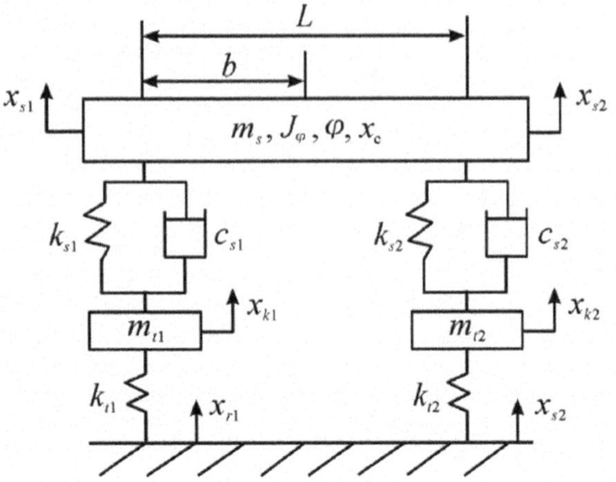

Fig. 9.58 Four-degree-of-freedom half- vehicle model

$$G_q(n) = G_q(n_0)\left(\frac{n}{n_0}\right)^{-w} \tag{9.61}$$

where n is the spatial frequency, n_0 is the reference spatial frequency,$n_0 = 0.1$, W is the frequency index, $G_q(n_0)$ is the road surface roughness coefficient at the reference spatial frequency. In this section, the C-level road surface is taken as the research object during optimization, $G_q(n_0)= 256 \times 10^{-6}$, and the vehicle speed is set as 20 m/s.

9.6.2 *Optimization of Design Parameters of Hydro-Pneumatic Suspension*

(1) Quantum Genetic Algorithm

The Genetic algorithm (GA) has extremely high robustness and is widely used in various research fields. However, GA is prone to problems such as many iterations or falling into local extreme values. Therefore, researchers have made a series of improvements to GA [33, 34].

The Quantum genetic algorithm (QGA) is a probabilistic evolutionary algorithm that combines quantum computer and GA. QGA introduces quantum state vector expression into genetic coding and uses quantum logic gates to realize the evolution of chromosomes. Since chromosomes encoded with qubits can express the superposition of multiple states, they have higher diversity than conventional GA and are not easily trapped in local extremum, and can be closer to the global extremum. QGA was first proposed by Chor in 1994. It is a new type of intelligent optimization algorithm and has been continuously improved in the application process. It has been used in signal processing, intelligent control, TSP problems, and other fields less [35–38].

(2) Objective function

The weighted acceleration root mean square value of the center of mass is the most commonly used index for evaluating vehicle ride comfort. The equation for calculating the root mean square value of acceleration is:

$$a_w = \left[\int_{0.5}^{80} W_k^2(f) G_a(f) df \right]^{\frac{1}{2}} \tag{9.62}$$

where $W_k(f)$ is the vertical vibration frequency weighting function, $G_a(f)$ is the power spectral density of the vertical vibration acceleration of the center of mass. This results in the objective function:

$$J_1 = a_w \tag{9.63}$$

Considering the ride comfort problem caused by the pitching motion of the vehicle body, it is expressed by the root mean square value of the weighted acceleration of the pitch angle vibration:

$$\beta_w = \left[\int_{0.5}^{80} W_e^2(f) G_\beta(f) df \right]^{\frac{1}{2}} \tag{9.64}$$

where $W_e(f)$ is the weighting function of pitch vibration frequency, $G_\beta(f)$ is the power spectral density of pitch angle vibration acceleration. This results in the objective function:

$$J_2 = \beta_w \tag{9.65}$$

Considering the separation of the wheels from the ground during driving on uneven roads, it is necessary to optimize the ground contact of the wheels. The dynamic load coefficient of the wheel is a common indicator for evaluating the ground contact of the wheel. Its physical meaning is the ratio of the root mean square value of the wheel dynamic load to the static load. The calculation equation is:

$$\xi = \left(\frac{1}{T} \int_0^T F_d^2 dt \right)^{\frac{1}{2}} \Big/ G \tag{9.66}$$

where F_d is the wheel dynamic load; G is the wheel static load. This results in the objective function:

$$\begin{cases} J_3 = \xi_f \\ J_4 = \xi_f \end{cases} \tag{9.67}$$

the above objective functions J_1, J_2, J_3 and J_4 are required to be as small as possible.

(3) Design variables

The parameters affecting the performance of the hydro-pneumatic suspension are many and complex, and it is necessary to take into account the optimization efficiency and the actual optimization effect. Considering the unification of actual production and processing, the inner cavity area A_g of the piston rod and the annular cavity area A_0 are kept unchanged, and the influence of other parameters on the performance of the oil–gas spring is studied by the control variable method.

It is easy to see from the simulation results combined with the relevant formulas that the check valve area has little influence on damping characteristics. Considering that the structure of the one-way valve of the oil and gas spring is relatively complex, for the convenience of production, the front and rear springs use the same size one-way valve. To sum up, the design variables can be determined to be the uniform equivalent one-way valve area, the initial inflation volume of the front and rear springs, and the equivalent spring damping hole area of the front and rear.

(4) Constraints

When designing the check valve, it is necessary to consider the mutual arrangement position of it and the piston. At the same time, since it is a built-in air chamber, the initial inflation volume should be smaller than the inner cavity volume of the piston rod. Since the small hole throttling formula is used, the selection of the equivalent orifice area should not be too large.

Intermediate variables express binding force in the algorithm in the form of penalty functions. 25 m/s^2. According to the acceleration root mean square value and human subjective feeling evaluation table, the maximum weighted acceleration root mean square value $a_{w\max}$ is set to 1.25 m/s^2. And it is generally believed that when the dynamic load coefficient of the wheel is lower than 1/3, the probability of the wheel jumping off the ground is below 0.15%. When the RMS value of the suspension's dynamic stroke S is lower than 1/3 of the suspension's maximum dynamic stroke $S_{\max}$, the impact probability of the limit block is below 0.3%.

9.6.3 Analysis of Optimization Results

In order to ensure the reliability of the results of the quantum genetic algorithm, three optimization operations were performed using the same initial parameters, and the optimization results are shown in Table 9.6.

It can be seen from Table 9.6 that the difference between the multiple optimization results of the quantum genetic algorithm is about 1%, indicating that the algorithm is relatively stable. The first optimization result with the best comprehensive optimization effect is selected for research, and the optimization process is shown in Fig. 9.59. It can be seen from Fig. 9.59 that the absolute value of the optimization target drops rapidly at the beginning of the calculation, reaches around the optimal value around 100 generations, and stabilizes after that.

Table 9.6 Quantum genetic algorithm optimization results

Objective function	Before optimization	First optimization	Second optimization	Third optimization
J	1	0.7172	0.7180	0.7173
J_1	1.0550	0.8303	0.8329	0.8351
J_2	0.8654	0.4187	0.4234	0.4212
J_3	0.3308	0.2531	0.2510	0.2520
J_4	0.3131	0.2975	0.2956	0.2931

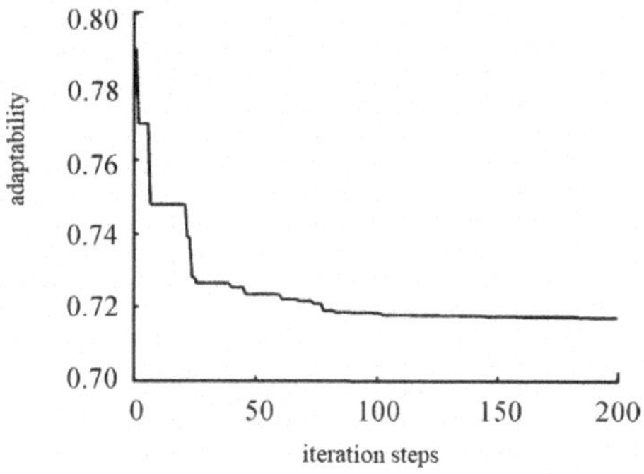

Fig. 9.59 Evolutionary process of quantum genetic algorithm

It can be seen from Table 9.6 that after the first optimization, the overall optimization objective, the vertical vibration weighted acceleration root mean square value, the pitch vibration weighted acceleration root mean square value, and the front and rear wheel dynamic loads have been improved by 29.3% and 21.3% %, 51.1%, 23.5%, 4.98%, respectively. The ride comfort of off-road vehicles has been greatly improved. The results are shown in Figs. 9.60, 9.61 and 9.62.

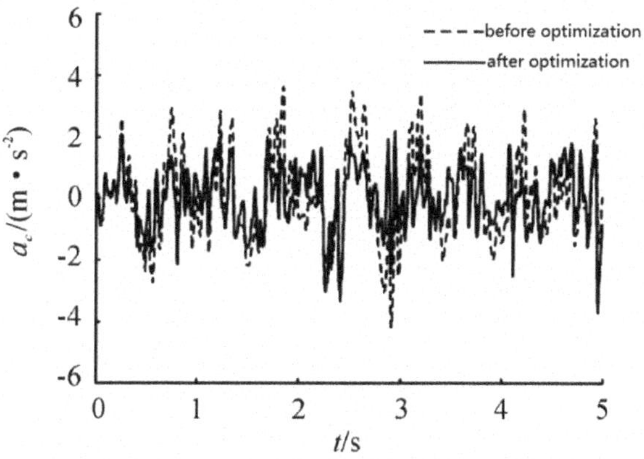

Fig. 9.60 Vertical vibration acceleration after optimization

Fig. 9.61 Front wheel dynamic load

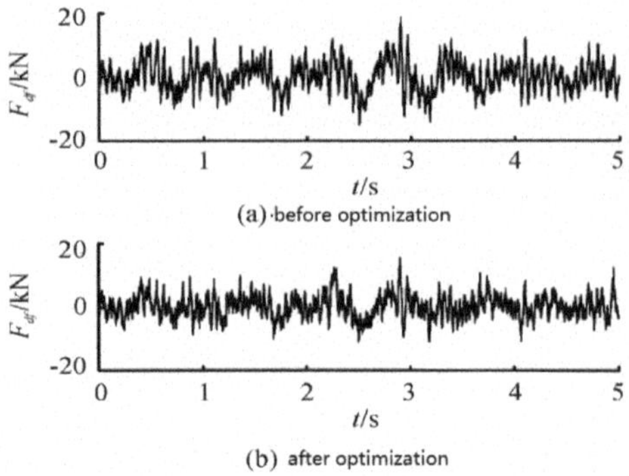

Fig. 9.62 Rear wheel dynamic load

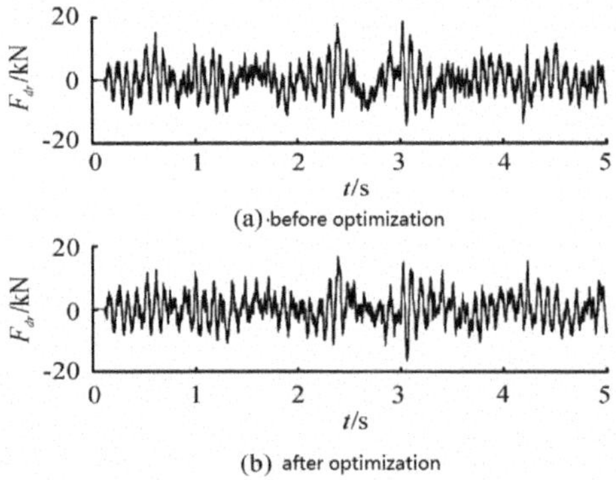

The large radial stiffness characteristic of mechanically elastic wheels leads to a reduction in vehicle ride. By establishing a vehicle ride comfort simulation model, the ride comfort of an off-road vehicle equipped with mechanically elastic wheels running at a speed of 20 m/s under the input of C-level road white noise is optimized. The optimization results show that all ride comfort evaluation indicators have been improved, it meets the requirements of ride comfort after optimization. It can be seen that the quantum genetic algorithm used in the multi-objective optimization design of the oil and gas suspension has successfully completed the task and has certain feasibility.

References

1. Liang HP. A simulation study on vehicle ride comfort with the combination of rigid-elastic coupling. Changchun: Jilin University; 2007.
2. Luo HY, Han LY, Liang SQ. The relationship between the parameters of automobile suspension system and the smoothness and operating stability of the whole. J Changchun Univ Technol. 2000;2:25–8.

3. Li CF. Research on evaluation method and experiment of automobile riding quality. Shenyang: Northeastern University; 2007.
4. Li ZC. Developments of simulation calculation procedures of the comfort of vehicles based on NJ6400GHA3. Nanning: Guangxi University; 2003.
5. Zhou YX. Non-stationary random vibration analysis for vehicle-bridge coupled systems. Dalian: Dalian University of Technology; 2009.
6. Xu ZX. Random vibration. Beijing: Higher Education Press; 1990.
7. Yu ZS. Automotive theory. 3rd ed. Beijing: China Machine Press; 2000.
8. Zhao JH. Application research report of pavement roughness spectrum analysis. Changchun: Changchun Automobile Research Institute Vehicle Research Room; 1985.
9. Au FTK, Cheng YS, Cheung YK. Effects of random road surface roughness and long-term deflection of prestressed concrete girder and cable-stayed bridges on impact due to moving vehicles. Comput Struct. 2001;79(8):853–72.
10. Zhang LX, Zhao YQ, Xu PM, et al. Simulation of road surface power spectrum density identification. Trans Chin Soc Agric Mach. 2007;38(5):15–8.
11. Zhang LX, Zhao YQ, Wu J, et al. A new method of road surface spectrum identification based on RBF neural network. J Highw Transp Res Dev. 2007;24(6):135–8.
12. Cebon D, Newland E. The artificial generation of road surface topography by the inverse FFT method. In: Proceedings of the 8th IAVSD Symposium on the Dynamics of Vehicles on Roads and Tracks, Cambridge;1982. p. 29–41.
13. Zhang HX. Computer prediction of vehicle ride comfort. Automot Eng. 1986;8(1):21–31.
14. Wu YJ, Xu BH, Chen BL. A new algorithm for the optimal design on vehicle suspension damping. Automot Eng. 1999;21(6):353–7.
15. Xie WD, Wang L, Yu YN, et al. Application of random signals to road roughness simulation. J Vibr Meas Diagn. 2005;25(2):126–30.
16. Zhao Y, Lu SF. A vehicle's time domain model with road input on four wheels. Automot Eng. 1999;21(6):112–7.
17. Tan RH, Chen Y, Lu YX. The mathematic models in time domain for the road disturbances and the simulation. China J Highw Transp. 1998;11(2):96–102.
18. Zhao YQ, He XM, Guo KH. Prediction of vehicle evolutionary random response excited by road. J Mech Eng. 2004;1:179–82.
19. Zhang LX. Research on inverse problem in vehicle handling stability and ride comfort. Nanjing: Nanjing University of Aeronautics And Astronautics; 2007.
20. Hu XZ. Ride comfort analysis and optimization for light truck. Changchun: Jilin University; 2007.
21. Song CZ, Zhao YQ, Wang L. Tri-objective co-evolutionary algorithm and application of suspension parameter design based on lizard behavior bionics. J Mech Sci Technol. 2014;28(12):4857–67.
22. Song CZ, Zhao YQ, Xie NG, et al. Multi objective optimization of eight-DOF vehicle suspension based on game theory. Trans Nanjing Univ Aeronaut Astronaut. 2010;27(2):138–47.
23. Song CZ, Zhao YQ, Wang L, et al. Multi-objective optimization design of passive suspension parameters based on competition-cooperation game model. J Mech Eng. 2014;12(1):13–24.
24. Song CZ, Zhao YQ. Fuzzy multi-objective optimization of passive suspension parameters. Fuzzy Inf Eng. 2010;1:87–100.
25. Tommaso C, Giulio P, Santosjo RGE, et al. Load leveling control for a hydro-pneumatic suspension of a tractor cabin: Modelling, identification and control. In: 2017 IEEE conference on control technology and applications, Mauna Lani; 2017. p. 1886–1891.
26. Tongjun O, Li W, Fu S. Analysis of the influence of the main parameters of the oil and gas spring on the performance of the suspension system. J Syst Simul 2008;20(9):2271–2274.
27. Xu W. Influence of oil and gas suspension on vehicle dynamics. Changchun: Jilin University; 2011.
28. Liu L. Research on parameter optimization of oil and gas spring based on vehicle ride comfort. Beijing: Beijing Institute of Technology; 2015.

29. Xu Y, Qin W, Ai T. Simulation analysis of the influence of vehicle suspension parameters on its performance based on Simulink. Sci Technol Innov. 2020;12:5–8.
30. Jiao N, Guo J, Liu S. Hydro-pneumatic suspension system hybrid reliability modeling considering the temperature influence. IEEE Access. 2017;5:19144–53.
31. Zhang H. Design and simulation of active suspension controller based on half-car model. Wuhu: Anhui Engineering University; 2012.
32. Yu Z. Automobile theory. 5th ed. Beijing: Machinery Industry Press; 2008. p. 206–12.
33. Back T. Evolutionary algorithms in theory and practice. Oxford: Oxford University Press; 1996.
34. Deb K, Agarwal S, Megarivan T. A fast and elitist multi-objective genetic algorithm: NSGA-II. Trans Evol Comput. 2002;6(2):182–97.
35. Shor P W. Algorithms for quantum computation: Discrete logarithms and factoring[C]. Proceedings 35th Annual Symposium on Foundations of Computer Science, Santa Fe, 1994: 124–134.
36. Han KH. Genetic quantum algorithm and its application to combinatorial optimization problem. In: Proceedings of the 2000 congress on evolutionary computation, La Jolla;2000. p. 1354–1361.
37. Lin M, Zhao J. Measurement data report of mechanical vibration road pavement spectrum (GB/T 7031–2005). Beijing: China Standard Press; 2005.
38. Zhou C. Vehicle ride comfort and suspension system design. Beijing: Machinery Industry Press; 2012. p. 192–206.

The manufacturer's authorised representative in the EU is Springer Nature Customer Service Centre GmbH, Europaplatz 3, 69115 Heidelberg, Germany. If you have any concerns regarding our products, please contact ProductSafety@springernature.com

Printed and bound by CPI Group (UK) Ltd, Croydon, CR0 4YY

01/10/2026

02225937-0013